COMPUTER TECHNOLOGY
Logic, Memory, and Microprocessors

A Bibliography

IFI DATA BASE LIBRARY

COMPUTER TECHNOLOGY:
Logic, Memory, and Microprocessors — A Bibliography

A. H. Agajanian

COMPUTER TECHNOLOGY
Logic, Memory, and Microprocessors

A Bibliography

COMPILED BY

A.H. Agajanian

IBM System Products Division
Hopewell Junction, New York

IFI/PLENUM • NEW YORK-WASHINGTON-LONDON

Library of Congress Cataloging in Publication Data

Agajanian, A   H
    Computer   technology: logic,   memory,   and   microprocessors—a   bibliography.

    (IF I data base library)
    Includes index.
    1. Computer engineering—Bibliography. 2. Microprocessors—Bibliography. I. Title.
Z5642.2.A35 [TK7885]   016.62138195   78-7369
ISBN 978-1-4684-6098-8   ISBN 978-1-4684-6096-4 (eBook)
DOI 10.1007/978-1-4684-6096-4

To my wife Else

Preface

During the past decade, dramatic advances have been made in
computer technology. These advances are primarily as a result of
the tremendous progress in semiconductor technology, paced by planar
processing and LSI. Great strides have also been made in mass
storage media, such as: magnetic recording, bubble domains,
charge-coupled devices, beam-addressed storage, and optical storage.
The mini- and microcomputer revolution has led to a major broadening
of the computer industry's horizons and applications. Many writers
have speculated on further progress in the coming years; other have
discussed the ultimate limits to further progress in integrated
electronic circuits towards higher complexity, smaller size, and
improved performance.

This book is a comprehensive bibliography of world literature
in computer technology. It is compiled to assist the workers in
the field to compare their work with that done by others. The book
is divided into four chapters designated A, B, C, and D. These
chapters contain the literature on logic, random-accessed memories,
serial and block-oriented memories (mass storage), and microprocessors/
microcomputers, respectively. Every chapter cites a number of books
and review articles; the review articles are selected to lead the
reader to the topic of that chapter. For easy access to the needed
references, each chapter is divided into many sections and subsections
(see Contents). A comprehensive subject index is also given to
assure easy access to the needed data. The literature on the fabrication technology and properties of semiconducting devices is the
subject of a previous book (A1) by the author.

The literature from January 1970 to September 1977 is covered.
Selected references prior to 1970 are also included. The main sources
searched for references are: Computer and Control Abstracts (CCA),
Electrical and Electronics Abstracts (EEA), Physics Abstracts (PA),
and Engineering Index. In addition to these sources, current
journals, current conference digests, Books in Print and Cumulative
Book Index are searched for material. The volumes and numbers of
the abstracts are given for access to the abstracts.

The author would like to thank Mr. R. B. Murphy, IBM East
Fishkill library manager, for encouragement and support, my wife
Else for encouragement and assistance, my colleagues Dr. H. Chang,
Mr. E. D. Councill, Dr. C. G. Jambotkar, and Dr. S. Singh for many
helpful discussions, Mr. L. A. Plaushin for editorial assistance,
and the IBM East Fishkill library staff for their valuable assis-

tance. Special thanks are due to Mrs. F. Sisson of Graphic
Techniques for typing the manuscript.

AVAILABILITY OF DOCUMENTS

The references in this book consist of technical articles presented in journals, books, symposia proceedings, technical meetings,
reports, patents and theses. If a reader does not have access to a
library to obtain copies of the references we suggest the following
sources:

1. Journal articles, books, symposia proceedings:
        Library of Congress
        Photoduplication Service
        Washington, D. C. 20540

        Engineering Societies Library
        345E 47th Street
        New York, New York 10017

2. U. S. government contract reports:
        National Technical Information Service
        Springfield, Virginia 22151

3. IAEC reports
        International Atomic Energy Agency
        Kaerntnerring A 1010
        Vienna, Austria

        National Lending Library
        Boston Spa, England

4. Theses listed in Dissertation Abstracts + number (U. S.)
        University Microfilms
        Dissertation Copies
        P.O. Box 1764
        Ann Arbor, Michigan 48106

    Others
        University Microfilms, Ltd.
        St. John's Road
        Penn, Buckinghamshire
        England

5. Patents: U. S. and foreign
        Patent Office
        Box 9
        Washington, D. C., 20231

Contents

A. LOGIC

I. BOOKS

1. Agajanian, A. H., "Semiconducting devices: A bibliography of fabrication technology, properties, and applications", IFI/Plenum, New York, 1976.
2. Altman, L., (Ed.), "Large scale integration", Electronics Book Series, McGraw-Hill Book Co., New York, 1976.
3. Becher, W. D., "Logical design using integrated circuits", Hayden Book Co., Inc., 50 Essex Street, Rochelle Park, N. J., 1977.
4. Biswas, N. N., "Introduction to logic and switching theory", Gordon & Breach, New York, 1975.
5. Blakeslee, T. R., "Digital design with standard MSI and LSI", Wiley, New York, 1975.
6. Breuer, M. A., "Diagnosis and reliable design of digital systems", Computer Science Press, Woodland Hills, California, 1976.
7. Carr, W. N. and Mize, J. P., "MOS/LSI design and applications", McGraw-Hill Book Co., New York, 1972.
8. Chirlian, P. M., "Analysis and design of digital circuits and computer systems", Matrix Publishers, Inc., Champaign, Ill., 1976.
9. Dempsey, J. A., "Basic digital electronics with MSI applications", Addison-Wesley Publishing Co., Reading, Mass., 1977.
10. Dokter, F. and Steinhauer, J., "Digital electronics", MacMillan Press, London, England, 1973.
11. Friedman, A. D., "Logical design of digital systems", Computer Science Press, Woodland Hills, California, 1975.
12. Greenfield, J. D., "Practical digital design using IC's", Wiley, New York, 1977.
13. Hunter, L. P., "Handbook of semiconductor electronics", McGraw-Hill Book Co., New York, 1970.
14. Hunter, W. L., "Master handbook of digital logic applications", Tab Books, Blue Ridge Summit, Pa., 1975.
15. Ilardi, F. A., "Computer circuit analysis: Theory and applications", Prentice-Hall, Inc., Englewood Cliffs, N. J., 1976.
16. Kostopoulos, G. K., "Digital engineering", Wiley, New York, 1975.
17. Lancaster, D. E., "TTL cookbook", H. W. Sams, Indianapolis, Indiana, 1974.
18. Langdon, G. G., "Logic design: A review of theory and practice", Academic Press, New York, 1974.
19. Lee, S. C., "Digital circuits and logic design", Prentice-Hall, Englewood Cliffs, New Jersey, 1976.
20. Lenk, J. D., "Handbook of logic circuits", Reston Publishing Co., Inc., Reston, Virginia, 1972.

21. Lenk, J. D., "Logic designer's manual", Reston Publishing Co.,
    Inc., Reston, Virginia, 1977.
22. Lewin, D., "Logic design of switching circuits", North-Holland,
    New York, 1974.
23. Marcus, M. P., "Switching circuits for engineers", Prentice-Hall,
    Englewood Cliffs, New Jersey, 1975.
24. Mazda, F. F., "Components of computers", Electrochemical Publi-
    cations, Ayr, Scotland, 1975.
25. Millman, J. and Halkias, C., "Electronic fundamentals and appli-
    cations for engineers and scientists", McGraw-Hill Book Co.,
    New York, 1976.
26. Millman, J. and Taub, H., "Pulse, digital, and switching wave-
    forms", McGraw-Hill Book Co., New York, 1965.
27. Mno, M. M., "Computer logic design", Prentice-Hall International,
    Englewood Cliffs, New Jersey, 1972.
28. Morris, R. L., "Designing with TTL integrated circuits", McGraw-
    Hill Book Co., New York, 1971.
29. Nagle, H. T., "An introduction to computer logic", Prentice-Hall,
    Englewood Cliffs, New Jersey, 1975.
30. Newhouse, V. L., (Ed.), "Applied superconductivity", Vols. 1 and
    2, Academic Press, New York, 1975.
31. Reeves, C. M., "Introduction to logical design of digital cir-
    cuits", Cambridge University Press, London, England, 1972.
32. Solymar, L., "Superconductive tunnelling and applications",
    Wiley-Interscience, New York, 1972.
33. Wakerly, J. F., "Logic design projects using standard integrated
    circuits", John Wiley & Sons, Inc., New York, 1976.
34. Wolfendale, E., (Ed.), "MOS integrated circuit design", Halsted
    Press, New York, 1973.
35. Zissos, D., "Problems and solutions in logic design", Oxford
    University Press, London, England, 1976.

II. REVIEW ARTICLES

36. Altman, L. and Mattera, L., "Several solid-state technologies
    show surprising new paces", Electronics, $\underline{49}$ (25), 90-5 (1976).
37. Anon., "Understanding simple logic circuits. I. The truth table",
    Electr. Rev., $\underline{199}$ (3), 43 (1976), CCA11-25077.
38. Anon., "Understanding simple logic circuits. II. Basic gates",
    Electr. Rev., $\underline{199}$ (7), 25 (1976), CCA11-31488.
39. Anon., "Understanding simple logic circuits. III. (Basic gates
    used in digital integrated circuits)", Electr. Rev., $\underline{199}$ (9),
    36 (1976), CCA11-31489.
40. Clarke, J., "The application of Josephson junctions to computer
    storage and logic elements and to magnetic measurements", AIP
    Conf. Proc., no. 29, 17-22 (1976), CCA11-22674.
41. Dabrowski, G., "Active electronic components. XX", Elektroniker,
    $\underline{14}$ (12), EL24-35 (1975), German, CCA11-12246.

42. Franson, P., "Logic family update - SSI/MSI still thrive in the world of LSI", EDN, 22 (4), 79-85 (1977), CCA12-15376.
43. Holton, W. C., "The large-scale integration of microelectronic circuits", Sci. Am., 237 (3), 82-94 (1977).
44. Kroeger, J. and Threewitt, B., "Review the basics of MOS logic", Electron. Des., 22 (6), 98-105 (1974), EEA78-8169.
45. Lutsch, A. G. K. and Verster, T. C., "Modern semiconductor logic families", Trans. S. Afr. Inst. Electr. Eng., 67 (2), 58-70 (1976), CCA11-17329.
46. Meindl, J. D., "Microelectronic circuit elements", Sci. Am., 237 (3), 70-81 (1977).
47. Noguchi, S. and Oizumi, J., "A survey of cellular logic", J. Inst. Electron. & Commun. Eng. Jap., 54 (2), 206-20 (1971), Japanese, CCA6-15133.
48. Ritchie, G. J., "A review of standard logic families and linear integrated circuits", Vacation School on Engineering Aspects of Microelectronics, 1969, Paper 9-1, CCA5-13667.
49. Wolff, M. F., "The genesis of the integrated circuit", IEEE Spectrum, 13 (8), 44-53 (1976), EEA80-9500.
50. Yanai, H. and Ikoma, T., "Recent development of transferred electron logic devices in Japan", IEEE MIT-S International Microwave Symposium, 1976, p. 161-3, CCA11-31484.

III. TECHNOLOGIES
  1. General

51. Anon., "Wire-wrap and ECL: Wiring problems in the changeover from TTL to ECL logic systems", Elektron Prax., 11 (4), 16, 18-20 (1976), German, CCA11-28120.
52. Bajenseco, T. I., "Is I^2L going to compete with C-MOS?", Electron. & Microelectron. Ind., no. 200, 49-51 (1975), French, EEA78-24456.
53. Bishop, A., "Which integrated circuit technology - I^2L or CMOS?", Elektron. Ind., 7 (11), 297-9 (1976), German, CCA12-8799.
54. Braeckelman, W., Fritzsche, H., Kroos, F. K., Trinki, W. and Wilhelm, W., "A masterslice LSI for subnanosecond random logic", 1977 International Solid-State Circuits Conference, Digest of the Technical Papers, p. 108-9.
55. Craney, P. M., "Relays and logic ICs can be working partners", Electronics, 49 (2), 107-10 (1976), EEA79-16873.
56. Crook, C., "Comparing the power of C-MOS with TTL", Electronics, 47 (10), 132-3 (1974), EEA77-22487.
57. Everist, J. A., "TTL MOS and all that", Electrotechnology, 1 (1), 15-18 (1973), CCA8-9479.
58. Fleisher, H., "Array logic and LSI", 1971 IEEE International Convention Digest, p. 372-3, CCA6-17552.
59. Garrett, L. S., "Integrated-circuit digital-logic families. I. Requirements and features of a logic family; RTL, DTL, and HTL devices", IEEE Spectrum, 7 (10), 46-58 (1970), CCA6-3722.

60. Garrett, L. S., "Integrated-circuit digital logic families. II.
    TTL devices", IEEE Spectrum, 7 (11), 63-72 (1970), CCA6-3723.
61. Garrett, L. S., "Integrated-circuit digital logic families. III.
    ECL and MOS devices", IEEE Spectrum, 7 (12), 30-42 (1970),
    CCA6-5590.
62. Gilder, J. H., "New bipolar technologies to compete with CMOS
    and ECL", Electron. Des., 24 (5), 18-19 (1976), CCA11-17341.
63. Grundy, D. and Peirce, C., "The uncommitted logic array", Micro-
    electronics, 5 (1), 12-17 (1973), CCA9-14285.
64. Henle, R. A., "How LSI is affecting logic design", 1971 IEEE
    International Convention Digest, p. 276-7, CCA6-15119.
65. Hodges, D. A., "Which logic technology for MSI and LSI systems",
    Digest of Papers of the Sixth Annual IEEE Computer Society Inter-
    national Conference, 1972, p. 135-7, CCA8-2552.
66. Hoppe, J. and Haase, H. D., "Problems of design and production of
    universal logic elements. I. Technological considerations", Radio
    Fernsehen Elektron., 21 (16), 518-9 (1972), German, CCA8-5234.
67. Joumard, R., "The integrated circuit logic", Report CEA-BIB-114,
    Comm. Energie Atomique, Centre d'Etudes Nucleaires, Grenoble,
    France, 1968, 47 pp.
68. Krivohlavek, J., "Orientation in the structures of present and
    prospective bipolar integrated circuits", Sdelovaci Tech., 25 (3),
    83-6 (1977), Czech, EEA80-28905.
69. McDermott, J., "I^{2}L and CMOS battle for lead as IC-technology
    race heats up", Electron. Des., 25 (7), 42, 44, 46 (1977).
70. McLay, A., "The kinks and kernels of I.P. Schottky and C.M.O.S.",
    New Electron, 8 (4), 42-3 (1975), EEA78-24485.
71. Malek, T. and Schwartz, W. B., "IC packaging panels for high-speed
    logic applications", Electron Packag. Prod., 16 (4), 35-40 (1976).
72. Miles, T. E., "Schottky TTL vs ECL for high speed logic", Comput.
    Des., 11 (10), 79-86 (1972), EEA76-4232.
73. Mouftah, H. T. and Jordan, I. B., "Integrated circuits for ternary
    logic", Proceedings of the 1974 International Symposium on Multi-
    Valued Logic, p. 285-302, EEA78-5095.
74. Nenfang, O., "New logic circuit symbols", Funk-Tech., 30 (20),
    647, 650 (1975), CCA11-6331.
75. Petri, M., "LSI integrated circuits; I^{2}L or C-MOS", Toute
    Electron, no. 415, 43-7 (1976), French, EEA80-3035.
76. Piening, J. and Wilhelm, W., "Integrated logic circuits",
    Elektronik, 19 (5), 171-4 (1970), German, CCA5-15513.
77. Rein, H. M., Worner, K. and Cluass, H., "Integrated subnanosecond
    circuits with low power dissipation and few components", Nach-
    richtentech Z., 25 (10), 465-70 (1972), German, CCA8-2490.
78. Secaze, G., "C-MOS or LPSTTL-which logic to choose?", Electron.
    & Microelectron. Ind., no. 200, 61-5 (1975), French, EEA78-24457.
79. Shagurin, M. I., Petrov, L. N. and Tat'yanin, V. I., "Comparison
    of TTL and DTL variants employing Schottky barrier diodes", Izv.
    VUZ Radioelektron., 14, 1365-9 (1971), Russian, EEA74-12666.

80. Sudo, T., Kamoto, T. and Mukai, H., "A monolithic 8 pJ/2 GHz logic family", IEEE J. Solid-State Circuits, 10 (6), 524-9 (1975), CCA11-4705.

2. Bipolar Logic
a. ECL

81. Alford, C., "System design considerations for high-speed ECL", New Electron, 9 (7), 40 (1976), EEA79-28128.
82. Anon., "ECL circuits offer optimum speed/power product", Comput. Des., 10 (5), 136-7 (1971), CCA6-17563.
83. Anon., "Wire-Wrap and Emitter-Coupled Logic (ECL)", Elektronik-praxis, 11 (4, 5), 16, 18-20 (1976).
84. Blood, W., "Building high speed logic systems using MECL circuitry", Control & Instrum, 6 (4), 42-3, 45 (1974), CCA9-20639.
85. Budinsky, J., "ECL integrated digital circuits", Slaboproudy Obzor, 32 (7), 302-10 (1971), Czech, CCA6-20072.
86. Crook, C., "High-speed low power e.c.1.", New Electron, 6 (4), 71-2 (1973), CCA8-11577.
87. Czajkowski, G., "Design principles for ECL-circuitry", Elecktronika, 16 (9), 385-8 (1975), Polish, CCA11-4701.
88. Fadrhons, J., "The emitter coupled logic", Sdelovaci Tech., 22 (2), 55-60 (1974), Czech, CCA9-16007.
89. Harding, J., "Development and application of picosecond logic", New Electron, 8 (17), 16, 18 (1975), EEA79-1067.
90. Hayasaka, A., Kaji, T., Honma, Y. and Harada, S., "High performance ECL gates made by SEL method", 1976 International Electron Devices Meeting, Technical Digest, p. 48-50, CCA12-10598.
91. McDonald, J. F., Harris, R. and Sustman, J., "Three ECL designs for microprogrammable writable control stores", Workshop on Microprogram, 6th Annual Conference Record, 1973, p. 36-43.
92. Marley, R. R., "Design considerations of temperature compensated emitter coupled logic", 1971 IEEE International Convention Digest, p. 138-9, CCA6-15116.
93. Narud, J. A., Meyer, C. S. and Lynn, D. K., "One nanosecond current mode logic circuits", Proceedings IEEE Electrical and Electronics Engineering Resources Roundup, 1969, p. 229-42, CCA5-2702.
94. Rein, H. M. and Ranfft, R., "Improved feedback ECL gate with low delay-power product for the subnanosecond region", IEEE J. Solid-State Circuits, 12 (1), 80-2 (1977), CCA12-3817.
95. Rein, H. M. and Straub, D., "SECL-an advanced high speed integrated circuit family", 1971 IEEE International Convention Digest, p. 140-1, CCA6-15117.
96. Rein, H. M. and Straub, D., "SECL digital integrated circuits with particularly fast switching time", Elektron. Ind., 2 (3), 26-31 (1971), German, CCA6-15128.

97. Russo, P. M. and Rohrer, R. A., "Computer optimization of the transient response of an ECL gate", IEEE Trans. Circuit Theory, CT-18 (1), 197-9 (1971), CCA6-9289.

98. Shagurin, I. J. and Strukov, V. N., "Feasibility of improving the noise immunity of emitter-coupled integrated transistor logic elements", Izv. VUZ Radioelektron., 13 (12), 1480-5 (1970), Russian, CCA6-7971.

99. Talbot, B. and James, J., "ECL-still the fastest", Elektroniker, 14 (12), EL14-16 (1975), German, CCA11-12251.

100. Tan, N. Q., "Emitter-coupled logic with dielectric insulation", Electron & Microelectron. Ind., no. 160, 44-50 (1972), French, CCA8-2505.

101. Thrift, P. R., "Using open emitter coupled logic gates", Comput. Des., 10 (5), 132-4 (1971), CCA6-17562.

102. Williams, K. R., "Emitter coupled logic to either long line or send receive mode bidirectional convert circuit", IBM Tech. Disclosure Bull., 18 (12), 4101-3 (1976), CCA11-28112.

b.   TTL

103. Aoki, S. J., Najmann, K. N. and Schettler, H., "Microwatt TTL circuits", IBM Tech. Disclosure Bull., 16 (10), 3272 (1974), CCA9-18328.

104. Arustamyan, V. E., "TTL circuits using multi-electrode transistors in large integrated circuits", Izv. VUZ Radioelektron., 19 (9), 66-9 (1976), Russian, CCA12-1407.

105. Baldey, R., "Schottky diode clamped TTL", Presented at Colloquium Digest on Very High Speed Bipolar Logic: ECL-Schottky TTL, London, England, October 1971, EEA75-452.

106. Beke, I. and Fejes, L., "The TTL gate, the high-speed element of the digital technique", Meres & Autom, 19 (2), 69-75 (1971), Hungarian, CCA6-12670.

107. Berthon, L. and Mick, J., "Digital systems: application of tri-state Schottky TTL", Electron. & Microelectron. Ind., no. 210, 53-7 (1975), French, EEA79-8382.

108. Bielser, E., "Experiments with 5 volt TTL circuits", Elektroniker, 15 (2), EL33-41 (1976), German, CCA11-14960.

109. Cavinaugh, C., "Using Schottky to upgrade performance of your TTL system", EDN, 18 (18), 38 (1973).

110. Cole, H. A., "A low-cost T-type binary using NAND gate t.t.l.", Des. Electron., 8 (7, 8), 80-5 (1971), CCA6-12642.

111. Fejes, L. and Beke, I., "Calculation of the operating points and characteristics of the TTL gate", Hiki, 11 (1), 30-56 (1971), Hungarian, CCA6-17550.

112. Fleischhammer, W., Schneider, G. and Koppe, G., "T^3L achieves C-MOS noise immunity while retaining TTL speed", Electronics, 47 (5), 94-7 (1974), EEA77-18278.

113. Fleischhammer, W., Schneider, G. and Koppe, G., "T^{3}L: Improved
     TTL circuits with increased noise immunity", Elektronik, 23 (7),
     235-7 (1974), German, CCA9-21952.
114. Fulkerson, D. E., "Direct-coupled transistor-transistor logic:
     A new high-performance LSI gate family", IEEE J. Solid-State
     Circuits, SC-10 (2), 110-17 (1975).
115. Gajewski, W., Pienkos, J. and Turczynski, J., "Noise in systems
     built from TTL integrated circuits. II. Causes, effects and
     remedies", Pomiary Autom. Kontr., 18 (9), 396-8 (1972), Polish,
     CCA8-2491.
116. Gary, P. A., Pedersen, R. A. and Soloway, B. H., "Designs of
     high performance TTL integrated circuits employing CDI component
     structures", IEEE International Solid State Circuits Conference,
     Digest of Technical Papers, 1970, p. 116-17, CCA5-13647.
117. Hintz, L., "Most Schottky TTL circuits use NPN inputs, but PNP
     is better - here's why", New Electron, 16, 28 (1973).
118. Kronlage, J. W., "Schottky clamped TTL", IEEE International
     Convention Digest, 1971, p. 144-5, CCA6-15118.
119. Kurz, B. and Barron, M. B., "Improved Schottky clamped (T^{2}L)
     circuits", IEEE J. Solid-State Circuits, SC-7, 175-9 (1972),
     EEA75-12660.
120. MacLeod, A. M., "An elementary introduction to TTL", Phys.
     Educ., 10 (6), 440-5 (1975), EEA79-12499.
121. MacLeod, A. M., "A second step to TTL circuitry", Phys. Educ.,
     11 (2), 111-6 (1976), CCA11-17354.
122. Manas, S., "Touch sensitive pulsator for TTL technology
     systems", Mundo Electron, no. 49, 89 (1976), Spanish, CCA11-14967.
123. Marinescu, N., "Analysis of TTL integrating circuits. Static
     characteristics", Autom. & Electron, 15 (1), 32-7 (1971),
     Rumanian, CCA6-;2637.
124. Nikolaeva, I. I. and Redina, S. F., "The principle of operation
     and the characteristics of integral schemes of transistor-
     transistor logic modules", Elektrotekhnika, 41 (10), 24-7
     (1970), Russian, CCA6-3716.
125. Oberman, R. M. M., "The J-K gate", IEEE Trans. Comput., C-25
     (11), 1156-9 (1976), CCA11-31532.
126. Palm, W. A., "Connect a 7400 gate as a Schmitt trigger", EDN,
     21 (15), 84 (1976).
127. Palm, W. A., "TTL gates drive twisted-pair lines", Electronics,
     50 (6), 111-12 (1977).
128. Rabbat, N. B., "Transient analysis of modified TTL gate",
     Electron. Lett., 7 (1), 22-4 (1971), CCA6-5596.
129. Ribenyi, A., "Schottky diode in TTL circuits", Proceedings of
     the International Conference on the Physics and Chemistry of
     Semi-conductor Heterojunctions and Layer Structures, 1970,
     p. V349-54, Publ. 1971, EEA74-26822.
130. Sechler, R. F., "Large scale integrated TTL circuit technology
     utilizing low voltage positive supplies", IBM Tech. Disclosure
     Bull., 18 (12), 4091-4 (1976), CCA11-28111.

131. Siemens, A. G., "Slow interference free TTL logic circuit",
     Patent UK 1406295, Publ. September 1975, EEA79-20063.
132. Varada Rajan, H. D., "Antisaturation TTL circuit", IBM Tech.
     Disclosure Bull., 14 (1), 335 (1971), CCA6-22196.
133. Wilamowski, B. M., "Influence of properties of inside elements
     on static characteristics and outside parameters of TTL gates",
     Arch. Elektrotech., 22 (1), 49-59 (1973), Polish, CCA8-13599.

c. I^2L/MTL

134. Altman, L., "New LSI bipolar chips are best buy for designers
     of fast systems", Electronics, 48 (14), 81-92 (1975).
135. Altstein, J., "I^2L: Today's versatile vehicle for tomorrow's
     custom LSI", EDN, 20 (4), 34-8 (1975).
136. Anon., "Integrated injection logic", Wireless World, 80 (1468),
     486 (1974).
137. Bast, M., "I^2L: a densely packed bipolar logic", Elektrotechnik,
     58 (18), 16-18 (1976), German, EEA79-46412.
138. Berger, H. H., "The injection model - a structure-oriented model
     for merged transistor logic (MTL)", IEEE J. Solid-State Circuits,
     SC-9 (5), 218-27 (1974), CCA9-23182.
139. Berger, H. H., "Bipolar devices for low power digital applica-
     tions - progress through new concepts", 3rd European Solid State
     Device Research Conference (Invited papers), 1973, p. 109-29,
     Publ. 1974, German, CCA9-16081.
140. Berger, H. H. and Wiedmann, S. K., "Terminal-oriented model for
     merged transistor logic (MTL)", IEEE J. Solid-State Circuits,
     SC-9 (5), 211-17 (1974), CCA9-23181.
141. Berger, H. H. and Wiedmann, S. K., "The bipolar LSI breakthrough,
     part 1; rethinking the problem", Electronics, 48 (18), 89-95
     (1975).
142. Berger, H. H. and Wiedmann, S. K., "The bipolar LSI breakthrough,
     part 2; extending the limits", Electronics, 48 (20), 99-103
     (1975).
143. Berger, H. H. and Wiedmann, S. K., "Advanced merged transistor
     logic by using Schottky junctions", Microelectronics, 7 (3),
     35-42 (1976), EEA79-32529.
144. Blackstone, S. C. and Mertens, R. P., "Schottky collector I^2L",
     IEEE J. Solid-State Circuits, SC-12 (3), 270-5 (1977),
     EEA80-24984.
145. Blatt, V., "Integrated injection logic", New Electron, 9 (4),
     27-9 (1976), CCA11-14951.
146. Blatt, V. and Sumerling, G. W., "Schottky I^2L (substrate fed
     logic) - an analysis of the implications of the vertical in-
     jector structure and Schottky collection", IEEE J. Solid-State
     Circuits, SC-12 (2), 128-35 (1977), CCA12-8820.

147. Blatt, V., Walsh, P. S. and Kennedy, L. W., "Substrate fed logic", IEEE J. Solid-State Circuits, SC-10 (5), 336-42 (1975).

148. Bruederle, S. and Smith, P., "Designing with I2L", 1975 WESCON Technical Papers, Vol. 19, Paper 19/2, 5 pp., CCA11-12268.

149. Cook, B. and McNally, S. H., "I^2L II", 1975 International Electron Devices Meeting, Technical Digest, p. 284-7, EEA79-46527.

150. Crippen, R. E., O'Brien, D., Rallapalli, K. and Verhofstadt, P. W. J., "High-performance integrated injection logic: a microprogram sequencer built with I^2L", IEEE J. Solid-State Circuits, SC-11 (5), 662-8 (1976), EEA79-46428.

151. Crooke, M. and Greyvenstein, R. F., "Experimental integrated injection logic circuits", Trans. S. Afr. Inst. Electr. Eng., 67 (4), 118-21 (1976), CCA11-22572.

152. Dao, T. T., "Threshold I^2L and its applications to binary symmetric functions and multivalued logic", IEEE J. Solid-State Circuits, SC-12 (5), 463-72 (1977).

153. Dao, T. T., Russell, L. K., Preedy, D. R. and McClosky, E. J., "Multilevel I^2L with threshold gates", 1977 International Solid-State Circuits Conference, Digest of Technical Papers, p. 110-11.

154. Davies, R. D., Estrench, D. B., Meindl, J. D. and Dutton, R. W., "I^2L DC functional requirements", IEEE J. Solid-State Circuits, SC-12 (2), 208-10 (1977), CCA12-8830.

155. DeTroye, N. C., "Proposer for bi-polar technology", Microelectronics, 6 (2), 9-13 (1974).

156. DeTroye, N. C., "Integrated injection logic - present and future", IEEE J. Solid-State Circuits, SC-9 (5), 206-11 (1974).

157. den Brinker, C. S. and Colman, D., "Whither I^2L?", New Electron, 9 (4), 32, 34, 37 (1976), CCA11-14952.

158. Elmasry, M. I., "Non-saturated integrated injection logic", Electron Lett., 11 (3), 63-4 (1975).

159. Elmasry, M. I., "Load-line analysis of I.I.L.", Electron Lett., 11 (3), 68-9 (1975).

160. Elmasry, M. I., "Folded-collector integrated injection logic", IEEE J. Solid-State Circuits, SC-11 (5), 644-7 (1976), EEA79-46426.

161. Estreich, D. B. and Dutton, R. W., "Modeling integrated injection logic (I^2L) performance and operational limits", IEEE J. Solid-State Circuits, SC-12 (5), 450-6 (1977).

162. Guetin, P. and Houssie, M., "I^2L from LSI to bipolar devices", Electron. & Microelectron. Ind., no. 218, 17-21 (1976), French, CCA11-17344.

163. Hennig, F., Hingarh, H. K., O'Brien, D. and Verhofstadt, P. W. J., "Isoplanar integrated logic: a high-performance bipolar technology", IEEE J. Solid-State Circuits, SC-12 (2), 101-9 (1977), CCA12-8816.

164. Herman, J. M., III, Evans, S. A. and Sloan, B. J., Jr., "Second generation I^2L/MTL: a 20 ns process/structure", IEEE J. Solid-State Circuits, SC-12 (2), 93-101 (1977), CCA12-8815.

165. Hewlett, F. W., Jr., "Schottky I^2L", IEEE J. Solid-State
     Circuits, SC-10 (5), 343-8 (1975), EEA78-37625.
166. Hewlett, F. W., Jr. and Ryden, W. D., "The Schottky I^2L tech-
     nology and its application in a 24 x 9 sequential access
     memory", IEEE J. Solid-State Circuits, SC-12 (2), 119-23 (1977).
167. Horton, R. L., Englade, J. and McGee, G., "I^2L takes bipolar
     integration a significant step forward", Electronics, 48 (3),
     83-90 (1975), CCA10-9502.
168. Jaeger, R. C., "Evaluation of injection modeling", Solid-State
     Electron, 19 (7), 639-43 (1976).
169. Kaneko, K., Okabe, T. and Nagata, M., "Stacked I^2L circuit",
     Electron. Lett., 12 (10), 249-50 (1976), EEA79-28102.
170. Kaneko, K., Okabe, T. and Nagata, M., "Stacked I^2L circuit",
     IEEE J. Solid-State Circuits, SC-12 (2), 210-12 (1977),
     CCA12-8831.
171. Kerns, D. V., Jr., "The effect of base contact position on the
     relative propagation delays of the multiple outputs of an I^2L
     gate", IEEE J. Solid-State Circuits, SC-11 (5), 712-17 (1976),
     CCA12-1395.
172. Klaassen, F. M., "Some considerations on high-speed injection
     logic", IEEE J. Solid-State Circuits, SC-12 (2), 150-4 (1977),
     CCA12-8822.
173. Kleitman, D., "I^2L: novel circuits and formats", IEEE Spectrum,
     14 (6), 35-6 (1977).
174. Lohstroh, J., "Dynamic behaviour of active charge in I^2L tran-
     sistors", 1976 IEEE International Solid-State Circuits Con-
     ference, Digest of Technical Papers, p. 94-5, CCA11-28125.
175. Lohstroh, J., "Integrated injection logic", Bull. Assoc. Suisse
     Electr., 68 (2), 53-9 (1977), CCA12-8804.
176. Maderbacher, F., "I^2L bipolar technology for LSI", Radio Elektron.
     Schau, 51 (21), 722-4 (1975), German, EEA77-16341.
177. Matthews, W., Merters, R. and Stulting, J., "Base current of
     I^2L gates at low-current levels", 1976 International Electron
     Devices Meeting, Technical Digest, p. 316-18, CCA12-15364.
178. Mulder, C. and Wulms, H. E. J., "High speed integrated injection
     logic (I^2L)", IEEE J. Solid-State Circuits, SC-11 (3), 379-85
     (1976), CCA11-22559.
179. Muller, R., "Current hogging injection logic - a new logic with
     high functional density", IEEE J. Solid-State Circuits, SC-10
     (5), 348-52 (1975).
180. Muller, R. and Graul, J., "CHIL and I^2L with passive isolation",
     1st European Solid State Circuits Conference - ESSCIRC, Extended
     Abstracts, 1975, p. 30-1, EEA79-28698.
181. Muller, R. and Graul, J., "I^2L and CHIL (current hogging injec-
     tion logic) in a high-speed technology with oxide isolation",
     Nachrichtentech Z., 29 (11), 825-7 (1976), EEA80-3019.
182. Nakano, T., Horiba, Y., Yasuoka, A., Tomisawa, O., Murakami, K.
     and Kato, S., "Vertical injection logic", 1975 International
     Electron Devices Meeting, Technical Digest, p. 555-8, CCA12-3806.

183. Nishizawa, J. and Wilamowski, B. M., "Integrated logic-static
     induction transistor logic", 1977 International Solid-State
     Circuits Conference, Digest of Technical Papers, p. 222-3.
184. Poorter, T., "Electrical parameters, static and dynamic response
     of I^2L", IEEE J. Solid-State Circuits, SC-12 (5), 440-9 (1977).
185. Ray, W., "I^2L: complementing Schottky TTL", IEEE Spectrum, 14
     (6), 35 (1977).
186. Refail, S. S., Elmasry, M. I. and Hersell, E. L., "Functional
     modeling of integrated injection logic - DC analysis", IEEE
     Trans. Electron Devices, ED-24 (3), 234-40 (1977).
187. Roesner, B. B. and McGreivy, D. J., "A new high speed I^2L
     structure", IEEE J. Solid-State Circuits, SC-12 (2), 114-18
     (1977), CCA12-8818.
188. Schwitz, A. and Slob, A., "Effect of isolation regions on the
     current gain of inverse NPN-transistors used in integrated
     injection logic (I^2L)", 1974 International Electron Devices
     Meeting, Technical Digest, p. 508-10.
189. Shinozaki, S., "High-speed I^2L gate with self-aligned double-
     diffusion injector", Electron. Lett., 13 (1), 6-7 (1977),
     CCA12-3803.
190. Shinozaki, S., Iizuka, T., Masuoka, F., Shunada, K. and Miyamoto,
     J. I., "Role of the external n-p-n base region on the switching
     speed of integrated injection logic (I^2L)", IEEE J. Solid-State
     Circuits, SC-12 (2), 185-91 (1977), CCA12-8827.
191. Sloan, B. J. and Stehin, R. A., "Advances in I^2L development",
     13th IEEE Computer Society International Conference, 1976,
     p. 56-8, CCA12-1414.
192. Stone, J. L. and Agraz-Guerena, J., "Application of ion implanta-
     tion techniques to the optimization of integrated injection/
     merged transistor logic circuits", International Microelectronic
     Conference, Proceedings of the Technical Program, 1975, p. 72-80.
193. Terada, H., Asada, T., Okuba T. and Kitamura, Z., "An integrated
     injection logic (I^2L) realization of phase locked loop", 1st
     European Solid State Circuits Conference - ESSCIRC, Extended
     Abstracts, 1975, p. 24-5, EEA79-28150.
194. Thornton, C. G., "Impact of I^2L", 1976 IEEE International Solid-
     State Circuits Conference, Digest of Technical Papers, p. 171,
     CCA11-28126.
195. Tokumaru, Y., Nakai, M., Shinozaki, S. and Nishi, Y., "I^2L with
     a self-aligned double-diffused injector", IEEE J. Solid-State
     Circuits, SC-12 (2), 109-14 (1977).
196. Tomisawa, O., Horiba, Y., Kato, S., Murakami, K., Yasuoka, A.
     and Nakano, T., "Vertical injection logic", IEEE J. Solid-State
     Circuits, SC-11 (5), 637-43 (1976), CCA11-31479.
197. Torrero, E. A., "The multifacets of I^2L", IEEE Spectrum, 14 (6),
     28-9 (1977).
198. Walsh, P. S., Blatt, V. and Kennedy, L. W., "Substrate fed logic
     - a novel form of injection logic", 1st European Solid State Cir-
     cuits Conference - ESSCIRC, Extended Abstracts, 1975, p. 122-3.

199. Walsh, P. S. and Summerling, G. W., "Schottky I^2L (substrate
     fed logic) - an optimum form of I^2L", IEEE J. Solid-State
     Circuits, SC-12 (2), 123-7 (1977), CCA12-8819.
200. Wulms, H. E. J., "Integrated injection logic (I^2L)", Polytech.
     Tijdschr. Elektrotech. Elektron, 80 (25), 827-38 (1975), Dutch,
     EEA79-12500.
201. Wulms, H. E. J., "Base current of I^2L transistors", 1976 Inter-
     national Solid-State Circuits Conference, Digest of Technical
     Papers, p. 92-3, CCA11-28124.

                    d. DTL

202. Auxin, V. Y. and Katunina, L. A., "Majority of logic circuit",
     Patent UK 1225196, Publ. March 1971, CCA6-17593.
203. Landau, A. D., "Stability of a DTL flip flop circuit with
     Miller integrators to perturbation signals", Autom & Electron.,
     15 (6), 259-67 (1971), Rumanian, CCA7-8236.
204. Lia Sonia, H. and Landau, A. D., "The design of a NAND circuit
     of the DTL type with Miller integrator, optimized from the
     point of view of the stability at disturbing signals", Posta
     & Telecommunicatii, 2 (12), 681-6 (1972), Rumanian, CCA8-11578.
205. Myk, A., "Logic circuits, III", Tech. Radia & Telew., 2/18 (3),
     22-8 (1973), Polish, CCA9-25016.
206. Palmonella, G. M., "Logic circuitry for the control of automatic
     digital measurements of resistance", RC Ruin. Ass. Electrotec.
     Ital., 44 (2), 2 pp. (1969), Italian, CCA6-13470.
207. Puncochar, J., "Extension of the function of the MZH 100 circuits",
     Sdelovaci Tech., 24 (8), 303 (1976), Czech, CCA12-3828.
208. Takahashi, M., Saito, M. and Suzuki, J., "Reliability evaluation
     of DTL type monolithic integrated circuits", Rev. Elec. Commun.
     Lab., 18 (9-10), 694-711 (1970), CCA6-9281.

                    e. Others

209. Dobronravov, O. E., Koloskov, L. A. and Ovchinnikov, V. V., "An
     investigation of the dynamics of the operation of threshold
     tunnel-diode-transistor logic networks", Conference of Tunnel
     Diodes in Computing and Measurement Technology, 1970, p. 59-68,
     Publ. 1972, CCA8-2530.
210. Fulkerson, D. E., "Direct-coupled transistor-transistor logic:
     A new high-performance LSI gate family", IEEE J. Solid-State
     Circuits, SC-10 (2), 110-17 (1975).
211. Graul, J., Kaiser, H., Wilhelm, W. J. and Ryssel, H., "Bi-polar
     high-speed low-power gates with double implanted transistors",
     IEEE J. Solid-State Circuits, SC-10 (4), 201-4 (1975).
212. Greer, D. L., "Associative logic for logic network implementa-
     tion", 1976 IEEE International Solid-State Circuits Conference,
     Digest of Technical Papers, p. 18-19, 222, CCA11-28122.

213. LeCan, C. J., "A new current-mode circuitry and process for fast logic and memory applications", International Solid-State Circuits Conference, Digest of Technical Papers, 1971, p. 66-7, CCA6-15269.

214. Lehning, H., "Current hogging logic (CHL) - a new bipolar logic for LSI", IEEE J. Solid-State Circuits, SC-9 (5), 228-33 (1974), CCA9-23183.

215. Meyer, F., "Base-coupled logic circuits for high-speed digital systems", Nachrichtentech. A. (NTZ), 29 (11), 828-30 (1976), CCA12-3825.

216. Rajashekhara, T. N. and Sonde, B. S., "New family of low-power CT^2L circuits", IEEE J. Solid-State Circuits, SC-10 (1), 77-9 (1975).

217. Sakai, T., Sunohara, Y., Nakamura, N. and Sudo, T., "A 100-ps bipolar logic", 1977 International Solid-State Circuits Conference, Digest of the Technical Papers, p. 196-7.

218. Takagaki, T. and Mukogawa, M., "Gold diffusion transistor logic: a new LSI gate family", 1975 International Electron Devices Meeting, Technical Digest, p. 559-62, EEA80-3536.

219. Wiedmann, S. K. and Berger, H. H., "Bipolar complementary transistor logic (CTL)", 1st European Solid State Circuits Conference - ESSCIRC, Extended Abstracts, 1975, p. 36-7, EEA79-28151.

3. FET Logic
    a. MOS

220. Asai, S., Masuhara, T. and Nakamura, T., "Back-gate-input MOS - a new low-power logic concept", 1976 International Electron Devices Meeting, Technical Digest, p. 185-7.

221. Baranyai, A., "Static and dynamic properties of MOS logic systems", Hiradastechnika, 23 (6), 161-6 (1972), Hungarian, EEA75-31718.

222. Bobenrieth, A., "Logic circuits - employing junction-type field-effect transistors", Patent USA 3969632, Publ. July 1976.

223. Borel, J., Bernard, J. and Suat, J. P., "A depletion load self-aligned technology", Solid-State Electron., 16 (12), 1377-81 (1973), CCA9-6181.

224. Bostock, D., "MNOS technology and non-volatile logic", Microelectron. & Reliab., 14 (4), 359 (1975), EEA79-13029.

225. Cook, P. W., Critchlow, D. L. and Terman, L. M., "Comparison of MOSFET logic circuits", IEEE J. Solid-State Circuits, SC-8, 348-55 (1973), EEA76-34918.

226. Dabrowski, G., "Active components in electronics. XIX", Elektroniker, 14 (10), EL33-8 (1975), CCA11-6342.

227. Decelereq, M. J. and Laurent, T., "A theoretical and experimental study of DMOS enhancement/depletion logic", IEEE J. Solid-State Circuits, SC-12 (3), 264-70 (1977), EEA80-24997.

228. Eichrodt, D., "Rewarding application of transmission gates in MOS-LSI circuits", Siemens Forsch. & Entwicklungsber, $\underline{5}$ (6), 324-6 (1976).

229. Ellul, J. P., Copeland, M. A. and Chan, C. H., "MOS capacitor pull-up circuits for high-speed dynamic logic", IEEE J. Solid-State Circuits, SC-$\underline{10}$ (5), 298-307 (1975), EEA78-37619.

230. Elmasry, M. I., "An optimal two-phase M.O.S.-LSI logic system", International Electrical, Electronics Conference and Exposition Digest, 1973, p. 136-7, EEA77-9715.

231. Gerlach, A., "MOS technology. II. Some distinctions and applications", Funkschau, $\underline{47}$ (25), 58-62 (1975), German, EEA79-16876.

232. Hayashi, Y., Koyanagi, T. and Taru, Y., "Propagation time and design optimization of complementary MOS logic circuits", Bull. Electrotech. Lab., $\underline{33}$ (6), 588-602 (1969), Japanese, CCA5-8293.

233. Hebenstreit, E., "Design of integrated, digital MOS-circuits for minimal transition times", Nachrichtentech. Z. (NTZ), $\underline{28}$ (12), 418-20 (1975), CCA11-6334.

234. Josephy, R. D., "Short channel MOS transistors for integrated logic circuits", Presented at International Electron Devices Meeting, Washington, D. C., October 1970, CCA6-7996.

235. Leuenberger, F., "Monolithic integrated MOS circuits", Tech. Mitt. PTT, $\underline{51}$ (2), 61-7 (1973), French, EEA76-12677.

236. Lilen, H., "Towards m.o.s. logic circuits as fast as the bipolar ones", Electron. & Microelectron. Ind., no. 146, 17-25 (1971), French, CCA6-22184.

237. Liu, T. K., "Synthesis of multilevel feed-forward MOS networks", IEEE Trans. Comput., C-$\underline{26}$ (6), 581-8 (1977), CCA12-15427.

238. Masaki, A., Kadono, S., Fujiwara, O. and Chiba, T., "Comparison of MOS basic logic circuits", Syst. Comput. Control, $\underline{6}$ (4), 11-19 (1975), CCA12-1411.

239. Naumov, Y. Y., "Pulse noise immunity of MOS logic circuits", Izv. VUZ Radioelektron., $\underline{14}$, 1382-4 (1971), Russian, EEA75-12667.

240. Odell, A. D. and Arton, K. A. M., "A system of synchronous MOS logic using a single phase clock", 1st European Solid State Circuits Conference - ESSCIRC, Extended Abstracts, 1975, p. 62-3, EEA79-28154.

241. Ohta, K., Morimoto, M., Saitoh, M., Fukuda, T., Morino, A., Shimizu, K., Hayashi, Y. and Tarui, Y., "A high-speed logic LSI using diffusion self-aligned enhancement depletion MOS IC", IEEE J. Solid-State Circuits, SC-$\underline{10}$ (5), 314-22 (1975), EEA78-37621.

242. Orton, D. W. R., "A simple MOS design example", In: MOS integrated circuit design, Wolfendale, E., (Ed.), Butterworth, London, England, p. 93-115 (1973), CCA8-13631.

243. Puri, Y., "A Monte Carlo based circuit-level methodology for algorithm design of MOS LSI static random logic circuits", IEEE J. Solid-State Circuits, SC-$\underline{12}$ (5), 560-5 (1977).

244. Rees, H., "Low-temperature FET for low-power high-speed logic", Electron. Lett., $\underline{13}$ (6), 156-8 (1977), CCA12-10595.

245. Reynolds, F. H. and Morton, W. D., "Metal-oxide-semiconductor (MOS) integrated circuits. II. Simple logic circuits", Post Off. Elec. Eng. J., _63_, Pt. 2, 105-12 (1970), EEA74-13260.

246. Rodgers, T. J. and Meindl, J. D., "VMOS: high speed TTL compatible MOS logic", IEEE J. Solid-State Circuits, SC-_9_ (5), 239-50 (1974), CCA9-23184.

247. Shenton, G., "Problems of optimum partitioning the logic in MOS systems", Contr. & Instrum., _4_, 41-3 (1972), EEA75-39089.

248. Soderman, L., "How MOS circuits function. V. Good MOS circuits are dynamic", Eltek. Aktuell Elektron., _77_ (4), 46-7 (1974), EEA77-26035.

249. Tozer, R. C., "Sample and hold gates using field effect transistors", Electron, _48_ (577), 47, 49 (1976).

250. Verjans, J. R. and van Overstraeten, R. J., "Nendep - a simple n-channel mos technology for logic circuits", IEEE J. Solid-State Circuits, SC-_10_ (4), 212-18 (1975).

251. Verjans, J. R. and van Overstraeten, R. J., "Electrical characteristics of boron-implanted n-channel mos transistors for use in logic circuits", IEEE Trans. Electron Devices, ED-_22_ (10), 862-8 (1975).

b. CMOS

252. Altman, L., "Logic, memory, I/O chips build computer systems (CMOS)", Electronics, _48_ (10), 85-8 (1975), EEA78-23709.

253. Anon., "CMOS prices plummet - will it topple TTL?", JEE, no. 105, 30-2 (1975), CCA11-4690.

254. Anon., "C-MOS Schmitt-trigger with adjustable hysteresis", Radio Elektron. Schau, _52_ (8), 34 (1976), German, EEA79-45628.

255. Atkinson, R. F., "C-MOS reset circuit ignores brief outages", Electronics, _49_ (16), 111-12 (1976), EEA79-37138.

256. Birchel, R., "CMOS - a logic family with (almost) ideal properties", Bauelem. Elektrotech., _8_ (5), 46-60 (1973), German, EEA76-31860.

257. Birchel, R., "The effects of interference on CMOS circuits compared to bipolar logic families", Bauelem. Elektrotech., _10_ (3), 46, 48-50, 52, 54-6 (1975), German, EEA78-28842.

258. Bishop, A., "C-MOS - past, present and future", New Electron., _7_ (4), 70-1, 73, 75 (1974), CCA9-14286.

259. Bishop, R. A., "Complementary m.o.s. offers many advantages to the digital-system designer", Electron. Engineering, _44_ (537), 67-70 (1972), CCA8-2503.

260. Bishop, R. A., "COS/MOS technology simplifies logic circuits", Elektron. Ind., _4_ (12), 288-9 (1973), German, EEA77-18281.

261. Blanford, D. and Bishop, A., "Power supplies for complementary m.o.s. circuitry", Electron Power, _21_ (4), 247-8 (1975).

262. Burgess, R. R. and Daniels, R. G., "C/MOS unites with silicon gate to yield micropower technology", Electronics, _44_ (18), 38-43 (1971), CCA6-22183.

263. Calebotta, S., "CMOS: the ideal family of logic units", Elettron.
     Oggi, no. 10, 1367-76 (1976), Italian, CCA12-12793.
264. Cergei, L., "Power dissipation in CMOS logic", Electron. Equip.
     News, 29-3 (1975), EEA79-1038.
265. Charles, F., "The contribution of CMOS transistors to majority
     logic circuits", Electron. & Microelectron. Ind., no. 214,
     45-9 (1976), French, CCA11-12254.
266. Choras, R., "CMOS logic systems I", Wiad. Telekomun., $\underline{15}$ (7-8),
     13-18 (1975), Polish, EEA79-4771.
267. Choras, R., "Logic systems with CMOS integrated circuits, II",
     Wiad. Telekomun., $\underline{15}$ (10), 37-44 (1975), Polish, EEA79-12718.
268. Compton, J., "CMOS an alternative to TTL?", Microelectron. &
     Reliab., $\underline{14}$ (4), 357 (1975), CCA11-9580.
269. Dingwall, A. G. F. and Stricker, R. E., "C^2L: A new high-speed,
     high density bulk CMOS technology", 1976 International Electron
     Devices Meeting, Technical Digest, p. 188-91.
270. Douglas, E. C. and Dingwall, A. G. F., "Surface doping using
     ion implantation for optimum guard layer design in COS/MOS
     structures", IEEE Trans. Electron Devices, ED-$\underline{22}$ (10), 849-57
     (1975).
271. Foltz, J. and Musa, F., "Computer-based design of complementary
     MOS-logic", Elektron. Ind., $\underline{6}$ (7-8), 148-51 (1975), German,
     CCA11-4702.
272. Griffin, R. T., "Buffered and unbuffered CMOS gate circuits -
     a comparison", Can. Electron. Eng., $\underline{19}$ (3), 26-9 (1975).
273. Griffin, R. T., "Are buffered CMOS gates right for you?",
     Electron. Des., $\underline{23}$ (15), 52-5 (1975).
274. Grimes, D. W., "Ternary CMOS logic device", IBM Tech. Disclosure
     Bull., $\underline{17}$, 1145-6 (1974), EEA78-5070.
275. Halligan, J., "CMOS: a new logic type for control systems",
     Contr. Eng., $\underline{19}$, 68-70 (1972), EEA75-31701.
276. Harrison, L., "CMOS: The tolerant logic", Electron, $\underline{27}$ (13-19),
     38 (1973), EEA76-35351.
277. Henderson, R. S., "CMOS in today's economic climate", Micro-
     electron. & Reliab., $\underline{14}$ (4), 355 (1975), CCA11-9579.
278. Kalin, W. F., "CMOS logic elements interface easily", Electron.
     Des., $\underline{21}$ (5), 66-70 (1973), CCA8-13603.
279. Kalthoff, M., "Dynamic CMOS techniques save space", Elektronik,
     $\underline{23}$ (7), 264-5 (1974), German, CCA9-21953.
280. Karstad, K., "CMOS for general-purpose logic design", Comput.
     Des., $\underline{12}$ (5), 99-106 (1973), EEA76-29308.
281. Mouftah, H. T. and Jordan, I. B., "Implementation for 3-valued
     logic with COS MOS integrated circuits", Electron. Lett., $\underline{10}$,
     441-2 (1974), EEA78-2505.
282. Muller, K. D., "Reducing the power consumption of CMOS circuits
     by logic structural means", 6th International Congress on
     Microelectronics, 1974, 14 pp., German, EEA78-24524.

283. Phillips, D. H., "CMOS/SOS NAND gate sapphire photocurrent compensation", IEEE Trans. Nucl. Sci., 22 (6), 2617-20 (1975), CCA11-17345.
284. Santoni, A. and Trolese, G., "Design ideas with COS/MOS", New Electron., 7 (9), 27, 29, 30, 33, 34 (1974), CCA9-18332.
285. Turinsky, G., "Connection of digital CMOS circuits with other logic families", Radio Fernsehen Electron., 26 (3), 76-8 (1977), German, EEA80-28404.
286. Venmans, F. A. C. M., "Operation and application of complementary metal oxide semi-conductor integrated circuits", Polytech. Tijdschr. Elektrotech. Elektron., 32 (1), 2-11 (1977), Dutch, EEA80-9525.
287. Vittoz, E. and Oguey, H., "Complementary dynamic M.O.S. logic circuits", Electron. Lett., 9 (4), 77-8 (1973), CCA8-11565.
288. Walker, W., "Examining worst-case fan-out of standard C-MOS buffers", Electronics, 47 (13), 125 (1974), CCA9-20641.
289. Wallace, J., "Designing control logic with c.m.o.s.", New Electron., 7 (16), 20-1, 25 (1974), EEA77-38496.
290. Wojslaw, C. F., "Use CMOS to simplify your integrator design", EDN, 21 (2), 84, 86 (1976).
291. Yates, V., "The current trend in CMOS technology", Electron. Equip. News, 15 (7), 37-9 (1973), CCA9-6175.

c. CCD/CTD

292. Mok, T. D. and Salama, C. A. T., "Logic array using charge-transfer devices", Electron. Lett., 8 (20), 495 (1972).
293. Mok, T. D. and Salama, C. A. T., "M.O.S. bucket-brigade AND/OR logic array", Electron. Lett., 9 (6), 135-6 (1973), EEA76-19442.
294. Mok, T. D. and Salama, C. A. T., "A charge-transfer-device logic cell", Solid-State Electron., 17 (11), 1147 (1974).
295. Upadhyayula, L. C., "Trigger sensitivity of transferred electron logic devices", IEEE Trans. Electron Devices, ED-23 (9), 1049-52 (1976), EEA79-41485.
296. Yanai, H., Suzuki, N. and Yonei, K., "Threshold logic using charge-coupled devices", Annu. Rep. Eng. Res. Inst. Fac. Eng. Univ. Tokyo, 32, 127-32 (1973), CCA9-15990.
297. Zimmerman, T. A., Allen, R. A. and Jacobs, R. W., "Digital charge-coupled logic (DCCL)", IEEE J. Solid-State Circuits, SC-12 (5), 473-84 (1977).

4. Josephson Logic

298. Anacker, W., "Superconducting tunneling circuits for computer applications", 1974 International Electron Devices Meeting, 20th, Technical Digest, p. 5-8.
299. Anon., "Ultra rapid switching using Josephson junctions", Rev. Polytech., no. 4, 331-3 (1976), French, CCA11-22677.

300. Baechtold, W., "Josephson junction complementary circuit", IBM Tech. Disclosure Bull., $\underline{18}$ (3), 919-20 (1975), CCA11-9561.

301. Baechtold, W. and Broom, R. F., "Complementary logic circuit", IBM Tech. Disclosure Bull., $\underline{18}$ (3), 921-2 (1975), CCA11-9573.

302. Basavaiah, S. and Broom, R. F., "Characteristics of in-line Josephson tunneling gates", IEEE Trans. Magn., MAG-$\underline{11}$ (2), 759-62 (1975).

303. Beha, H., "Asymmetric 2-Josephson-junction interferometer as a logic gate", Electron. Lett., $\underline{13}$ (7), 218-20 (1977), CCA12-15365.

304. Chan, H. W. and Duzer, T. V., "Josephson nonlatching logic circuits", IEEE J. Solid-State Circuits, SC-$\underline{12}$ (1), 73-9 (1977), CCA12-3816.

305. Fang, F. and Matisoo, J., "Self-resetting scheme for terminated Josephson junction circuits", IBM Tech. Disclosure Bull., $\underline{16}$ (9), 3068-9 (1974), CCA9-16004.

306. Fulton, T. A., "Some aspects of the dynamics of Josephson junction circuits and devices", IEEE Trans. Magn., MAG-$\underline{11}$ (2), 749-50 (1975).

307. Fulton, T. A., Magerleing, J. H. and Dunkleberger, L. N., "A Josephson logic design employing current switched junctions", IEEE Trans. Magn., MAG-$\underline{13}$ (1), 56-8 (1977), CCA12-10602.

308. Gueret, P., Mohr, T. O. and Wolf, P., "Single flux-quantum memory cells", IEEE Trans. Magn., MAG-$\underline{13}$ (1), 52-5 (1977).

309. Herrell, D. J., "Femtojoule Josephson tunneling logic gates", IEEE J. Solid-State Circuits, SC-$\underline{9}$ (5), 277-82 (1974), CCA9-23187.

310. Herrell, D. J., Landman, B. S. and Zappe, H. H., "Memory compatible logic scheme for Josephson tunneling memories", $\underline{\text{IBM}}$ Tech. Disclosure Bull., $\underline{18}$ (5), 1607-9 (1975), CCA11-4823.

311. Jutzi, W. W., "Self-resetting Josephson junction", IBM Tech. Disclosure Bull., $\underline{16}$ (9), 3031-2 (1974), CCA9-18304.

312. Klein, M., "Josephson tunneling logic gate with thin electrodes", IEEE Trans. Magn. MAG-$\underline{13}$ (1), 59-62 (1977), CCA12-10603.

313. Lofstrom, K. H. and Van Duzer, T., "Josephson logic circuit with a sinusoidal current supply", IEEE Trans. Magn., MAG-$\underline{13}$ (1), 597-600 (1977), CCA12-10613.

314. Lum, W. Y., Chan, H. W. and Van Duzer, T., "Memory and logic circuits using semiconductor-barrier Josephson junctions", IEEE Trans. Magn., MAG-$\underline{13}$ (1), 48-51 (1977).

315. Nakajima, K. and Onodera, Y., "Logic circuits using Josephson junction lines", Oyo Buturi, $\underline{45}$ (8), 779-85 (1976), Japanese, EEA80-3025.

316. Schlig, E. S., "Terminated line logic circuits for Josephson cryotrons", IBM Tech. Disclosure Bull., $\underline{16}$ (10), 3435-6 (1974), CCA9-21959.

317. Schlig, E. S., "A proposed distributed Josephson logic circuit", IEEE J. Solid-State Circuits, SC-$\underline{11}$ (3), 424-6 (1976), CCA11-22550.

318. Yao, Y. L., "Electrical characteristics of a two-level single
     ground-plane interconnection in Josephson tunneling logic", IEEE
     Trans. Parts, Hybrids & Packag., PHP-12 (13), 236-40 (1976),
     CCA12-1398.
319. Zappe, H. H., "Quantum interference Josephson logic devices",
     Appl. Phys. Lett., 27 (8), 432-4 (1975), CCA11-4688.
320. Zappe, H. H., "Josephson quantum interference computer devices",
     IEEE Trans. Magn., MAG-13 (1), 41-7 (1977), CCA12-10600.

          5.   Schottky Logic

321. Barnes, J. J., Lomax, R. J. and Haddad, G. I., "Finite-element
     simulation of GaAs MESFET's with lateral doping profiles and
     submicron gates", IEEE Trans. Electron Devices, ED-23 (9), 1042-8
     (1976), EEA79-41557.
322. Bertin, C. L. and Williams, T. A., "FET-Schottky diode logic",
     IBM Tech. Disclosure Bull., 13, 1276 (1970).
323. Bhatia, H. S., Calhoun, H. C., Genin, D. J. and Renbeck, R. B.,
     "Elimination of parasitic currents in transistor-Schottky barrier
     diode circuit", IBM Tech. Disclosure Bull., 19 (7), 2522 (1976).
324. Chang, A. W., "Transistor-diode clamping and biasing circuit",
     IBM Tech. Disclosure Bull., 15, 1085-6 (1972), EEA76-8179.
325. Dhaka, V. A., "Integrated circuits with surface barrier diodes",
     Patent USA 3506893, Publ. April 1970.
326. Dorler, J. and Swietek, D., "Integrated Schottky diode and tran-
     sistor circuit", IBM Tech. Disclosure Bull. 14, 3214-5 (1972),
     EEA75-27742.
327. Fawcette, J., "MSI logic reaches 4.5 GHz can application be far
     off?", Microwave Syst. News, 7 (2), 13-14, 16, 18 (1977),
     EEA80-28389.
328. Heald, R. A. and Hodges, D. A., "Design of Schottky-barrier
     diode clamped transistor layouts", IEEE J. Solid-State Circuits,
     SC-8, 269-75 (1973), EEA76-29312.
329. Ishikawa, H., Kusakawa, H., Suyama, K. and Fukuta, M., "Normally-
     off type GaAs MESFET for low-power, high-speed logic circuits",
     1977 International Solid-State Circuits Conference, Digest of
     Technical Papers, p. 200-1.
330. Jones, R., "Interconnection Schottky", New Electron., 10 (3),
     34-9 (1977), EEA80-28395.
331. Kan, D. T., "Shunt-feedback Schottky clamped logic gates", IEEE
     J. Solid-State Circuits, SC-7, 404-11 (1972), EEA75-35294.
332. Kataoka, S., Hashizume, N., Kawashima, M. and Komamiya, Y.,
     "High field domain functional logic devices with multiple control
     electrodes", Cornell Electronic Engineering Conference, 4th
     Bienn., Proc., 1973, p. 225-34.
333. Mathews, K. F., "Schottky direct-coupled transistor logic circuit",
     6th Asilomar Conference on Circuits and Systems, 1972, p. 410-14,
     Publ. 1973, EEA76-23049.

334. Mich, J. R., "A Schottky m.s.i. four-bit shifter", New Electron.,
     7 (17), 42, 44, 49, 52 (1974), EEA78-378.
335. Mick, J. R., "Using Schottky 3-state outputs in bus-organized
     systems", EDN, 19 (23), 35-9 (1974).
336. Muta, H., Suzuki, S., Yamada, K., Nagahashi, Y., Tanaka, T.,
     Okabayashi, H. and Kawamura, N., "Femto joule logic circuit
     with enhancement-type Schottky barrier gate FET", IEEE Trans.
     Electron Devices, ED-23 (9), 1023-7 (1976), CCA11-28102.
337. Nakamura, M., Kodera, H. and Migitaka, M., "Computer study on
     GaAs Schottky barrier IMPATT diodes", Solid-State Electron.,
     16 (6), 63-7 (1973), EEA76-22927.
338. Nawata, K., Ikeda, M. and Ishii, Y., "Millimeter-wave GaAs
     Schottky-barrier IMPATT diodes", IEEE Trans. Electron Devices,
     ED-21 (1), 128-30 (1974), EEA77-13858.
339. Picquendar, J. E., "Nanoelectronic project and ASTEC logic",
     J. Vac. Sci. & Technol., 10 (6), 1132 (1973), CCA9-15992.
340. Sugera, T. and Yanai, H., "Logic and memory applications of
     the Schottky-gate Gunn-effect digital device", Proc. IEEE, 60
     (2), 238-40 (1972), CCA7-8229.
341. Suzuki, S., Muta, H., Tanaka, T., Yamada, K., Nagahashi, Y.,
     Okabayashi, H. and Kawanura, N., "Logic circuits with 2 μm gate
     Schottky barrier FETs", Oyo Buturi, 44 (Suppl.), 219-24 (1975),
     EEA78-41684.
342. Van Tuyl, R. L. and Liechti, C. A., "High-speed integrated logic
     with GaAs MESFET's", IEEE J. Solid-State Circuits, SC-9 (5),
     269-76 (1974), EEA77-38469.
343. Van Tuyl, R. and Liechti, C., "Gallium arsenide spawns speed",
     IEEE Spectrum, 14 (3), 41-7 (1977).
344. Van Tuyl, R. L., Liechti, C. A., Lee, R. E. and Gowen, E.,
     "GaAs MESFET logic with 4-GHz clock rate", IEEE J. Solid-State
     Circuits, SC-12 (5), 485-96 (1977).
345. Yanagisawa, S., Wada, O. and Toyama, Y., "Integrated Gunn-effect
     logic circuits for gigabit rate operation", Fujitsu Sci. & Tech.
     J., 12 (4), 129-42 (1976), CCA12-8801.

6. Others

346. Bennett, L. A. M., "Improved forms of threshold-logic function
     implementation", Electron. Lett., 13 (12), 368-70 (1977).
347. Goto, G., Nakamura, T., Hasuo, S., Kazetani, K. and Isobe, T.,
     "Gunn-effect logic device using transverse extension of a high
     field domain", IEEE Trans. Electron Devices, ED-23 (1), 21-7
     (1976), CCA11-6333.
348. Herrell, D. J., "High-speed, wide-margin, logic family", IBM
     Tech. Disclosure Bull., 16 (9), 3052-5 (1974), CCA9-16003.
349. Kazetani, K., Goto, G. and Nakamura, T., "Logic gates using
     Gunn-effect devices", Fujitsu Sci. & Tech. J., 12 (4), 99-188
     (1976), CCA12-8800.

350. Kostsov, E. G. and Mishin, A. I., "Photoelectro-optical logic elements", Avtometriya, no. 4, 28-34 (1976), Russian, CCA12-1397.
351. Prosser, F. and Winkel, D., "Mixed logic leads to maximum clarity with minimum hardware", Comput. Des., $\underline{16}$ (5), 111-17 (1977).
352. Schlegel, E. and Page, D., "Gate-assisted turn-off thyristor with cathode shunts and dynamic gate", 1976 International Electron Devices Meeting, Technical Digest, p. 487-90.
353. Sitnikov, L. S. and Utyakov, L. L., "Multistable logic simplifies man-machine interface", Electronics, $\underline{42}$ (17), 105-8 (1969), CCA5-1353.
354. Upadhyayula, L. C., "Quasienhancement-mode operation of transferred-electron logic devices (T.E.L.D.S)", Electron. Lett., $\underline{12}$ (10), 262-3 (1976), CCA11-17326.
355. Upadhyayula, L. C., Smith, R. E., Wilhelm, J. F., Jolly, S. T. and Paczkowski, J. P., "Transferred electron logic devices (TELDs) for gigabit rate signal processing", 1976 IEEE MTT-S International Microwave Symposium, p. 164-5, CCA11-31485.
356. Whitehead, D. G., "Subnanosecond exclusive-OR logic circuits", Int. J. Electron, $\underline{40}$ (1), 93-6 (1976).

IV.   LOGIC DESIGN

357. Agusa, K. and Kambayashi, Y., "Synthesis of a hazardless NAND network", Syst. Comput. Control, $\underline{6}$ (2), 13-21 (1975), CCA11-28116.
358. Allen, C. M., "Prospects for the future of non-binary digital design", Proceedings of 1975 IEEE Southeastern Region 3 Conference on Electricity and Expanding Technology, Pt. 11, p. 4E-3/1-7.
359. Anon., "Circuit design using AEG Telefunken CNY 36 and CNY 37 modules", Antenna, $\underline{48}$ (7-8), 280-1 (1976), Italian, CCA12-8803.
360. Banks, W. and Majithia, J. C., "Design and application of a modified universal logic module", Proc. IEEE, $\underline{61}$ (5), 687-8 (1973), CCA8-13627.
361. Baugh, C. R. and Wooley, B. A., "Statistical analysis of a differential threshold logic circuit configuration", IEEE Trans. Comput., C-$\underline{25}$ (7), 745-54 (1976).
362. Bell, C. G. and Grason, J., "The register transfer module design concept", Comput. Des., $\underline{10}$ (5), 87-94 (1971), CCA6-17595.
363. Beloiu, D., "Comparison methods of two binary numbers", Autom. & Electron., $\underline{14}$ (5), 224-7 (1970), Rumanian, CCA6-7982.
364. Bennett, L. A. M., "Threshold-logic functions and the exclusive-OR", Electron. Lett., $\underline{13}$ (7), 195-6 (1977), EEA80-16811.
365. Bilous, O. and Turnbull, J., "An approach to the design of custom LSI logic modules", International Conference on Large Scale Integrated Circuits, 1974, p. 199-206, EEA78-11755.
366. Blacksher, R., "Arithmetic logic unit design for an LSI minicomputer", Comput. Des., $\underline{74}$ (4), 96-9 (1975).

367. Blyumin, S. L., Ignatenko, A. D., Mashlykin, V. G. and Cherni-
     kov, Y. V., "Analysis of a typical thyristor logical element",
     Autom. & Remote Control, 33 (4), Pt. 2, 669-73 (1972),
     CCA8-2500.
368. Bonisch, W. and Beuter, R., "Logic circuits with integrated
     NAND gate circuit", Grundig Tech. Inf., 21 (2), 299-301 (1974),
     CCA9-21957.
369. Borisenko, L. I., "Logic integrated microcircuits", Avtom
     Telemekh. Svyaz, no. 12, 16-21 (1972), Russian, CCA8-11558.
370. Braddock, R., Yamanaka, H. and Epstein, G., "Multiple-valued
     logic design and applications in binary computers", In:
     Computer Science and multiple-valued logic. Theory and
     applications, Rine, D. C., (Ed.), North-Holland, Amsterdam,
     Netherlands, p. 35-44 (1977), CCA12-10640.
371. Bradshaw, F. T., "Directed graph models for hardware/software
     design", International Symposium on Computer Hardware Descrip-
     tion Languages and their Applications, 1975, p. 7-15, CCA11-31543.
372. Butyl'skii, Y. T. and Brunchenko, A. V., "An algorithm for cut-
     ting a two-dimensional graph and its use for designing digital
     devices in LSI systems", Avtom. & Vychisl. Tekh., no. 4, 72-6
     (1976), Russian, CCA11-31549.
373. Buzzelli, G. E., "Logical systems, IV. Fundamental elements of
     combinatory logics and their representation by equivalent cir-
     cuits", Industr. Ital. Elettrotec. Elettronica, 22 (4), 333-46
     (1969), Italian, CCA5-2692.
374. Cagliano, D. and Taddei, R., "MOS-D: a program for the place-
     ment of integrated logic gates", Proceedings of the 2nd Inter-
     national Computing Symposium, 1972, p. 335-44, EEA75-31825.
375. Caplener, H. D. and Janku, J. A., "Top-down approach to LSI
     system design", Comput. Des., 13 (8), 143-8 (1974).
376. Cavaliere, J. R., "Logic gate", IBM Tech. Disclosure Bull.,
     13 (5), 1364-5 (1970), CCA6-3729.
377. Chin, W. B., Gruodis, A. J., Kraft, W. R. and Tai, E., "Inte-
     grated Schottky diode-transistor design and layout", IBM Tech.
     Disclosure Bull., 17 (12), 3580-1 (1975), EEA78-28861.
378. Conruyt, P. and Serrand, J. P., "NOR logical gate", Patent UK
     1257116, Publ. December 1971, CCA7-14667.
379. Crowther, G. O. and Emms, E. T., "Inter-relation between LSI
     technology and electronic system design", Tijdschr. Ned.
     Elektron. - & Radiogenoot., 41 (5), 165-9 (1976), Dutch,
     EEA80-21006.
380. Dandapani, R. and Reddy, S. M., "On the design of logic net-
     works with redundancy and testability considerations", IEEE
     Trans. Comput., C-23 (11), 1139-49 (1974), CCA9-23221.
381. Darwood, N., "Evolving an arithmetical aid for sequential-
     logic design", Electron. Eng., 43 (519), 60-3 (1971), CCA6-12644.

382. Davidesko, I., "Designing a microprogrammable control unit for the Remez 4000 digital minicomputer", 9th Convention of Electrical and Electronic Engineers in Israel, 1975, p. B-1-3/1-6, Hebrew, CCA11-31565.

383. De Jong, M. L., Dart, L. and Stokstad, P., "PASCO scientific digital mini-modules, an evaluation (teaching apparatus)", Phys. Teach., 15 (4), 248-51 (1977), EEA80-28401.

384. Dubus, D. and Tosser, A., "Duality of Boolean functions and logic schemes", Automatisme, 21 (11), 354-65 (1976), French, CCA12-15415.

385. Dubus, E. and Tosser, A., "Synthesis of multiple output logic circuits - an algebraic analysis", Int. J. Electron., 40 (6), 11-16 (1976), CCA11-22562.

386. Dunbridge, B., Miller, C. S. and Tsou, H. S., "Building block approach to digital signal processing", IEEE Electron and Aerospace System Convention Record, 1974, p. 469-76.

387. Dunderdale, H., Majithia, J. C. and Pugh, A., "Designing asynchronous counters", Radio & Electron. Eng., 43 (3), 227-8 (1973), CCA8-13615.

388. D'yakonov, V. P. and Sterlyagov, A. A., "High-speed logical circuits based on avalanche transistors", Instrum. & Exp. Tech., 16 (5), Pt. 1, 1389-91 (1973), CCA9-25004.

389. Edwards, C. R. and Hurst, S. L., "An analysis of universal logic modules", Int. J. Electron., 41 (6), 625-8 (1976), EEA80-3014.

390. Elmasry, M. I. and Thompson, P. M., "Logic partition for multi-emitter two-level structures", IEEE Trans. Circuits & Syst., CAS-21 (3), 354-9 (1974), CCA9-14283.

391. Elmasry, M. I., "Logic design using EFL structures", IEEE Trans. Comput., C-25 (9), 952-6 (1976).

392. Feller, A., "Design automation techniques for custom LSI arrays", AGARD Lecture Series, 1975, Paper 7, 15 pp.

393. Fellera, A. and Lombardi, T., "Design, development, fabrication and delivery of register and multiplexer units", Report NASA-CR-123562, RCA, Van Nuys, Calif., 1971, 51 pp., EEA75-35425.

394. Frenzel, L. E., Jr., "Don't overlook compatible NAND and NOR gates in reducing logic", Contr. Eng., 17 (10), 70-2 (1970), CCA6-5592.

395. Frisiani, A. L., "Design of optimal two-level logic networks using NOR, AND, NAND, OR gates", Int. J. Electr. Eng. Educ., 12 (4), 318-21 (1975).

396. Fulcomer, E. J. and Verma, S. P., "Modelling and test generation for a custom LSI chip", 1976 Semiconductor Test Symposium, Digest of Papers, p. 41-4, CCA12-3835.

397. Gramolin, V. V. and Shamrov, M. I., "Design of fast adders using integrated components", Avtom. & Vychisl. Tekh., no. 3, 82-6 (1972), Russian, CCA8-11557.

398. Hanratty, J. J., "Model your ROM with NAND gates when you are using computer-aided design to prove out a digital system", Electron. Des., 24 (20), 88-94 (1976), CCA12-1511.

399. Heath, C., "Design of programmable sequential logic circuits", Electron. Engineering, $\underline{49}$ (588), 45-7, 49 (1977), CCA12-10634.

400. Hemming, C. W., Jr. and Hemphill, J. M., "Digital logic simulation models and evolving technology", 12th Design Automation Conference, Proceedings, 1975, p. 85-94.

401. Henderson, R., "Let the designer beware", Electron Power, $\underline{21}$ (10), 629-30 (1975).

402. Herring, S., Scalf, H. L. and Ford, D. W., "LSI design choices for minimum cost", Electron. Packag. & Prod., $\underline{17}$ (7), 171-7 (1977).

403. Hnatek, E. R., "Switching power supplies: design considerations", Comput. Des., $\underline{16}$ (2), 89-94 (1977), CCA12-10621.

404. Ho, H. H., "Logic model for a thyristor and its application in invertor circuit simulation", Inst. Eng. Aust. Electr. Eng. Trans., EE$\underline{11}$ (2), 68-72 (1975), CCA11-14963.

405. Holdsworth, B. and Zissos, L., "Logic design. I. Boolean algebra and Karnaugh maps", Wireless World, $\underline{83}$ (1493), 51-4 (1977), EEA80-9033.

406. Holdsworth, B., "Logic design II. Combinational logic", Wireless World, $\underline{83}$ (1494), 49-53 (1977), CCA12-10638.

407. Holdsworth, B. and Zissos, D., "Logic design III. Event-driven circuits", Wireless World, $\underline{83}$ (1495), 49-53 (1977), CCA12-10639.

408. Hu, S. C., "Probabilistic approach of designing more reliable logic gates with asymmetric input faults", IEEE Trans. Comput., C-$\underline{24}$ (10), 1012-14 (1975).

409. Hurst, S. L., "Improvements in circuit realisation of threshold-logic gates", Electron. Lett., $\underline{9}$ (6), 123-4 (1973), CCA8-11566.

410. Hurst, S. L., "Logic network synthesis using digital-summation threshold-logic gates", Microelectronics, $\underline{6}$ (4), 42-8 (1975).

411. Hurst, S. L., "Application of multioutput threshold-logic gates to digital-network design", Proc. Inst. Electr. Eng., $\underline{123}$ (2), 128-34 (1976).

412. Hurst, S. L., "More powerful logic will change design philosophy", Electron. Engineering, $\underline{48}$ (576), 53-5 (1976), EEA79-12463.

413. Iser, D. A., Jurison, J. and Russell, B. J., "Design considerations for aerospace digital computers", Comput. Des., $\underline{13}$ (8), 113-23 (1974).

414. Ishizuka, O., "A procedure for the design of Esaki-diode-transistor multithreshold logic circuits", Electron. & Commun. Jap., $\underline{54}$ (8), 143-4 (1971), CCA8-2504.

415. John, S., "A graphical analysis of the transients in logic interconnections", Thesis, Univ. Coll. Swansea, Wales, 1976, EEA80-3038.

416. Johnson, D. W., "Go from flow chart to hardware", Electron. Des., $\underline{24}$ (18), 90-5 (1976), CCA12-1510.

417. Kamayashi, Y., "Optimum logic design using memory-type arrays", Proceedings of the Symposium on Uniformly Structured Automata and Logic, 1975, p. 199-205, CCA11-31547.

418. Katz, D., "Logic gate arrays: 'catalog' IC's with custom advantages", Bell Lab. Record, 55 (7), 187-91 (1977).

419. Kellerman, Y. I. and Cherkashenko, M. V., "Analysis of logic circuits in control systems for automated machines", Mekh & Avtom. Proiz., no. 7, 39-41 (1976), Russian, CCA11-31494.

420. Kohn, K. S., "A study on the synthesis of logic networks with negative gates", J. Korean Inst. Electron Eng., 9 (6), 297-304 (1972), Korean, CCA8-13613.

421. Knutrud, T., Kusko, A. and Cain, J. J., "Designing reliability into equipment having power semiconductors", Electronics, 49 (5), 111-15 (1976).

422. Lagovier, B. A. and Lapina, T. A., "Design algorithm for automata with a complex register serving as memory", Mekh. & Avtom. Upr., no. 2, 58-61 (1976), Russian, CCA11-22643.

423. Laleuf, M., "Positive and negative logic and duality", Onde Electr., 57 (4), 312-15 (1977), French, EEA80-28399.

424. Lallier, K. W. and Savkar, A. D., "Relating logic design to physical geometry in LSI chip", IBM Tech. Disclosure Bull., 19 (6), 2140-3 (1976), EEA80-28899.

425. Langheld, E., "Introduction to threshold and majority gate logic II.", Elektronik, 25 (2), 73-8 (1976), German, CCA11-14959.

426. Launee, R., "High-speed current-mode threshold gate with the clamps and current comparison", Electron. Lett., 13 (4), 99-100 (1977), EEA80-16810.

427. Leck, R. P., "Logic gates and LED indicate phase lock", Electronics, 47 (11), 106 (1975).

428. Levin, V. A. and Nisnevich, D. G., "Some problems in digital frequency synthesizer design", Telecommun. & Radio Eng. Pt. 1, 30 (4), 44-50 (1976), EEA80-16769.

429. Lewin, D. W., "Advanced aspects of asynchronous logic design", Comput. J., 14 (3), 254-9 (1971), CCA6-22182.

430. Liu, T. K., "Synthesis of logic networks with (metal oxide semiconductors) complex cells", Report PB-210108, Univ. Illinois, Urbana, 1972, 232 pp., EEA76-8132.

431. Lokos, S., "Only one line is true detector for digital parallel lines", Electron. Engineering, 49 (587), 29 (1977), CCA12-6577.

432. Majithia, J. C., "Some comments concerning design of pipeline arithmetic arrays", IEEE Trans. Comput., C-25 (11), 1132-4 (1976), CCA11-31531.

433. Marcelius, A., "The connection of inductive loads to logic circuits", Und-oder-Nor & Steurungstech., no. 1-2, 27-9 (1977), German, EEA80-28410.

434. Masset, G. J. and Tosser, A. J., "Minimization of NAND-NOR networks by factoring", Int. J. Electron., 41 (6), 525-56 (1976), EEA80-3013.

435. Meggitt, G. C., "TTL system programmer for cyclic activation analysis", Nucl. Instrum. Methods, 124 (2), 455-9 (1975).

436. Moto-Oka, T., Kurachi, T., Shiino, T. and Sugimoto, M., "Logic design system in Japan", 12th Design and Automation Conference, Proceedings, 1975, p. 241-50.

437. Mouftah, H. T., "A study on the implementation of three-valued logic", 6th International Symposium on Multi-Valued Logic, Technical Digest, 1976, p. 123-6.

438. Muehldorf, E. I., "Designing LSI logic for testability", 1976 Semiconductor Test Symposium, Digest of Papers, 1976, p. 45-9.

439. Murugesan, S., "Universal logic gate and its applications", Int. J. Electron., 42 (1), 55-63 (1977), CCA12-3819.

440. Nordmann, B. J., Jr., "Modular asynchronous control design", IEEE Trans. Comput., C-26 (3), 196-207 (1977), CCA12-8858.

441. Ogus, R. C., "Design and evaluation of ultra-reliable hybrid redundant digital systems", Thesis, Stanford University, California, 1975, 200 pp., Order No. 75-13569.

442. Oswald, R. S., Laurance, N. L. and Devlin, S. S., "Design considerations for an onboard computer system", SAE Special Publication, No. 393, 1975, p. 79-84.

443. Parhami, B., "Design of self-monitoring logic circuits for application in fault-tolerant digital systems", IEEE International Symposium on Circuits and Systems, Proceedings, 1976, p. 57-60, CCA12-2884.

444. Pedersen, R. J., "Overlapped next addressing in a control storage system", IBM Tech. Disclosure Bull., 19 (2), 602-4 (1976), CCA12-1453.

445. Pessen, D., "Design method for near-optimal NOR logic circuits", ASME Pap., Paper 76-WA/Flcs-4, 1976, 9 pp.

446. Philips Electronic & Assoc. Ind. Ltd., "Logic circuit building block", Patent UK 1283623, Publ. August 1972, CCA8-2523.

447. Pratapa Reddy, V. C. V. and Neelakantaswamy, P. S., "Note on digital-summation threshold-logic gates", Proc. Inst. Electr. Eng., 121 (10), 1085-6 (1974), CCA9-25009.

448. Procek, E., "Multifunction gate", New Electron., 9 (5), 14 (1976), CCA11-17351.

449. Puncochar, J., "Realisation of logic functions IF and EXCLUSIVE-OR by a diode bridge and transistors", Sdelovaci Tech., 24 (12), 453-4 (1976), Czech, CCA12-6580.

450. Rallapalli, K. and Wilnai, D., "Programmed logic - A valuable hardware design tool", 12th IEEE Computer Society International Conference, COMPCON 76, p. 158-9.

451. Rony, P. R., Titus, J. A. and Larsen, D. G., "What is a logical instruction", Am. Lab., 9 (2), 158-60, 62-3 (1977), CCA12-15417.

452. Russo, R. L., "Partitioning for LSI: on the tradeoff between logic performance and circuit-to-pin ratio", 1971 IEEE International Convention Digest, p. 376-7, CCA6-17554.

453. Santoni, A., "Digital systems spawn new tasks in measurement", Electronics, 49 (22), 100-2, 105-6 (1976), CCA12-1402.

454. Schulz, J., "Describing digital component groups using ortho-
     gonal function transforms", Z. Elektr. Inf.- & Energietech.,
     6 (2), 171-6 (1976), German, CCA11-22548.
455. Semenov, A. T. and Yakubovich, S. D., "Calculation of steady-
     state characteristics of a many-resonator injection laser
     [logic system]", Sov. J. Quantum Electron., 4 (1), 100-2 (1974),
     CCA9-25015.
456. Sholl, H. A. and Yang, S., "Design of asynchronous sequential
     networks using read-only memories", IEEE Trans. Comput., C-24
     (2), 196-206 (1975).
457. Solter, T. W. and Carr, W. N., "Automated breadboard approach
     to prototype circuit design", Proceedings of the 27th Annual
     ISA Conference, 1972, Paper 601, 4 pp., CCA8-11581.
458. Stewart, A., "Thermal considerations in digital circuit design",
     New Electron., 9 (20), 56, 58, 64, 68 (1976), EEA80-285.
459. Suprun, V. P., "Synthesis of logic circuits with linear com-
     plexity", Avtom. & Vychisl. Tekh., no. 5, 22-5 (1976),
     Russian, CCA12-3865.
460. Thomas, T., Rombeck, H., Gilmour, W. and Caughey, M., "Logi-
     flex - faster design for integrated circuits", Telesis, 3 (3),
     72-8 (1973), EEA77-14605.
461. Thompson, R. A., "Computer aided design instruction of combina-
     tional logic", Proceedings of the 1975 IEEE Southeastern Region
     3 Conference on Electricity and Expanding Technology, Pt. I,
     p. 21E-4/1-3, CCA11-31540.
462. Tolmacheva, A. Y., "Efficiency of design algorithms for digital
     devices", Autom. & Remote Control, 37 (9), 1437-44 (1976),
     CCA12-15416.
463. Tose, D., "Digital logic board design with test needs in mind",
     Electron. Engineering, 48 (586), 73-5 (1976), EEA80-6248.
464. Tose, D., "Digital logic board design with test needs in mind.
     II.", Electron. Engineering, 49 (878), 46, 49 (1976), EEA80-9004.
465. Tose, D., "Design circuits for testability to save time and cut
     bottlenecks", EDN, 22 (10), 95-8 (1977), EEA80-28375.
466. Tosser, A. J., "Opportunity of inhibition to optimize NAND-OR
     networks", Comput. & Electr. Eng., 3 (3), 281-92 (1976),
     CCA12-3867.
467. Toy, W. N., "Modular LSI control logic design with error detec-
     tion", IEEE Trans. Comput., C-20 (2), 161-6 (1971), CCA6-8006.
468. Ustyuzhaninov, V. N., "Estimating the effects of redundancies
     on the technical characteristics of digital integrated elec-
     tronics", Izv. VUZ Priborostr., 20 (4), 122-7 (1977), Russian,
     EEA80-28385.
469. Veshkurtsev, Y. M. and Bronshtein, B. G., "Distortion of phase
     distribution laws by a digital delay line", Izv. VUZ Radio-
     elektron., 19 (12), 55-8 (1976), Russian, EEA80-24397.

470. Volkogon, V. P., Korncichuk, V. I., Molchanov, A. A. and Teslenko, A. K., "Logical modelling of large integrated circuits using algorithmic FORTRAN language", Isv. VUZ Radioelektron., 19 (6), 44-50 (1976), Russian, CCA11-31550.

471. Von Puttkamer, E., "Simple hardware buddy system memory allocator", IEEE Trans. Comput., C-24 (20), 953-7 (1975).

472. Wada, K., "LSI gate placement program ALPS", Rev. Electr. Commun. Lab., 24 (5-6), 383-9 (1976).

473. Waggener, W. N., "Designer's guide to: digital synchronization circuits. I.", EDN, 21 (14), 56-61 (1976), CCA12-3840.

474. Waggener, W. N., "Designer's guide to: digital synchronization circuits. II.", EDN, 21 (15), 75-82 (1976), CCA12-6583.

475. Waggener, W. N., "Designer's guide to digital synchronization circuits. III.", EDN, 21 (16), 99-105 (1976), CCA12-6603.

476. Wakerly, J. F., "Documentation standards clarify design", Comput. Des., 16 (2), 75-85 (1977), CCA12-10633.

477. Warn, P., "Logic design to layout", In: MOS integrated circuit design, Wolfendale, E., (Ed.), Butterworth, London, England, p. 49-64 (1973), CCA8-13630.

478. Weller, D. R., "A coordinated building block for systems design", 9th Annual Connector Symposium, 1976, p. 357-68, CCA12-10642.

479. Winkel, D. and Prosser, E., "Hardware design style: the vital element", Behav. Res. Methods & Instrum., 8 (2), 113-7 (1976), CCA11-31523.

480. Winn, G. C. E., "Digital approach to the efficient synthesis of threshold gates", Comput. J., 18 (3), 239-42 (1975).

481. Wittenzeliner, E., "Computer-aided design of large-scale integrated I^{2}L logic circuits", IEEE J. Solid-State Circuits, SC-12 (2), 199-204 (1977), EEA80-9019.

V.   TESTING

482. Adham, M. and Friedman, A., "Digital system fault diagnosis", J. Des. Autom. & Fault-Tolerant Comput., 1 (2), 115-32 (1977), CCA12-8863.

483. Agrawal, P. and Agrawal, V. D., "On Monte Carlo testing of logic tree networks", IEEE Trans. Comput., C-25 (6), 664-7 (1976).

484. Akers, S. B., "Partitioning for testability", J. Des. Autom. & Fault-Tolerant Comput., 1 (2), 133-46 (1977), CCA12-8864.

485. Allen, D. P., "Logic tester uses single trial procedure to troubleshoot digital IC boards", Electronics, 46 (23), 89-93 (1973), EEA77-3750.

486. Anderson, A. G., "Function tester for TTL components", Elektronik, 24 (10), 106 (1975), German, CCA11-4700.

487. Anon., "Constructing a logic analyser", Electron. & Microelectron. Ind., no. 227, 29-31 (1976), French, CCA12-6598.

488. Anon., "Automatic switching network tester", Electron. & Microelectron. Ind., no. 227, 55 (1976), French, CCA12-6599.

489. Anon., "Novel aids for detecting faults in digital circuits", Sdelovaci Tech., 24 (12), 450 (1976), Czech, CCA12-8842.

490. Anon., "An automatic testing system MB2410S", Electron. & Appl. Ind., no. 231, 42-3 (1977), French, EEA80-28377.

491. Anon., "TAU1: a tool for logic circuit testing", Rev. Tech. Thomson-CSF, 9 (1), 127-48 (1977), EEA80-28407.

492. Aucouturier, J. L., Dom, J. P., Lacroix, G. and Riviere, J. J., "Characterisation by energy of short duration noise in logic systems. I. Method of measuring the parasitic energy on a connection between logic circuits", Onde Electr., 56 (10), 397-402 (1976), French, CCA12-1410.

493. Barbasov, V. M., "Measurement of the average delay time of signal propagation in high-speed logic elements", Meas. Tech., 16 (8), 1125-7 (1973), CCA9-18301.

494. Beckwith, J. F., "Current tracer: A new way to find low-impedance logic-circuit faults", Hewlett-Packard J., 28 (4), 2-8 (1976), CCA12-8810.

495. Beckwith, J. and Bronson, B., "Tracing current by inductive pickup tracks logic faults precisely", Electronics, 49 (24), 106-8 (1976), EEA80-6250.

496. Belforte, P., Colonnelli, U. and Guaschino, G., "Use of digital wave filters to simulate the interconnections between high speed logical devices", Alta Freq., 45 (11), 649-50 (1976), Italian, CCA12-8802.

497. Beville, B., "Troubleshoot parallel logic with word recognizers", Electron Prod., 17 (9), 19-20 (1975).

498. Blunden, D. F., Boyce, A. H. and Lawson, D. J., "Some aspects of testing logic circuits", Digital Processes, 1 (2), 171-6 (1975).

499. Bonar, R. G. and Millham, E. H., "Dynamic logic tester for large-scale integration", IBM Tech. Disclosure Bull., 27, 1426-7 (1974), EEA78-8616.

500. Borchert, M., "A 16-channel logic analyzer", Tekscope, 7 (4), 2-7 (1975), CCA11-17355.

501. Bouricius, W. G., Carter, W. C., Jessep, D. C., Jr. and Wadia, A. B., "Self-testing logic functions for control of encoder & self-repairing computer subunits", IBM Tech. Disclosure Bull., 14 (1), 276-80 (1971), CCA6-22194.

502. Bowman, J. A., "Short and long term stability measurements using automatic data recording system", Symposium on Frequency Control, 27th Annual, Proceedings, Paper, 1973, p. 440-5.

503. Bray, D. W., "Digital network test generation using t-expressions", Digital Processes, 2 (3), 181-207 (1976), CCA12-3870.

504. Bray, D. W., "Partitioning for test generation of large synchronous sequential circuits", 1976 Semiconductor Test Symposium, Digest of Papers, p. 50-3, CCA12-3885.

505. Brovarnik, V. V. and Pepelyaev, V. A., "Algorithm for generating
     tests for detecting faults of type $\sim$ of logical elements in
     combinational circuits", Cybernetics, 11 (3), 394-9 (1975),
     CCA12-1400.
506. Brown, E. T., "Count up/count down error-detect circuit", IBM
     Tech. Disclosure Bull., 19 (7), 2424 (1976), EEA80-28383.
507. Buhler, O. R., "Stop counter errors", Electron Des., 25 (1),
     104-6 (1977).
508. Chen, I. and Carroll, B. D., "Test pattern generation for
     iterative arrays and sequential circuits", Southeast Symposium
     on System Theory, 6th Annual, Proceedings, 1974, 3 pp.
509. Choras, R., "Diagnostics of logic circuits", Wlad. Telekomun.,
     16 (5), 133-6 (1976), Polish, CCA11-28119.
510. Citrin, D. A., "Electrical characterization of complex micro-
     circuits", Report AD-748242, General Electric Co., Pittsfield,
     Mass., 1972, 306 pp., CCA8-13711.
511. Coles, D., "Digital logic checker", Pract. Electron., 12 (1),
     52-4 (1976), CCA11-9611.
512. Control Data Corp., "Computer-controlled logic circuit tester",
     Patent UK 1255441, Publ. December 1971, CCA7-14664.
513. Cook, R., "Analyze I^2L accurately", Electron. Des., 25 (13),
     90-2 (1977).
514. Curlander, P. J., "Logic array checking", IBM Tech. Disclosure
     Bull., 18 (7), 2044-6 (1975), CCA11-9610.
515. Davis, W. A. and Berka, V. F., "Fault diagnosis in sequential
     circuits", Microelectron. & Reliab., 15, Suppl., 15-24 (1976),
     EEA80-24403.
516. DesMarais, P. and Krieger, M., "Fault simulation and digital
     circuit testability", Microelectron. & Reliab., 15, Suppl., 5-13
     (1976), EEA80-24402.
517. Dhillon, B. S., "Modification to fault tree 'AND' gate", Micro-
     electron & Reliab., 15 (6), 625-6 (1976).
518. Down, R. L., "Understanding logic analyzers", Comput. Des.,
     16 (6), 188-91 (1977).
519. Ebel, B., Geisselhardt, W. and Schlaper, J., "Computation of
     fault-diagnosis test sets for logic boards using the equivalence-
     method", 1975 International Symposium on Fault-Tolerant Com-
     puting, Digest of Papers, p. 240, CCA11-28128.
520. Edrington, J., "Battery-powered logic probe needs no on/off
     switch", EDN, 20 (10), 72, 74 (1975).
521. English, W., "Digital IC tester", Pract. Wireless, 52 (10),
     830-4 (1977), EEA80-28403.
522. Farnbach, W. A., "Logic state analyzers - a new instrument
     for analyzing sequential digital processes", IEEE Trans.
     Instrum. Meas., IM-24 (4), 353-6 (1975).
523. Farr, D. L. and Turer, M., "Static Reed test switch", Relay
     Conference, 3rd International and 22nd National Proceedings,
     Papers, 1974, Paper 25, 4 pp.

524. Ferrero, R. O., "A logic circuit probe", Rev. Telegr. Electron.,
     64 (762), 301-2, 306, 310 (1976), Spanish, EEA80-24410.
525. Ferrie, R. G., "In-circuit IC tester checks TTL and C-MOS",
     Electronics, 47 (11), 120-1 (1974), CCA9-18310.
526. Fike, J. L., Smith, R. J., II and Szygenda, S. A., "Diagnostic
     test generation for digital logic - a review", Automation Sup-
     port System for Advanced Maintainability, Symposium, ASSC
     Record, 1973, p. 126-30.
527. Forrest, C., "How to improve testability of logic design",
     Electron. Ind., 2 (10), 28-9 (1976), CCA12-3879.
528. Fowler, F. H., "Salvaging diagnostic fault dictionaries and
     fault tallies after a minor logic change in the tested device",
     Sixth Annual IEEE Computer Society International Conference,
     Digest of Papers, 1972, p. 251-4, CCA8-2554.
529. Fowler, F. H., "Analysis of fault behavior in digital systems",
     Advanced Technqiues in Failure Analysis Symposium, Paper, 1976,
     p. 110-14, CCA11-31520.
530. Fox, J. R., "Test-point condensation in the diagnosis of digital
     circuits", Proc. Inst. Electr. Eng., 124 (2), 89-94 (1977),
     CCA12-8839.
531. Froggatt, J., "Logic probe", Pract. Electron., 13 (3), 222
     (1977), EEA80-28402.
532. Gienapp, A., Hubner, R. and Witte, C., "Test-probe for TTL-
     circuits", Radio Fernsehen Elektron., 23 (2), 62-3 (1974),
     German, CCA9-16005.
533. Gobzemis, A. Y., Sklyarevich, A. N. and Udalov, V. I., "Diag-
     nostic approach to the analysis of a multifunctional logic
     module", Avtom. & Vychisl. Tekh., no. 5, 1-4 (1976), Russian,
     CCA12-3864.
534. Gorajek, A., "TTL logic tester displays H or L", Electronics,
     49 (15), 121 (1976), EEA79-37136.
535. Gossi, R., "Rational testing and error detection of digital
     electronic devices", Feingeraete Tech., 25 (11), 497-9 (1976),
     CCA12-8809.
536. Graham, S., "Logic circuit tests wiring assemblies",
     Electronics, 48 (26), 86 (1975).
537. Gucky, T., "Functional testing of logic integrated circuits",
     Sdelovaci Tech., 19 (3), 100-2 (1971), Czech, CCA6-22200.
538. Hanson, B., "Do you have a 'digital screwdriver' in your tool
     box?", Instrum. & Control Syst., 50 (1), 51-3 (1977), CCA12-15379.
539. Harikumar, M. and Gopalan Nair, N., "Jitter generator tests
     bit synchronizer", Electronics, 47 (11), 127-9 (1975).
540. Hayes, J. P., "Transition count testing of combinational logic
     circuits", IEEE Trans. Comput., C-25 (6), 613-20 (1976).
541. Hellman, H. I., "Testgen: an interactive test generation and
     logic simulation program", RCA Eng., 21 (1), 35-9 (1975).
542. Hill, J. C. and Fiedler, C., "Logic analyzers in system debug-
     ging make time run backward", Comput. Des., 14 (12), 67-72
     (1975).

543. Hlavicka, J. and Kottek, E., "Fault model for TTL circuits", Digital Processes, $\underline{2}$ (3), 169-80 (1976), CCA12-3869.

544. Hnatek, E. R., "User's tests, not data sheets, assure IC performance [digital LSI]", Electronics, $\underline{48}$ (24), 108-13 (1975), CCA11-4689.

545. Huey, B. M. and Hill, F. J., "Fault test generation using a design language", International Symposium on Computer Hardware Description Languages and Their Applications, 1975, p. 91-5, CCA11-31551.

546. Jakubaschk, H., "MOS logic tester for negative logic", Radio Fernsehen Elektron., $\underline{25}$ (3), 101-2 (1976), German, EEA79-24077.

547. Jayasimha Prasad, S. and Muralidharan, M. R., "Logic tester has unambiguous display", Electronics, $\underline{50}$ (5), 117 (1977), CCA12-8854.

548. Jolivet, J. P., "Logic analyzer", Mes. Regul. Autom., Spec. Issue, $\underline{1}$, 29-31 (1977), French, EEA80-28386.

549. Kan, D. T., "Programmable interface drivers simplify logic testing", Comput. Des., $\underline{15}$ (7), 106-8 (1976), CCA11-31524.

550. Karchevskii, V. P., "A device for short circuit detection", Mekh. & Avtom. Proiz., no. 7, 32-3 (1976), Russian, CCA11-31518.

551. Kemp. J. C. and Patzer, W. J., "Test access arrangement for LSI computers", IBM Tech. Disclosure Bull., $\underline{16}$ (3), 942-4 (1973).

552. Kinsky, I., "Clock burst testing of digital circuits", 9th Convention of Electrical and Electronic Engineers in Israel, 1975, p. C-4-4, CCA12-1413.

553. Kirschner, N., "A method to evaluate the electrical stability of TTL gates", Frequenz, $\underline{27}$ (2), 41-4 (1973), German, CCA8-11567.

554. Klinger, A. R., "Logic probe built from IC timer is compatible with TTL, HTL and CMOS", Electron. Des., $\underline{24}$ (12), 156 (1976), EEA79-37147.

555. Knapp, J., "Momentary power failure indicator", Radio & Electron. Constructor, $\underline{30}$ (5), 309-11 (1976), EEA80-24409.

556. Kodandapani, K. L., "Fault location in Reed-Muller canonic networks", Proc. Inst. Electr. Eng., $\underline{124}$ (4), 345-8 (1977), EEA80-20394.

557. Kodanadapani, K. L. and Swamy, S., "Fault location in cutpoint cellular arrays", Proc. Inst. Electr. Eng., $\underline{123}$ (6), 515-16 (1976), CCA11-22571.

558. Koga, Y., "Diagnosis of fail-safe logic systems", Syst. Comput. Control, $\underline{6}$ (3), 25-31 (1975), CCA12-6608.

559. Koga, Y. and Sasaki, I., "Methods of wiring check and evaluation of their validity", Inf. Process. Soc. Jap., $\underline{17}$ (8), 711-19 (1976), Japanese, CCA12-6606.

560. Koren, I. and Kohavi, Z., "Adaptive fault-locating tests for digital systems", Digital Processes, $\underline{2}$ (3), 209-23 (1976), EEA80-2989.

561. Krajewski, R. A., "A circuit for the detection of the off-state of silicon controller rectifier", Przegl. Elektrotech., $\underline{52}$ (11), 430-2 (1976), Polish, EEA80-6268.

562. Kreuwels, W. G. J., "Structural testing of digital circuits", Philips Tech. Rev., $\underline{35}$ (10), 261-70 (1975).

563. Kulikovskii, K. L., Aliev, A. G. and Alyshev, K. R., "Method for testing cells of digital circuits and systems in integrated design", Izv. VUZ Priborostr., $\underline{18}$ (12), 53-4 (1975), Russian, CCA11-12248.

564. Lee, P. M., "Practical aspects of LSI testing", 1971 IEEE International Convention Digest, p. 556-7, CCA6-17555.

565. Lemaigre, P., "LOGTEST: generation and simulation of test-vectors for logic boards", Proceedings of the 8th AICA Congress on Simulation of Systems, 1976, p. 979-83, Publ. by North-Holland, 1977, CCA12-15433.

566. Lowry, D. A., "Skillfully used logic analyzers solve problems large and small", EDN, $\underline{22}$ (8), 86-90 (1977).

567. Manoharan, L. C. and Neelakantan, S., "Online checking of digital integrated circuits for 0's and 1's", J. Inst. Eng., $\underline{56}$ pt. (ET), 21-2 (1975).

568. Martin, G. W., "Versatile logic probe displays four modes", Electronics, $\underline{49}$ (24), 130-1 (1976), CCA12-3810.

569. Mazundar, H. S. and Bandyopadhyay, K. K., "Low-cost data generator uses only two IC's", Electron. Eng., $\underline{47}$ (573), 17 (1975).

570. Mercier, J. J., "Self-testing comparator", Electron. Lett., $\underline{12}$ (19), 489-90 (1976), CCA11-31526.

571. Miguel, B., "A probe for logic circuits", Rev. Esp. Electron., $\underline{24}$ (267), 48-9 (1977), Spanish, EEA80-28408.

572. Minhinnett, P., "Logic analysers - the digital test tool", Commun. Int., $\underline{4}$ (1), 28, 31-2 (1977), EEA80-24382.

573. Morrill, J. S., Jr., "Interfacing a parallel-mode logic state analyzer to serial data", Hewlett-Packard J., $\underline{28}$ (4), 21-4 (1976), CCA12-8814.

574. Morrill, J. S., "Adaptation of a logic state analyser for operation with serial and parallel data systems", Electron. & Appl. Ind., no. 231, 36-9 (1977), French, EEA80-28376.

575. Muchldorf, E. I., "Test pattern generator", IBM Tech. Disclosure Bull., $\underline{15}$ (8), 2409-12 (1973), CCA8-13610.

576. Muller, J., "Logic analysis in the data domain", Elektronik, $\underline{26}$ (4), 72-8 (1977), German, EEA80-28371.

577. Nelson, G. F., "Parametric tests meet the challenge of high-density ICs", Electronics, $\underline{48}$ (25), 108-11 (1975).

578. Osman, M. Y. and Weiss, C. D., "Shared logic realizations of dynamically self-checked and fault-tolerant logic", IEEE Trans. Comput., C-$\underline{22}$ (3), 298-306 (1973), CCA8-7291.

579. Ostrego, M. A., "Diagnosing defects in complex digital circuit packs", Western Electric Eng., $\underline{20}$ (4), 3-7 (1976), CCA12-3832.

580. Pal, M., Palit, A., Pal, S., Basu, M. S. and Choudhury, A. K.,
     "Fault detection in a two level negative gate network", J.
     Inst. Electron. Telecommun. Eng., 21 (2), 58-65 (1975).
581. Palmquist, S. and Chapman, D., "Expanding the boundaries of
     LSI testing with an advanced pattern controller", 1976 Semi-
     conductor Test Symposium, Digest of Papers, p. 70-5, CCA12-3837.
582. Papaioannou, S. G., "Test generation in combinational networks
     by decomposition", Comput. & Electr. Eng., 3 (4), 387-94 (1976),
     EEA80-24381.
583. Patton, R. L., "Automatic diagnosis of logic faults in ceramic
     circuit packs", West. Electr. Eng., 20 (1), 16-19 (1976).
584. Pines, K., "What do logic analyzers do?", Digital Des., 7 (9),
     55-70 (1977).
585. Prasad, S. J., "Designing your own logic clip for testing",
     Electron. Eng., 49 (592), 19 (1977), EEA80-24383.
586. Prasad, S. J. and Muralidharan, M. R., "Logic tester has unam-
     biguous display", Electronics, 50 (5), 117 (1977).
587. Puri, Y., "Stress testing FET gates without the use of test
     patterns", IEEE J. Solid-State Circuits, SC-10 (5), 294-8 (1975).
588. Quayle, B. R., "Stress testing for improved power electronics
     reliability", IEEE Power Electron Special Conference, Pro-
     ceedings, 1974, p. 116-17.
589. Quenelle, R. C., "New logic probe troubleshoots many logic
     families", Hewlett-Packard J., 28 (4), 9-11 (1976), CCA12-8811.
590. Rault, J. C., Bastin, D. and Girard, E., "Fault detection and
     location in logic circuits: general principles", Revue Tech.
     Thomson-CSF, 4 (1), 49-88 (1972), French, CCA7-14656.
591. Riggins, C., "Automated testing for logic circuits-benefits
     and drawbacks", Presented at the Semiconductor Integrated
     Circuit Processing and Production Conference, Anaheim, Califor-
     nia, February 1971, CCA6-15146.
592. Robinson, J. P. and Hoffner, C. W., II, "Easily tested three-
     level gate networks for T or more of N symmetric functions",
     IEEE Trans. Comput., C-24 (3), 331-5 (1975).
593. Russell, G., "Fault diagnosis in logic systems using complex
     gates", IEEE Conference, Publication, No. 411, 1974, p. 137-42.
594. Rutter, G. G. R., "Digital circuit tester", Pract. Electron.,
     12 (10), 826 (1976), CCA11-31497.
595. Santoni, A., "Needed for logic testing: a new breed of instru-
     ments", Electronics, 48 (19), 88-93 (1975).
596. Sawyer, D. E. and Berning, D. W., "Laser scanning of MOS IC's
     reveals internal logic states nondestructively", Proc. IEEE,
     64 (3), 393-4 (1976).
597. Schaaf, R. L., "Tachometer checking", IBM Tech. Disclosure
     Bull., 19 (7), 2462-3 (1976), EEA80-24398.
598. Schleinitz, E., "Tester for TTL circuits", Radio Fernsehen
     Elektron., 24 (23), 758-71 (1975), German, EEA79-16825.
599. Schneider, D., "Designing logic boards for automatic testing",
     Electronics, 47 (15), 100-4 (1974), CCA9-20642.

600. Schuler, D. M., Baker, T. E., Bryant, S. P. and Ulrich, E. G., "A computer program for logic simulation, fault simulation, and the generation of tests for digital circuits", Proceedings of the 8th AICA Congress on Simulation of Systems, 1976, p. 453-9, Publ. by North-Holland, 1977, CCA12-15432.

601. Schulz, J., "Application of Walsh functions to electronic function testing of digital systems", Nachrichtentech. Elektron., 26 (12), 454-7 (1976), German, CCA12-10637.

602. Seth, S. C., "Data compression techniques in logic testing: an extension of transition counts", J. Des. Autom. & FAult-Tolerant Comput., 1 (2), 99-114 (1977), CCA12-8862.

603. Seth, S. C. and Kodandapani, K. L., "Diagnosis of faults in linear tree networks", IEEE Trans. Comput., C-26 (1), 29-33 (1977).

604. Shulman, S. E., "Current tests ensure IC-package orientation", Electronics, 48 (25), 117, 119 (1975).

605. Siwek, K., "Minicomputer simulator for digital logic testing", Electron. Packag. Prod., 15 (9), 91-2, 95 (1975).

606. Sjogren, A., "The logic testers hand instruments: Details, performance and price", Eltek. Aktuell Elektron., 17 (5-6), 38-47 (1974), Swedish, EEA77-36052.

607. Sotomayor, J. M. and Garcia, F. H., "Evaluation of logic families", Mundo Electron., no. 56, 93-6 (1976), CCA12-3821.

608. Srimani, P. K., "A programmable fault testing equipment for combinational networks", J. Inst. Electron. & Telecommun. Eng., 22 (7), 486-91 (1976), EEA80-283.

609. Stamm, K., "Tester for logic integrated circuits", Automatis-ierungspraxis, 15 (12), 269 (1972), CCA8-11560.

610. Szmolnik, J. and Garai, G., "Devices for detecting faults in digital logic networks", Hiradastechnika, 25 (4), 107-11 (1974), Hungarian, CCA9-18318.

611. van Dalen, J., "Logic probe", Radio Electron., 24 (17), 563 (1976), Dutch, CCA12-12791.

612. Vranesic, Z. G., "Multi-valued circuits in fault detection of binary logic circuits", Microelectron. & Reliab., 15, Suppl., 25-33 (1976), EEA80-24404.

613. Wan, C. J. and Mo, H. W., "Test pattern generation for the functional test of logic networks", J. Korean Inst. Electron. Eng., 13 (3), 74-9 (1976), Korean, EEA80-282.

614. Williams, M. J. Y. and Angell, J. B., "Enhancing testability of large-scale integrated circuits via test points and addi-tional logic", IEEE Trans. Comput., C-22 (1), 46-60 (1973), CCA8-5195.

615. Woodward, F. I., "Portable logic tester and frequency counter", New Electron., 10 (2), 36 (1977), EEA80-28393.

616. Yamada, A., Wakatsuki, N. and Tomita, K., "Parallel fault simulator based on circuit delay and selective tracing of asynchronous logic circuit", Inf. Process. Soc. Jap., 17 (7), 569-76 (1976), Japanese, EEA80-9022.

617. Zdrazil, Z., "Logic state analysers", Automatizace, 19 (10),
     260-3 (1976), Czech, CCA12-3838.
618. Zobniw, L. M., "Realtime diagnosis (probe algorithm) using
     single pin probe", IBM Tech. Disclosure Bull., 18 (8), 2471-4
     (1976), CCA11-12281.
619. Zobniw, L. M., "Multidefect real-time diagnosis using single-
     pin probe", IBM Tech. Disclosure Bull., 19 (5), 1656-61 (1976),
     CCA12-15428.

VI.   RELIABILITY

620. Abdel-Latif, M. and Strutt, M. J. O., "Dynamic malfunction
     limits in high-speed TTL and ECL integrated digital gates",
     Microelectron. & Reliab., 12 (1), 57-60 (1973), CCA8-11576.
621. Abraham, J. A., "Combinatorial solution to the reliability of
     interwoven redundant logic networks", IEEE Trans. Comput., C-24
     (5), 578-84 (1975).
622. Aggarwal, K. K., "Reliability of logic circuits with random
     inputs", Int. J. Electron., 39 (4), 459-64 (1975).
623. Aggarwal, K. K., "Reliability of probabilistic logic circuits
     with random inputs", Microelectron. & Reliab., 15 (6), 627-8
     (1976), CCA12-8838.
624. Aggarwal, K. K., "Distribution function and reliability of
     complex digital circuits with random inputs", Int. J. Electron.,
     42 (1), 91-5 (1977), CCA12-8836.
625. Aggarwal, K. K., Misra, K. B. and Gupta, J. S., "Reliability
     evaluation. A comparative study of different techniques",
     Microelectron. & Reliab., 14 (1), 49-56 (1975).
626. Barela, K., "Method of automatic reliability analysis in logical
     combinational networks", Rozpr. Electrotech., 22 (4), 925-38
     (1976), Polish, CCA12-6612.
627. Chirkov, M. K., "Reliability analysis of combined logic networks
     containing OR-NOT elements", Vychisl. Tehh. & Vopr. Kibera,
     no. 10, 61-6 (1974), Russian, CCA9-21965.
628. DesMarais, P. and Krieger, M., "Reliability analysis of logic
     circuits", IEEE Trans. Reliab., R-24 (1), 46-52 (1975).
629. DesMarais, P. and Krieger, M., "Reliability analysis of logic
     circuits", Microelectron. & Reliab., 16 (1), 29-33 (1977),
     EEA80-20388.
630. Doring, H., "Computation and application of position systems
     in circuit technology", Nachrichtentech. Elektron., 26 (12),
     461-3 (1976), German, CCA12-10617.
631. Fox, J. L., "Availability design of the system/370 model 168
     multi processor", USA-Japan Computer Conference, 2nd,
     Proceedings, 1975, p. 52-7.
632. Foxhall, G. F., "Bipolar linear integrated circuit reliability",
     1976 International Electron Devices Meeting, Technical Digest,
     p. 41.

633. Fussell, J. B., Aber, E. F. and Rahl, R. G., "On the quantitative analysis of priority-AND failure logic", IEEE Trans. Reliab., R-25 (5), 324-6 (1976), CCA12-10596.

634. Genis, Y. G., Nechaev, B. Y. and Yakobson, G. R., "Reliability analysis for a redundant recoverable system with a switch with an arbitrary MTBF function", Avtom. & Vychisl. Tekh., no. 6, 31-2 (1976), Russian, CCA12-15414.

635. Greenwood, W. T., Jr., "Reliability testing for industrial use", Computer, 10 (7), 26-30 (1977).

636. Heavey, P. P. and Simpson, V. E., "Functional and level fail detection for register array testing", IBM Tech. Disclosure Bull., 15 (4), 1135-6 (1972), CCA8-5272.

637. Hewlett, F. W., Jr. and Pedersen, R. A., "Reliability of integrated injection logic circuits for the Bell System", 14th Annual Proceedings of the Reliability Physics Symposium, 1976, p. 5-10.

638. Hlavicka, J. and Kottek, E., "Fault model for TTL circuits", Digital Processes, 2 (3), 169-80 (1976), EEA80-2987.

639. Kats, B. Z., "The errors of a pulse-phase detector", Telecommun. & Radio Eng. Pt. 2, 31 (1), 128-30 (1976), EEA80-2978.

640. Kemeny, A. and Kalmar, G., "New results concerning a 100 million device-hr reliability assessment of TTL SSI integrated circuits. I. Test methods and testing conditions", Hiradastechnika, 26 (10), 289-99 (1975), Hungarian, CCA11-9556.

641. Kemeny, A. and Kalmar, G., "New results concerning the 100 million device-hr reliability assessment of TTL SSI integrated circuits. II. Test results and their interpretation from a failure physics aspect", Hiradastechnika, 26 (11), 324-44 (1975), Hungarian, CCA11-9557.

642. Kemeny, A. P., Kalmar, G. and Stefaniay, V., "Results of a 160 x 10^6 device-hour reliability assessment and failure analysis of TTL SSI integrated circuits-1, 2", Microelectron. & Reliab., 14 (5-6), 469-526 (1975).

643. Klein, M. R., "Microcircuit device reliability: Digital detailed data", Report MDR-4, Rome Air Development Center, Griffiss Air Force Base, N. Y., 1976, 225 pp.

644. Krieger, M., Zisapel, Y. and Kella, J., "Hazards in combinational circuits", Microelectron. & Reliab., 15, Suppl., 35-46 (1976), EEA80-24405.

645. Lehtonen, D. E., "Microcircuit reliability assessment through accelerated testing", Electron. Packag. & Prod., 17 (7), 288-93 (1977).

646. Nakamichi, M. and Fujikura, N., "Composite construction methods for ternary 3-level logic elements and their classification", Syst. Comput. Controls, 5 (6), 1-12 (1974).

647. Patrick, J. L., "Integrated circuits and the reliability of numerical control", Inter. Electronique, 23 (12), 32, 39 (1968), French, CCA5-2694.

648. Platteter, D., "Basic integrated circuit failure analysis techniques", 14th Annual Proceedings Reliability Physics Conference, 1976, p. 248-55, EEA80-17260.

649. Popov, V. A., Sitnijkov, V. O. and Tikhotskii, A. I., "One method of reliability analysis for simple logic circuits", Izv. VUZ Priborostr., 17 (5), 66-72 (1974), Russian, CCA9-21960.

650. Roe, J. M., "Microwave interference effects in integrated circuits", IEEE Electromagnetic Comp. Symp. Records, 1975, Pap. Ib, 6 pp.

651. Sapiecha, K., "Detection and location of multiple failures to fan-out free combinational circuits", Arch. Autom. & Telemech., 21 (3), 371-83 (1976), Polish, EEA80-266.

652. Sarekha, K., "Application of circuit discriminant for fault diagnosis", Arch. Autom. & Telemech., 21 (3), 385-99 (1976), Polish, EEA80-267.

653. Sirvent, J. M., "Introduction to fail-safe circuits. II. Logic circuits", Mundo Electron., no. 60, 55-8 (1977), Spanish, EEA80-28388.

654. Viele, A. A., "Failure analysis technique for locating the fail site in MOSFET (LSI) logic chips with sputtered SiO2 passivation", Proceedings of the 12th Annual Reliability Physics Conference, 1974, p. 16-21.

655. Zirphile, J., "An aid for failure diagnosis in logic integrated circuits", Onde Electr., 57 (3), 231-8 (1977), EEA80-20391.

VII. MISCELLANEOUS CIRCUITS
1. General

656. Berges, G. A., "Use of postfix notation for the description of microcellular arrays", IEEE Trans. Comput., C-26 (2), 185-91 (1977).

657. Dane, R. J., "MOS logic circuits", In: MOS integrated circuit design, Wolfendale, E., (Ed.), Butterworth, London, England, p. 23-48 (1973), CCA8-13619.

658. Duchene, J., "Simplification and money-save in logic-circuits by use of three state logic", Onde Electr., 57 (2), 126-9 (1977), French, CCA12-15429.

659. Fet, Y. I., "Functional possibilities of simple computing media", Autom. Control Comput. Sci., 8 (3), 42-7 (1974).

660. Inokuchi, S. and Suzuki, Y., "Application techniques of integrated circuits. III", Syst. & Control, 20 (9), 547-56 (1976), Japanese, CCA12-15411.

661. Logunova, N. L. and Norkin, K. B., "Discrete operations with the use of multilevel representation of variables", Autom. Remote Control, 35 (11), 1825-9 (1974).

662. Muller, R., Pfleiderer, H. J. and Stein, K. U., "Energy per logic operation in integrated circuits: definition and determination", IEEE J. Solid-State Circuits, SC-11 (5), 657-61 (1976), EEA80-277.

663. Murray, J., "Digital computation basic principles of automatic
     computation", Eng. Mater. & Des., 20 (11), 47-52 (1976),
     CCA12-6600.
664. Smilde, J. G., "A universal counter/clock component. I", Radio
     Electron., 24 (17), 564-8 (1976), Dutch, EEA80-20403.
665. Soderman, L., "How MOS circuits function. IV. Bipolar and static
     logic are similar", Eltek. Aktuell Elektron, 17 (3), 28-9
     (1974), Swedish, CCA9-15996.
666. Solovei, N. I. and Artyukhov, V. L., "Logic elements with mag-
     netically controlled contacts", Prib. & Sist. Upr., no. 9,
     42-3 (1976), Russian, EEA80-6267.
667. Sotomayor, J. M. and Garcia, F. H., "Evaluation of logic
     families", Undo Electron, no. 56, 93-6 (1976), EEA80-3018.
668. Wolfgarten, W., "Digital circuits: the realisation of optimal
     structures", Funk-Tech., 31 (7), 192-8 (1976), German, CCA11-22556.
669. Yamamoto, M., "Some consideration on the identification of com-
     binational cellular logic arrays", Syst. Comput. Control, 6 (4),
     56-62 (1975), EEA80-3033.
670. Yanai, H., Sugeta, T., Tamaki, Y. and Tanimoto, M., "Signal-
     processing of a Schottky-gate Gunn-effect digital functional
     device", Annu. Rep. Eng. Res. Inst. Fac. Eng. Univ., 31, 139-44
     (1972), Japanese, CCA8-7264.

2.   Programmable Logic Arrays (PLA)

671. Anon., "Designing with programmable logic arrays", Digital
     Des., 6 (10), 70-2 (1976), EEA80-28367.
672. Balasubramanian, P. S. and Venkataraman, K. V., "Programmable
     logic array using charge-coupled devices", IBM Tech. Disclosure
     Bull., 19 (3), 993-5 (1976), CCA12-3820.
673. Bennett, L. A. M., "Threshold logic functions in programmable
     logic arrays", Electron. Lett., 12 (11), 279-80 (1976),
     CCA11-22554.
674. Bennett, L. A. M., "Modified programmable logic array with more
     product terms", Electron. Lett., 13 (15), 443-5 (1977).
675. Birkner, J., "PLA's: When and how to use them", 12th IEEE
     Computer Society International Conference, COMPCON 76, p. 160-4.
676. Cavlan, N., "Structure and applications of field programmable
     logic arrays", Microelectron. & Reliab., 15 (4), 285-95 (1976),
     CCA11-31495.
677. Cavlan, N. and Cline, R., "Field-PLA simplify logic designs",
     Electron. Des., 23 (18), 84-90 (1975).
678. Cavlan, N. and Cline, R., "FPLA applications - exploring
     design problems and solutions", EDN, 21 (9), 63-9 (1976).
679. Cavaliere, J. R., Eardley, D. B. and Weinberger, A., "Current
     switch read-only programmable logic array", IBM Tech. Dis-
     closure Bull., 18 (10), 3245-8 (1976), CCA11-601.

680. Cha, C. W., "Prime faults in a double-bit partition programmable logic array", IBM Tech. Disclosure Bull., 18 (8), 2715-17 (1976), CCA11-603.

681. Cook, P. W., "Array logic", Proceedings of the 1974 IEEE International Symposium on Circuits and Systems, p. 251-19, CCA9-25017.

682. Fleisher, H. and Maissel, L. J., "An introduction to array logic", IBM J. Res. & Develop., 19 (2), 98-109 (1975).

683. Frankenberg, R. J., "Programmable logic: a user's viewpoint", 1975 WESCON Technical Papers, Vol. 19, Paper 26/1, 13 pp., CCA11-12269.

684. Graff, M. H., "Boundary conditions for programmable logic replacement", 1975 WESCON Technical Papers, Vol. 19, Paper 26/3, 4 pp., CCA11-12271.

685. Hebenstreit, E. and Horninger, K., "High-speed programmable logic arrays in ESFI SOS technology", IEEE J. Solid-State Circuits, SC-11 (3), 370-4 (1976), CCA11-22558.

686. Hemel, A., "The PLA: a 'different kind' of ROM", Electron. Des., 24 (1), 78-84 (1976), CCA11-15062.

687. Henderson, R. and Brown, J. T., "How to design with programmable logic arrays", New Electron, 7 (13), 14-16, 18 (1974), CCA9-23227.

688. Hwang, K., "TTL programmable logic array and its applications", Proc. IEEE, 64 (3), 368-9 (1976).

689. Jones, J. W., "Array logic macros", IBM J. Res. & Develop., 19 (2), 120-6 (1975).

690. Kambayashi, Y., "Optimum logic design using memory-type arrays", International Symposium of Unif. Struct. Auto. and Logic, Proc., 1975, p. 199-205.

691. Lebech, F., "PLA: A programmable logic circuit", Elektronik, no. 3, 22-3 (1976), Danish, CCA11-14961.

692. Lilen, H., "A programmable logic array (PLA)", Electron. & Microelectron. Ind., no. 215, 19-24 (1976), French, CCA11-17343.

693. Logue, J. C., Brickman, N. F., Howley, F., Jones, J. W. and Wu, W. W., "Hardware implementation of a small system in programmable logic arrays", IBM J. Res. & Develop., 19 (2), 110-19 (1975).

694. May, P. and Schiereck, F. C., "High-speed static programmable logic array in LOCMOS", IEEE J. Solid-State Circuits, SC-11 (3), 365-9 (1976), CCA11-22557.

695. Miles, G., "Field programmable logic arrays for next generation designs", 1975 WESCON Technical Papers, Paper 26/2, 6 pp., CCA11-12270.

696. Miles, G., "FPLA's offer a design alternative for development of system logic", EDN, 20 (20), 85-9 (1975).

697. Miles, G., "Increased flexibility for digital circuits using field programmable logic arrays", Elektron. Ind., 7 (10), 243-4 (1976), German, CCA12-3877.

698. Mitchell, T. W., "Programmable logic arrays", Electron. Des., 24 (15), 98-101 (1976).

699. Pittman, P., "The f.p.l.a. is a sequential circuit!", New
     Electron., 9 (4), 51-2 (1976), CCA11-14969.
700. Scrupski, S. E., "High-density bipolars spur advances in
     computer design", Electronics, 48 (22), 82-6 (1975).
701. Shoup, R. G., "Programmable cellular logic arrays", Report
     AD-706891, Carnegie Mellon Univ., Pittsburgh, Pa., 1970,
     242 pp., CCA6-10709.
702. Vodovoz, E., "Second look at the world of PLA's", Electron.
     Prod., 18 (9), 39-42 (1976).
703. Wood, R. A., "High-speed dynamic programmable logic array
     chip", IBM J. Res. & Dev., 19 (4), 379-83 (1975), CCA11-6344.
704. Yakubaitis, E. A., "Structure synthesis of a programmable logic
     array", Avtom. & Vychisl. Tekh., no. 4, 1-10 (1976), Russian,
     CCA11-31522.
705. Zilko, P. S., "New generation of programmable logic controllers",
     Proceedings of the 9th IEEE Ind. Appl. Soc. Meeting, 1974,
     p. 613-19.

              3.   Associative Memories

706. Armstrong, C. V. W., "Functional memory techniques applied to
     the microprogrammed control of an associative processor", 2nd
     Annual Symposium on Computer Architecture, Conference Proceed-
     ings, 1975, p. 34-40.
707. Berkovich, S. Y. and Kochin, Y. Y., "Search for numbers that
     are nearest to a given number", Autom. Remote Control, 36 (2),
     343-5 (1975).
708. Berkovich, S. Y., Kochin, Y. Y. and Lapir, G. M., "Algorithms
     for group word processing in an associative memory and their
     realization by homogeneous computing arrays", Autom. Remote
     Control, 35 (8), 1342-9 (1974).
709. Berra, P. B. and Singhania, A. K., "A multiple associative
     memory organization for pipe-lining a directory to a very large
     data base", 12th IEEE Computer Society International Con-
     ference, COMPCON 76, p. 109-12.
710. Brundage, R. E. and Batson, A. P., "Performance enhancement
     of descriptor-based virtual memory systems through the use of
     associative registers", 2nd Annual Symposium on Computer
     Architecture, Conference Proceedings, 1975, p. 85-90.
711. Brundage, R. E. and Batson, A. P., "Use of associative
     memory in symbolically-segmented virtual memory systems",
     Rev. Fr. Autom. Inf. Rech. Oper., 10 (1), 47-60 (1976).
712. Chakrabarti, K. and Kolp, O., "Dynamic storage allocation
     algorithm and its hardware implementation", Int. J. Syst. Sci.,
     6 (3), 289-300 (1975).
713. Davis, D. E. and Hannaford, C. W., "Associative-memory system
     using latchable search drivers", IBM Tech. Disclosure Bull.,
     15 (3), 719-20 (1972), CCA8-2665.

714. Dyke, J. G. and Lea, R. M., "An associative parallel processor for local editing applications", Digital Processes, 1 (2), 89-101 (1975), CCA11-9858.

715. Fairhurst, M. C., "Economical associative storage with universal sequential logic networks", Electron. Lett., 12 (23), 612-13 (1976), CCA12-1491.

716. Finnila, C. A. and Love, H. H., Jr., "The associative linear array processor", IEEE Trans. Comput., C-26 (2), 112-25 (1977), CCA12-10651.

717. Gershuny, E. S. and Lamb, O. L., "Code translation using multiple associative memory searches", IBM Tech. Disclosure Bull., 15 (4), 1109-10 (1972), CCA8-5196.

718. Greer, D. L., "An associative logic matrix", IEEE J. Solid-State Circuits, SC-11 (5), 679-91 (1976), EEA80-3010.

719. Handler, W., "A unified associative and VonNeumann processor EGPP and EGPP-array", Parallel Processing, 1974, p. 97-9, Publ. 1975 by Springer-Verlag, New York, CCA10-25986.

720. Houston, G. B. and Simonsen, R. H., "Associative memory device", Patent USA 3906455, Publ. September 1973, CCA11-4778.

721. IBM Corp., "Associative memory", Patent UK 1289249, Publ. September 1972, CCA8-7354.

722. Landis, D., "Multiple-response resolution in associative systems", IEEE Trans. Comput., C-26 (3), 230-5 (1977), CCA12-8884.

723. Lea, R. M., "Design for a high-speed m.o.s. associative memory", Electron. Lett., 8, 391-3 (1972), EEA75-32707.

724. Lea, R. M., "Low-cost high-speed associative memory", IEEE J. Solid-State Circuits, SC-10 (3), 179-81 (1975).

725. Lea, R. M., "Design for a low-cost high-speed m.o.s. associative memory", Radio Electron. Eng., 45 (4), 177-82 (1975).

726. Lea, R. M., "A NAND-gate implementation for high-speed associative memory", Digital Processes, 2 (1), 83-8 (1976), CCA11-22640.

727. Lea, R. M., "The comparative cost of associative memory", Radio & Electron. Eng., 46 (10), 487-96 (1976).

728. Leibrand, R., "Associative memories", Presented at Man Made Memories, Conference Summaries, London, England, March 1971, CCA6-12755.

729. Linton, R. H. and Sonoda, G. Y., "MOSFET associative cell", IBM Tech. Disclosure Bull., 17, 3340-1 (1975), EEA78-28302.

730. Matsue, S., "High speed associative memory circuits", Patent USA 3703709, Publ. November 1972, CCA8-13707.

731. Meyers, B. F., "Data structuring for associative memory", IEEE Conference on Computer Graphics, Pattern Recognition, and Data Structure, Proceedings, 1975, p. 94-7.

732. Mundy, J. L., Burgess, J. F., Joynson, R. E. and Neugebauer, C., "Low-cost associative memory", IEEE J. Solid-State Circuits, SC-7, 364-9 (1972), EEA75-36285.

733. Naito, S. and Nagano, T., "New model of associative memory with backward lateral connections", Syst. Comput. Controls, 5 (2), 17-25 (1974).

734. Pao, Y. H. and Altman, J., "Associative memories as multipath logic switches", Proceedings of the 1975 International Symposium on Multi-Valued Logic, p. 242-5, CCA11-9621.

735. Pao, Y. H. and Merat, F. L., "Distributed associative memory for patterns", IEEE Trans. Syst. Man. Cybern., SMC-5 (6), 620-5 (1975).

736. Pfahlbusch, H. and Rettelbusch, L., "A coupling element for the transmission of analog and digital information [associative storage control]", Fernmeldetechnik, 16 (3), 97-9 (1976), German, CCA11-22641.

737. Pronina, V. A. and Chudin, A. A., "Syntax-analysis implementation in an associative parallel processor", Autom. Remote Control, 36 (8), pt. 2, 1303-8 (1975).

738. Shead, C. J., "The associative memory: a versatile circuit element", GEC J. Sci. & Technol., 40 (3), 119-25 (1973), CCA9-6240.

739. Wolf, G., "Associative memories and processors", Elektron. Rechenanlagen, 17 (6), 264-71 (1974), German, CCA11-9859.

740. Wyland, W. C., "Associative read-only memory technique: A key to LSI control logic", Comput. Des., 10 (9), 61-8 (1971).

4.   Decoders/Encoders

741. Askin, H. O., Boss, C. D. and Lee, J. M., "Gated output word decoder", IBM Tech. Disclosure Bull., 27 (5), 1359-60 (1974), EEA78-8569.

742. Blacksher, R., "PROM decoder replaces chip enabling logic", Electronics, 49 (18), 100-2 (1976).

743. Buzunov, Y. A. and Knyazev, A. Y., "A method for two-dimensional encoding of microinstructions", Avtom. & Vychisl. Tekh., no. 5, 81-8 (1976), Russian, CCA12-3888.

744. Chen, C. L. and Chien, R. T., "Recent developments in majority-logic decoding", Report AD-712683, Univ. Illinois, Urbana, 1970, 17 pp., CCA6-17597.

745. Dockerty, R. C. and Knepper, R. W., "Enhancement/depletion decoder circuit", IBM Tech. Disclosure Bull., 19 (5), 1681-2 (1976), EEA80-20378.

746. Duttweiler, D. L., "A µ255 encoder requiring no precision components", IEEE Trans. Commun., COM-25 (3), 297-303 (1977), CCA12-8853.

747. Hsieh, J. C. and Sonoda, G., "MOSFET storage array addressing system", IBM Tech. Disclosure Bull., 13 (8), 2383-4 (1971), CCA6-12733.

748. Huffman, D. R., Rossi, F. R. and Shea, D. J., "Memory address decode circuit", IBM Tech. Disclosure Bull., 19 (1), 28-9 (1976), CCA11-22564.

749. Kammerer, J., "Application of digital circuits. V.", Elektro-
     meister & Dtsch. Elektrohandwerk, 51 (21), 1367-9 (1976),
     German, EEA80-9010.
750. Lavrinchuk, V. M. and Khorunzhii, A. J., "Accuracy of angle
     encoders based on a rotating transformer", Izv. VUZ Radio-
     elektron, 19 (11), 115-18 (1976), Russian, EEA80-16903.
751. Leefsma, E. H., "Morse decoder with DDLL (dot-dash length
     logic)", Elektro, 3 (3), 48-51 (1977), EEA80-24393.
752. Neal, T., "Logic circuit ensures definite break-before-make
     action for relay drive", Electron. Des., 24 (16), 80 (1976),
     CCA11-28109.
753. Oliver, J. P., "Priority encoder simplifies clock-to-computer
     interfaces", Electronics, 49 (26), 70-1 (1976), CCA12-3839.
754. Overington, W. J. G., "BCD-addressed memory with 1000 addresses",
     Electron. Eng., 48 (578), 19 (1976).
755. Regitz, W. M. and Reed, J. A., "MOS memory decoder circuit",
     Patent USA 3863230, Publ. January 1975, EEA78-33160.
756. Schuenemann, C. and Wolf, P., "Interleaved Josephson junction
     tree decoder", IBM Tech. Disclosure Bull., 18 (12), 4168-70
     (1976), CCA11-28268.
757. Starr, R. F., "Decoders convert binary code for hexadecimal
     display", Electronics, 49 (14), 107 (1976).
758. Valil'eva, N. P. and Dubrovin, V. I., "Magnetic film memory
     matrices with internal decoding", Autom. & Remote Control, 37
     (2) pt. 2, 272-8 (1976), CCA11-31594.
759. Yiu, K. P., "Feedback polynomial of a majority logic decoder",
     Electron. Lett., 12 (8), 195-6 (1976), CCA11-17339.

5.   Switches

760. Anderson, S. D., "High-speed switch handles $\pm$ 400-V peak and
     is controlled by logic levels", Electron. Des., 23 (2), 74 (1975).
761. Anon., "Form factor of switching power supply meets user re-
     quirements for diverse designs", Comput. Des., 15 (7), 142-3
     (1976), CCA11-31504.
762. Anon., "The operation and application of threshold-value de-
     tecting switches", Elektromeister & Dtsch. Elektrohandwerk,
     52 (6), 388-90 (1977), German, EEA80-24388.
763. Armgarth, D., "Operation and characteristics of some MSI
     switching circuits", Radio Fernsehen Elektron., 23 (4),
     135-8 (1974), German, CCA9-16006.
764. Balthasar, P. P. and Reimers, E., "Integrated power switch",
     IEEE Trans. Ind. Appl., IA-12 (2), 179-91 (1976).
765. Barna, A., "Delay and rise time in current-mode switching cir-
     cuits", Arch. Elektron. & Uebertragungstech., 30 (3), 112-16
     (1976), CCA11-14954.

766. Bejasov, J., Zwetkov, G., Nenov, S., Schwartz, A., Schrepel, D., Rockstroh, M. and Konig, G., "Improved signal transmission by coupled fluidic switching elements", Mess. Steuern Regeln Mit Automatisierungprax., 19 (6), 212-16 (1976), German, CCA11-22551.
767. Benichou, C., "Path test system for digital switching network", IBM Tech. Disclosure Bull., 18 (8), 2541-2 (1976), CCA11-12282.
768. Birke, W., "The switching matrix", Elektrotechnik, 58 (11), 20-2 (1976), German, CCA11-22555.
769. Bonisch, W. and Beuter, R., "Logic switching with integrated circuits", Grundig Tech. Inf., 23 (4), 777 (1976), German, CCA12-15378.
770. Bredeson, J. G., "Synthesis of multiple input-change hazard-free combinational switching circuits without feedback", Int. J. Electron., 39 (6), 615-24 (1975).
771. Brockman, D. E. and Nelson, R. W., "Hall-effect sensors - magnetic switches that have no contacts", Mach. Des., 47 (25), 123-7 (1975).
772. Brown, F. M., "Equational realizations of switching functions", IEEE Trans. Comput., C-24 (11), 1054-66 (1975).
773. Brown, F. M., "Weighted realizations of switching functions", IEEE Trans. Comput., C-24 (12), 1217-21 (1975).
774. Burens, J. H., "Switching power supplies: specification criteria", Comput. Des., 16 (3), 91-7 (1977), CCA12-15387.
775. Catto, C. J. D. and Webdell, D. S., "Debouncing a multi-way switch with hysteresis", Electron. Eng., 47 (569), 9 (1975).
776. Chan, H. W., Lum, W. Y. and Van Duzer, T., "High-speed switching and logic circuits using Josephson devices", IEEE Trans. Magn., MAG-11 (2), 770-3 (1975).
777. Chang, A. and Swietek, D., "Current switch logic circuit", IBM Tech. Disclosure Bull., 13 (5), 1101-2 (1970), CCA6-3725.
778. Chauhan, A. S. and Maheshwari, L. K., "Turn-off switching analysis of a saturation-controlled transistor circuit [TTL]", Int. J. Electron., 42 (4), 405-7 (1977), CCA12-10611.
779. Chin, W. B. and Robbins, G. J., "MTL-compatible current switch circuit", IBM Tech. Disclosure Bull., 19 (6), 2081-3 (1976), EEA80-20384.
780. Cole, T. G. and Wrona, S., "Current switch logic circuit", IBM Tech. Disclosure Bull., 14 (1) 336 (1971), CCA6-22197.
781. Diklic, M. and Mihailovic, P., "Ledik - highly prospective magnetic element for the realization of switch logic functions", Tehnika, 31 (11), 1734-40 (1976), Croatian, CCA12-3831.
782. Dittman, C. D., "Sonar transmit-receive switch is compatible with most logic levels", Electron. Des., 24 (4), 172, 174 (1976), EEA79-24402.
783. Embley, R. W., "Solid-state analog switch matrix replaces relay and crossbar selectors", Comput. Des., 14 (11), 112-14 (1975).
784. Estvander, R. A., "Survey of the application of redundancy in switching processors", Proceedings National Electron Conference, Vol. 29, 1974, p. 55-6.

785. Franson, P., "Today's monolithic switching circuits greatly simplify power-supply designs", EDN, 22 (6), 47-8, 51, 53 (1977), CCA12-15391.

786. Forrest, S., "Multiple switch debouncer", New Electron, 9 (19), 17 (1976), CCA11-31496.

787. Fukumura, T., "Recent theories of switching circuits. I. Many-valued logic", J. Inst. Electron. Commun. Eng. Japan, 53 (7), 925-34 (1970), Japanese, CCA6-3730.

788. Gensel, J., "The modem-matrix calculus in switching networks. II.", Mon. Tech. Rev., 15 (8), 153-6 (1971), CCA6-24413.

789. Givens, S., "Field-effect transistors as analog switches", Comput. Des., 13 (6), 106, 108-10 (1974), CCA9-21950.

790. Goedecke, W. D., "Sequential switching circuits for various switching requirements", Ind.-Anz., 98 (78), 1379-82 (1976), German, CCA12-1408.

791. Gold, D. E., "Applications of some switching network results to dynamic allocation of memories in a hierarchy", Digest of Papers of the Six Annual IEEE Computer Society International Conference, 1972, p. 127-9, CCA8-2626.

792. Goser, K., "MOS transistors as high-speed switches in a selection matrix", Nachrichtentech. Z. (NTZ), 23 (10), 512-21 (1970), English and German, CCA6-3761.

793. Gratzer, D. and Schlenk, K. W., "Magnetic components for switching circuits", Elektro-Anz., 29 (22), 540 (1976), German, EEA80-9066.

794. Gray, J., "Power switch ROM's and PROM's quickly", Electron. Des., 25 (9), 102-4 (1977).

795. Green, C. W., "Checking out semiconductor memories for electronic switching system", Bell Lab. Record, 55 (5), 131-6 (1977).

796. Hage, D. and Givens, S., "Switching high-frequency signals with FET integrated circuits", Microelectronics, 5 (4), 32-44 (1974).

797. Hahner, K. W., "Polarity reversing switch without mechanical switching elements", Elektronik, 26 (1), 52 (1977), German, EEA80-16895.

798. Honeywell Inc., "Magnetic memory integrated switching circuit", Patent UK 1184722, Publ. March 1970, CCA6-3735.

799. Kamionka-Mikula, H., "Synthesis of two-level minimal switching circuits used to control flip-flops", Arch. Autom. & Telemech, 19 (1), 73-88 (1974), Polish, CCA9-18306.

800. Kandel, A., "Inexact switching logic", IEEE Trans. Syst. Man Cybern, SMC-6 (3), 215-19 (1976).

801. Kopperschmidt, G., "Majority logic in practical circuitry", Elektronik, 25 (10), 71-8 (1976), German, EEA80-274.

802. Kuhn, E., "Low distortion line equaliser for TTL switching circuits", Fernmeldetechnik, 13 (4), 198-202 (1973), German, CCA9-18317.

803. McLeod, N. W., "An analogue switch using a digital integrated circuit", Proc. Inst. Radio Electron. Eng. Aust., 32 (2), 63-5 (1971), CCA6-15138.

804. Mehlhor, K. and Galil, Z., "Monotone switching circuits and boolean matrix product", Comput. <u>16</u> (1-2), 99-111 (1976).
805. Morley, D., "Touch operated on-off switch using CMOS nor gates", Electron. Eng., <u>47</u> (569), 14-15 (1975).
806. Nishimoto, Y., "P-N-P-N switching performance of N-P-N transistors in integrated circuits", Proc. IEEE, <u>63</u> (4), 723-4 (1975).
807. Nuzillat, G., Arnodo, C. and Puron, J. P., "A subnanosecond integrated switching circuit with MESFET's for LSI", IEEE J. Solid-State Circuits, SC-<u>11</u> (3), 385-94 (1975), EEA79-33121.
808. Ogawa, T. and Wada, T., "A high-speed chrome permalloy switching core", IEEE Trans. Magn., MAG-<u>9</u> (3), 511-14 (1973), CCA9-3418.
809. Pfannschmidt, H., "Limitation of transmission rate in high-speed TDM-switching networks using Schottky-TTL circuit technology", 1976 International Conference on Digital Communications, p. C3/1-6, CCA11-13542.
810. Shaw, H. H., "Digitally controlled electronic switch", Electron, no. 24, 37-9 (1973), CCA8-13605.
811. Stitt, M., "CMOS switch inverts analog signal under control of digital logic", Electron. Des., <u>23</u> (25), 92 (1975).
812. Su, S. Y. H. and Cheung, P. T., "Computer simplification of multi-valued switching functions", In: Computer science and multiple-valued logic. Theory and applications, Rine, D. C., (Ed.), North-Holland, New York, p. 189-220 (1977), CCA12-10645.
813. Thayse, A., "Applications of discrete functions-3. The use of functions of functions in switching circuits", Philips Res. Rep., <u>29</u> (5), 429-52 (1974).
814. Walsh, J. L., "Binary switching circuit", IBM Tech. Disclosure Bull., <u>14</u> (1), 193 (1971), CCA6-22192.
815. Winch, N. R. and Hyde, P. J., "Preprocessor for SPC switching systems", International Switching Symposium, Proceedings, Paper 147, 1974, 6 pp.

6. Memory Peripheral Circuits

816. Arzubi, L. M., "Twin cell, high performance write circuit", IBM Tech. Disclosure Bull., <u>19</u> (7), 2482-3 (1976), EEA80-25012.
817. Bloom, G. E., "Saturable core transformers harden latch memories", Electronics, <u>50</u> (7), 104-5 (1977), EEA80-20361.
818. Choras, R., "Semiconductor memory circuits", Wiad. Telekomun., <u>16</u> (7-8), 198-203 (1976), Polish, EEA79-46448.
819. Christensen, N. T., Gannon, P. M., Lee, A. E. and Wright, L. W., "Simplified control of memory arrays", IBM Tech. Disclosure Bull., <u>16</u> (8), 2459-60 (1974), CCA9-14349.
820. Copeland, M. A., Ellul, J. P. and Chan, C. H., "High-speed dynamic MOS capacitor pull-up circuits", 1974 International Electron Devices Meeting, 20th, Technical Digest, p. 112-14.
821. Critchlow, D. L., "Sense amplifier for IGFET memory", IBM Tech. Disclosure Bull., <u>13</u> (6), 1720-2 (1970), CCA6-5686.

822. Dilatush, E., "Power supplies and testers - new markets and shifting priorities have upset many a tradition", EDN, $\underline{21}$ (22), 214-16 (1976), CCA12-15390.

823. Eardley, D. B., "Driver latch circuit", IBM Tech. Disclosure Bull., $\underline{19}$ (7), 2528-9 (1976), EEA80-24400.

824. Fuss, R. C. and Harland, R., "Peripheral circuits for one-transistor cell MOS RAM's", IEEE J. Solid-State Circuits, SC-$\underline{10}$ (5), 255-61 (1975).

825. General Instrument Corp., "Memory circuit", Patent UK 1296068, Publ. November 1972, CCA8-13699.

826. Guidry, M. R., Amelio, G. F. and Early, J. M., "A sense amplifier for a low clock capacitance 16K CCD memory", 1976 IEEE International Solid-State Circuits Conference, Digest of Technical Paper, p. 190-1, CCA11-28304.

827. Heller, L. G., Spampinato, D. P. and Yao, Y. L., "High sensitivity charge-transfer sense amplifier", IEEE J. Solid-State Circuits, SC-$\underline{11}$ (5), 596-601 (1976).

828. Hemingway, T. K., "Regulator gives over-voltage protection for TTL", Electronics, $\underline{43}$ (26), 54 (1970), CCA6-5593.

829. Johnson, R., Feldman, P. and Fisher, E., "Inductor simplifies memory-driver circuit", Electronics, $\underline{48}$ (2), 100 (1975).

830. Kruggel, R. H., "Voltage regulator sense circuit", IBM Tech. Disclosure Bull., $\underline{16}$ (9), 2784-5 (1974), CCA9-16074.

831. Lang, T. and Stone, H., "Shuffle-exchange network with simplified control", IEEE Trans. Comput., C-$\underline{25}$ (1), 55-65 (1976).

832. LeBlanc, A. R., "Field-effect transistor memory circuit", IBM Tech. Disclosure Bull., $\underline{15}$ (4), 1292-3 (1972), CCA8-5274.

833. Lum, W. Y., Chan, H. W. and VanDuzer, T., "Memory and logic circuits using semiconductor-barrier Josephson junctions", IEEE Trans. Magn., MAG-$\underline{13}$ (1), 48-51 (1977), CCA12-10601.

834. Marda, H., Sakuma, T. and Kobayashi, K., "Driving technique for thin film memory elements", Patent USA 3704457, Publ. November 1972, CCA8-13686.

835. Moore, S. F., "Voltage regulator protects logic pull-up transistors", Electronics, $\underline{47}$ (11), 106 (1974), CCA9-18339.

836. Novak, L., "DC constant voltage sources", Mereni A Regulace, $\underline{21}$ (6), 189-92 (1973), Czech, CCA9-18341.

837. Sehitoglu, H. and Klein, R. E., "Identification of multivalued and memory nonlinearities in dynamic processes", Simulation, $\underline{25}$ (3), 86-92 (1975).

838. Shapiro, S. D., "Dynamic memory storage allocation", 10th IEEE Computer Society International Conference, COMPCON 75, Digest of Papers, p. 105-8.

839. Smith, C. H. and Wittie, L. D., "Memory hardware for high speed job selection", IEEE Trans. Comput., C-$\underline{25}$ (2), 148-56 (1976), CCA11-6364.

840. Townend, M., "Memory-mapped I/O", New Electron, $\underline{9}$ (22), 19, 22 (1976), EEA80-6264.

841. Vishnyakov, V. A., Taranov, G. V., Sukhomlinov, V. B. and Bruk,
     A. I., "Synthesis and realization of a memory controller with
     electrical rewriting on MOSFETs", Izv. VUZ Priborostr, 19 (3),
     65-9 (1976), Russian, EEA79-28589.
842. Wiedmann, S. K., "Pinch load resistors shrink bipolar memory
     cells", Electronics, 47 (5), 130-3 (1974), CCA9-11961.

7.   Flip-Flops

843. Baechtold, W., Forster, T., Heuberger, W. and Mohr, T. O.,
     "Complementary Josephson-junction circuit: A fast flip-flop
     and logic gate", Electron Lett., 11 (10), 203-4 (1975).
844. Balph, T. and Gnauden, H., "Build a clock bias circuit for ECL
     flip-flops", EDN, 21 (9), 116 (1976).
845. Budinsky, J., "Integrated TTL flip-flop circuits", Slaboproudy
     Obzor, 30 (8), 364-73 (1969), Czech, CCA5-2696.
846. DasGupta, K. S. and Kumar, K. M. C., "Spare J-K flip-flops as
     two input nand gate", Electron. Eng., 46 (561), 21 (1974).
847. DeFord, D. I., "Gates convert J-K flip-flop to edge-triggered
     set/reset", Electronics, 50 (1), 106 (1977), CCA12-3812.
848. Fischer, K., "A trigger circuit for monostable flip-flops",
     Elektronik, 26 (2), 73 (1977), German, EEA80-16807.
849. Gimer, G., "A COS/MOS edge-triggered R-S flip-flop", New
     Electron, 9 (23), 39 (1976), EEA80-3024.
850. Halijak, C. A., "Characteristic equation and stable states of
     flipflops", Comput. & Electr. Eng., 3 (3), 339-43 (1976),
     CCA12-3868.
851. Hinch, P. G., "Low power CMOS switch debouncer", Electron.
     Engineering, 48 (586), 17 (1976), CCA12-3808.
852. Irving, T. A., Jr., Shiva, S. G. and Nagle, H. T., Jr., "Flip-
     flops for multiple-valued logic", IEEE Trans. Comput., C-25
     (3), 237-46 (1976).
853. Kammerer, J., "Application of digital circuits", Elektromeister
     & Dtsch. Elektrohandwerk, 51 (10), 647-8 (1976), German,
     CCA11-22664.
854. Kling, A., "TTL-RS-flip-flop with high static disturbing voltage
     security", Elektroniker, 12 (2), EL13-14 (1973), German,
     CCA8-13602.
855. Kuhne, H., "Tests of pulse-edge triggered flip-flops with TTL
     gates", Radio Fernsehen Elektron., 26 (3), 98-9 (1977),
     German, EEA80-28406.
856. Kunzel, R., "The state diagram", Elektronik, 22 (3), 97-100
     (1973), German, CCA8-13601.
857. Lal, M. and Parshad, R., "Single trigger source using binary
     flip-flop", J. Inst. Electron. Telecommun. Eng., 22 (4),
     191-2 (1976).
858. Levin, V. I., "Transients in elementary asynchronous automata
     with memory", Autom. Control Comput. Sci., 8 (2), 19-23 (1974).

859. Liu, B. and Gallagher, N. C., "On the 'metastable region' of
     flip-flop circuits", Proc. IEEE, 65 (4), 58o-3 (1977),
     EEA80-20392.

860. Mageswaran, A., "Flip-flop helps in multiplexing control bits",
     Electron. Engineering, 48 (586), 25 (1976), EEA80-6245.

861. Maholick, A. W., "Gated J-K flip-flop", IBM Tech. Disclosure
     Bull., 14 (2), 491-2 (1971), CCA6-24404.

862. Mange, D., "Definition, analysis and modes of representation
     of flip-flops", Syst. Logiques, no. 3, 171-83 (1971), French,
     CCA6-15142.

863. Martin, D. P., "Another way to build a two-gate flip-flop",
     Electronics, 47 (12), 124-5 (1974), CCA9-18311.

864. Mor, M., "On three valued elements simulation", International
     Symposium on Computer Hardware Description Languages and
     Their Applications, 1975, p. 179, CCA11-31553.

865. Pomeranz, J. N., "Complementary metal-oxide semiconductor J-K
     flip-flop and shift cell", IBM Tech. Disclosure Bull., 16 (11),
     3613-14 (1974), EEA78-366.

866. Pucknell, D. A., "Application equations, transition equations
     and the realisation of sequential networks of JK flip-flops",
     Monitor, 37 (11), 339-42 (1976), EEA80-28390.

867. Rath, S. B., "Ternary flip-flop circuit", Int. J. Electron, 38
     (1), 41-7 (1975).

868. Ropplinger, R. E., "Flip-flop", Patent USA 3437827, Publ. April
     1969, CCA5-2704.

869. Sintonen, L., "A clocked multivalued flip-flop", IEEE Trans.
     Comput., C-26 (3), 292-4 (1977), CCA12-8860.

870. Sudò, T. and Kindo, H., "New circuit for ultra high speed flip-
     flops", Rev. Electr. Commun. Lab., 22 (11-12), 1026-34 (1974).

871. Tohma, Y., "Design technique of fail-safe sequential circuits
     using flip-flops for internal memory", IEEE Trans. Comput.,
     C-23 (11), 1149-54 (1974), CCA9-23222.

872. Vancura, B., "A monostable flip-flop with a long period",
     Sdelovaci Tech., 25 (2), 67 (1977), Czech, EEA80-24411.

873. Volkov, B. A., Garipov, F. G. and Gorozhanin, V. A., "Flipflop
     based on solid-state logic circuits", Instrum. & Exp. Tech.,
     15 (4), 1102-3 (1972), CCA8-13611.

874. Wolf, G., "The incorrect operation of asynchronously driven
     flipflops", Frequenz, 31 (3), 71-6 (1977), German, EEA80-20374.

875. Yen, T. T., "C-MOS flip-flop can do more than logic tasks",
     Electronics, 48 (6), 123-6 (1975), EEA78-14897.

876. Zalevka, Z., "An accurate monostable flip-flop", Sdelovaci
     Tech., 24 (11), 421-2 (1976), Czech, EEA80-6271.

8.   Multivibrators

877. Anon., "Detection and storage of words supplied as serial bits
     in increasing word order", Mundo Electron., no. 52, 94 (1976),
     Spanish, CCA11-22569.

878. Bakonin, E. V. and Ostrovskiy, A. I., "A driven multivibrator
     with pulse-width modulation", Telecommun. & Radio Eng. Pt. 1,
     29 (11), 61-3 (1975), EEA80-255.
879. Bowes, R. C., "Improved crosscoupled multivibrator controllable
     in frequency over a wide range", Electron. Lett., 7 (8), 180-2
     (1971), CCA6-12632.
880. Coufal, J., "Compensated adjustment of the operating modes of
     gates of integrated logic circuits used in multivibrators",
     Sdelovaci Tech., 21 (2), 53-4 (1973), Czech, CCA8-13617.
881. D'yakonov, V. P., "A broadband self-oscillatory multivibrator
     based on integrated transistor-transistor-logic microcircuits",
     Instrum. & Exp. Tech., 19 (2), 435-9 (1976), EEA80-16793.
882. D'yakonov, V. P., "Driven multivibrators based on integrated
     circuits", Instrum. Exp. Tech., 19 (3), 785-8 (1976), EEA80-20385.
883. Hajek, J., "A multivibrator with thermal coupling (lighting and
     temperature control)", Sdelovaci Tech., 25 (1), 31 (1977),
     Czech, EEA80-16821.
884. Jung, W. G., "Hybrid one-shot logic input and bipolar output",
     Electron. Eng., 30 (4), 66 (1971), CCA6-17568.
885. Ochlerking, B., "An astable multivibrator with a pulse duty
     factor of 1:1", Elektronik, 26 (1), 52 (1977), German,
     EEA80-16801.
886. Palanichamy, S., "Transistorized astable multivibrator for pulse
     width modulation", J. Inst. Eng. (India) Electron. & Tele-
     commun. Eng. Div., 56 (pt. ET-3), 101-3 (1976), EEA80-2965.
887. Pogson, I., "Circuit and design ideas", Electron. Aust., 38 (6),
     75, 77 (1976), EEA80-16897.
888. Roberts, J. A., "T.T.L./D.T.L. compatible astable multivibrators",
     Des. Electron., 8 (7, 8), 74-8 (1971), CCA6-12641.
889. Sayanarayana, P., "Multivibrator suppresses switch contact
     bounce", Electron. Eng., 48 (577), 15 (1976).
890. Spasoje, T. and Vasiljevic, D., "Quartz multivibrator using
     two regenerative switches", Proc. IEEE, 65 (1), 166-7 (1977).
891. Trussen, C. T. B., "Gated CMOS multivibrator", New Electron,
     9 (23), 36, 39 (1976), EEA80-3023.
892. Yoshida, B., Takayanagi, M. and Wada, A., "Analysis of MOSFET
     astable multivibrator", Bull. Nagoya Inst. Technol., no. 27,
     455-60 (1975), Japanese, EEA80-16759.

       9.   Shift Registers
          a.   General

893. Anon., "Shift registers", Funkschau, 49 (1), 19-22 (1976),
     German, EEA80-16818.
894. Grodzki, Z., "New class of the controlled shift-registers-1",
     Probl. Control Inf. Theory, 5 (2), 181-8 (1976).
895. Grodzki, Z., "New class of the controlled shift-registers-2",
     Probl. Control Inf. Theory, 5 (3), 263-70 (1976).

896. Hsiao, M. Y., "Theories and applications of parallel linear
     feedback shift register", Thesis, University of Florida,
     Gainesville, 1967, 126 pp., Order No. 70-20632.
897. Kaspenkovitz, D., "Analysis and improvement of a static shift
     register", Microelectron. & Reliab., 13 (6), 501-15 (1974).
898. Zagurskii, V. Y., "Analysis of the dynamic operating regime
     of a fast shift register with intermediate diode memory",
     Avtom. & Vychisl. Tekh., no. 5, 55-64 (1971), Russian,
     CCA7-8234.

                    b.   Technologies
                    i.   Bipolar

899. Anon., "Bipolar LSI shift registers offer design flexibility",
     Comput. Design, 8 (11), 186-7 (1969), CCA5-8306.
900. Berger, H. and Wiedmann, S., "Dynamic shift register", IBM
     Tech. Disclosure Bull., 14 (1), 301 (1971), CCA6-22195.
901. Dunn, R. and Hartsell, G., "At last, a bipolar shift register
     with the same bit capacity as MOS", Electronics, 42 (25), 84-7
     (1969), CCA5-8295.
902. Ho, I. T., "Microwatt bipolar dynamic shift register cell",
     1970 International Hybrid Microelectronic Symposium, 1970,
     3 pp., CCA6-17592.
903. Lagerberg, J. P. L., "A high density static master slave shift
     register in I^2L", 1st European Solid State Circuits Conference -
     ESSCIRC, Extended Abstracts, 1975, p. 16-17, EEA79-28146.

                    ii. MOS

904. Bernstein, H., "MOS dynamic shift registers", Elektron. Ind.,
     7 (12), 351-4 (1976), German, EEA80-24389.
905. Cernoch, M., "An integrated MOS shift register", Sdelovaci
     Tech., 19 (2), 47-50 (1971), Czech, CCA6-22199.
906. Cooke, I. J., "MOS charge coupled shift registers for memory
     applications", Presented at Colloquium Digest on Computer
     Memories, London, England, February 1973, CCA8-11652.
907. Gaensslen, F. H. and Krick, P. J., "Noncomplementary static
     shift register", IBM Tech. Disclosure Bull., 13 (5), 1349-50
     (1970), CCA6-3728.
908. Ho, A. P. and Ting, Y. M., "Using a random-access memory as
     a shift register", IBM Tech. Disclosure Bull., 16 (11), 3567
     (1974), CCA9-25000.
909. Hoff, M. F., Jr. and Mazor, S., "Operation and application of
     MOS shift registers", Comput. Des., 10 (2), 57-62 (1971),
     CCA6-12639.
910. Kalisz, J., "Five-transistor delay cell of the dynamic two-
     phase MOS shift register", Elektronika, 16 (12), 511-12 (1975),
     CCA11-12252.

911. Kasperkovitz, D. and van Santen, J. G., "Integration density
     and power dissipation of MOS and bipolar shift registers - a
     comparison", Microelectron. & Rel., 9 (6), 497-501 (1970),
     CCA6-5591.
912. Milet, M. M., "Dynamic shift registers and their applications",
     Toute Electron, no. 365, 35-40 (1972), French, CCA7-14661.
913. Oliva, I., "Mos shift registers - application and new tech-
     nology", Ind. Ital. Elettrotec. & Elettron., 26 (4), 290-4
     (1973), Italian, EEA76-34916.
914. Ordermann, F., "Semiconductor storage unit with MOS shift
     register", Presented at Semiconductor Integrated and Produc-
     tion Conference and Exhibition, Anaheim, California, February
     1971, CCA6-15252.
915. Owczarek, A., "MOS-shift register as an example of sequence
     system", Elektronika, 16 (12), 513-16 (1975), CCA11-12253.
916. Pallum, C., "N-channel MOS registers and stores", Electron. &
     Microelectron. Ind., no. 162, 52-5 (1972), French, EEA76-4235.
917. Popov, V. P., Soldatenko, V. A., Chervan, B. A. and Antonov,
     M. A., "Signal distortion in a dynamic MOS-transistor shift
     register", Izv. VUZ Radioelektron., 17 (12), 31-4 (1974),
     Russian, EEA78-14915.
918. RCA, "Data store for dynamic shift register", Patent UK
     1401487, Publ. July 1975, CCA11-6387.
919. Seewann, E., "Dynamic MOSFET shift register array clock driver",
     IBM Tech. Disclosure Bull., 16 (9), 2767-8 (1974), CCA9-15997.
920. Shum, M. N. Y., "Using RAM as a shift register", Electron.
     Engineering, 48 (583), 23 (1976), CCA11-28106.
921. Soderman, L., "How MOS circuits function. V. Good MOS circuits
     are dynamic", Eltek. Aktuell Elektron, 17 (4), 46-7 (1974),
     Swedish, CCA9-18315.

iii.   CCD

922. Anon., "CCD's simplify complex electronics", Mach. Des., 47
     (7), 89 (1975).
923. Barsan, R. M., "Charge transfer and signal degradation in the
     presence of interface traps in charge-coupled shift registers",
     Rev. Roum. Phys., 22 (1), 57-71 (1977), CCA12-15386.
924. Berglund, C. N., Powell, R. J., Nicollian, E. H. and Clemens,
     J. T., "Two-phase stepped oxide CCD shift register using under-
     cut isolation", Appl. Phys. Lett., 20 (11), 413-4 (1972),
     EEA75-23383.
925. Chan, C. H. and Chamberlain, S. G., "A CCD serial to parallel
     shift register", IEEE J. Solid-State Circuits, SC-8 (5), 388-41
     (1973), EEA76-34920.
926. Chen, T. C. and Ho, I. T., "Flow-steering switches in semi-
     conductor shift register technology", IBM Tech. Disclosure
     Bull., 19 (5), 1686-9 (1976), EEA80-20380.

927. Collins, D. R., Rhines, W. C., Barton, J. B., Shortes, S. R.,
     Brodersen, R. W. and Tasch, A. F., Jr., "Electrical character-
     istics of 500-bit Al-Al$_2$-O3-Al CCD shift registers", Proc. IEEE,
     $\underline{62}$ (2), 282-4 (1974), EEA77-9706.
928. Dessertenne, B. T., Lakhoua, N. and Poirier, R., "Two-phase
     128 bits, charge-coupled shift register", Onde Electr., $\underline{54}$ (7),
     325-30 (1974), French.
929. Dickson, J. F., "Charge coupled device shift registers", Pro-
     ceedings of the Conference on Video and Data Recording, 1976,
     p. 49-56, CCA11-31626.
930. Kosonocky, W. and Carnes, J. E., "Charge-coupled digital cir-
     cuits", IEEE J. Solid-State Circuits, SC-$\underline{6}$ (5), 314-22 (1971),
     CCA6-22185.
931. Krupp, R. S. and Tomko, L. A., "Switching networks of planar
     shifting arrays", Bell Syst. Tech. J., $\underline{52}$ (6), 991-1007 (1973),
     EEA76-36167.
932. Linnenbrink, T. E., Monahan, M. J. and Rea, J. L., "CCD based
     transient data recorder", International Conference on the
     Applications of Charge-Coupled Devices, Proceedings, 1975,
     p. 443-53.
933. Mavor, J. and Davie, M. C., "Techniques for increasing the
     effective charge-transfer efficiency of tapped CDD registers",
     Electron. Lett., $\underline{13}$ (1), 31-3 (1977), EEA80-8994.
934. Moore, G. E., "A perspective of charge transfer devices and
     other shift registers", 1971 Digests of the International Mag-
     netics (Intermag.) Conference, 1 pp., CCA6-17589.
935. Siemens, K., Robinson, C. R. and Mastonardi, J., "Fast access
     bulk memory using CCD's", International Conference on the
     Applications of Charge-Coupled Devices, Proceedings, 1975,
     p. 429-34.
936. Singh, M. P., Lamb, D. R. and Roberts, P. C. T., "Effect of
     signal and fat-zero size on the performance of 3-phase CCD
     OC shift registers", Proceedings Inst. Electr. Eng., $\underline{122}$ (7),
     693-6 (1975).
937. Smith, G. E., "Charge transfer devices", 1971 Digests of the
     International Magnetics (Intermag.) Conference, 1 pp.,
     CCA6-17588.
938. Tiemann, J. J., Baertsch, R. D., Engeler, W. E. and Brown,
     D. M., "A surface-charge shift register with digital refresh",
     IEEE J. Solid-State Circuits, SC-$\underline{8}$ (2), 146-51 (1973),
     CCA8-11570.
939. Tompsett, M. F., Amelio, G. F. and Smith, G. E., "Charge
     coupled 8-bit shift register", Appl. Phys. Lett., $\underline{17}$ (3),
     111-5 (1970), EEA73-34343.
940. Williams, E. W., Brookes, R. and Windle, D., "Optimum operating
     conditions for an aluminum-gate CCD shift register", J. Phys.
     D, $\underline{7}$ (15), 2063-76 (1974), CCA9-25006.

iv. Bucket-Brigade

941. Berglund, C. N., "The bipolar transistor bucket-brigade shift register", IEEE J. Solid-State Circuits, SC-7 (2), 180-5 (1972), EEA75-12661.
942. Berglund, C. N. and Boll, H. J., "Performance limitations of the IGFET bucket-brigade shift register", IEEE Trans. Electron Devices, ED-19, 852-60 (1972), EEA75-23283.
943. Brunn, E. and Olesen, O., "Analog capacitance ROM with IGFET bucket-brigade shift register", IEEE J. Solid-State Circuits, SC-10 (1), 55-9 (1975), EEA78-8609.
944. Herrmann, F., "The calculation of MOS bucket-chain (bucket-brigade) circuits", Radio Fernsehen Elektron., 25 (24), 808-11 (1976), German, EEA80-25016.
945. Hinkle, F. E., "C-MOS decade divider clocks bucket-brigade delay line", Electronics, 46 (16), 117-18 (1975).
946. Pricer, W. D., "Pseudo bucket brigade operation in bipolar technology", IBM Tech. Disclosure Bull., 19 (1), 24-5 (1976), CCA11-22563.
947. Sangster, F. L. J., "The 'bucket-brigade delay line', a shift register for analogue signals", Philips Tech. Rev., 31 (4), 97-110 (1970), CCA6-15135.
948. Thornber, K. K., "Incomplete charge transfer in IGFET bucket-brigade shift registers", IEEE Trans. Electron. Devices, ED-18 (10), 941-50 (1971).
949. Vogl, N. G., "Bucket-brigade shift register tutorial", Electrochemical Society Fall Meeting, Extended abstracts, 1974, p. 477-8, EEA78-37685.
950. Yao, Y. L., "Bucket-brigade circuits", IBM Tech. Disclosure Bull., 15 (4), 1165-6 (1972), CCA8-5198.

v. Magnetic

951. Hughes Aircraft Co., "Magnetic shift register", Patent UK 1287967, Publ. July 1970, CCA8-2657.
952. Ooba, Y., Fukuoka, O. and Miyamoto, E., "A thin magnetic film shift register", IEEE Trans. Magn., MAG-8 (3), 311 (1972), CCA8-13607.
953. Popa, I., "Calculation of a single-ended shift register with ferritic cores and diodes", Bul. Stiint. & Teh. Inst. Politeh. 'Traian Vuia' Timisoara, 21 (1), 73-4 (1976), Rumanian, EEA80-28365.
954. Reach, R. W., Shapiro, D. and Barry Wright Corp., "Shift register having main and auxiliary shift windings", Patent USA 3438011, Publ. April 1969, CCA5-1087.

vi.   Others

955. Amirkhanov, A. V., Rybal'chenko, V. I., Stafeyev, V. I. and
     Fursin, G. I., "A shift register employing n-p-i structures
     with negative resistance", Radio Eng. & Electron. Phys., $\underline{21}$
     (3), 150-1 (1976), EEA80-9029.
956. Fulton, T. A., Dynes, R. C. and Anderson, P. W., "The flux
     shuttle-a Josephson junction shift register employing single
     flux quanta", Proc. IEEE, $\underline{61}$ (1), 28-35 (1973), CCA8-5203.
957. Ipri, A. C. and Sarace, J. C., "CMOS/SOS semi-static shift
     registers", IEEE J. Solid-State Circuits, SC-$\underline{11}$ (2), 337-8
     (1976).
958. Kan, Y. S. and Rakhubovskiy, V. A., "A two-cycle shift register
     using cryotrons", Sov. Autom. Control, $\underline{9}$ (3), 66-8 (1976),
     CCA12-8843.
959. Kasperkovitz, D., "An integrated 11 MHz p-n-p-n shift register",
     Solid-State Electron, $\underline{15}$ (5), 501-4 (1972), CCA7-14657.
960. Kasperkovitz, D., "20-MHz p-n-p-n shift register with current
     mirror coupling", IEEE J. Solid-State Circuits, SC-$\underline{10}$ (3),
     125-9 (1975).
961. Mause, K., Hesse, E. and Schlachetzki, A., "Shift register with
     Gunn devices for multiplexing techniques in the gigabit-per-
     second range", IEE J. Solid-State & Electron. Devices, $\underline{1}$ (1),
     17-23 (1976), CCA11-31492.
962. Meyer, J. E., Burns, J. R. and Scott, J. H., "High speed sili-
     con-on-sapphire 50-stage shift register", 1970 International
     Solid State Circuits Conference, Digest of Technical Papers,
     p. 200-1, CCA5-13674.
963. Mick, J. R., "A Schottky m.s.i. four-bit shifter", New
     Electron, $\underline{7}$ (17), 42, 44, 49, 52 (1974), CCA9-25008.
964. Yanagisawa, S., Wada, O. and Takanashi, H., "Gigabit rate Gunn-
     effect shift register", 1975 International Electron Devices
     Meeting, Technical Digest, p. 317-19, CCA11-31501.
965. Yao, Y. L. and Herrell, D. J., "Experimental Josephson junction
     shift register element", 1974 International Electron Devices
     Meeting, 20th, Technical Digest, p. 145-8.

c.   Properties

966. Anon., "Shift registers", Funkschau, $\underline{49}$ (1), 19-22 (1976),
     German, CCA12-15377.
967. Arzubi, L. M., "Shift register circuit for regeneration of
     dynamic flip-flops", IBM Tech. Disclosure Bull., $\underline{18}$ (8),
     2448-9 (1976), CCA11-12257.
968. Bluethman, R. G. and Southworth, R. A., "Synchronization of
     two dynamic shift registers", IBM Tech. Disclosure Bull.,
     $\underline{14}$ (2), 566-7 (1971), CCA6-24406.

969. Chen, T. C. and Tung, C., "Storage management operations in linked shift-register loops", 11th IEEE Computer Society Conference, COMPCON 75, Digest of Papers, p. 16-18.
970. Damashek, M., "Shift register with feedback generates white noise", Electronics, 49 (11), 107, 109 (1976).
971. Dujardin, J. P., "One NOR gate starts shift-register loop", Electronics, 48 (7), 103 (1975).
972. Fichtenbaum, M. L., "Scope-triggered register freezes serial data", Electronics, 49 (16), 108 (1976).
973. Fredricsson, S. A., "Pseudo-randomness properties of binary shift register sequences", IEEE Trans. Inf. Theory, IT-21 (1), 115-20 (1975).
974. Golunkov, Y. V., "Realization of microprogram bases on shift registers", Cybernetics, 11 (3), 379-85 (1975), CCA12-1399.
975. Harvey, J. T., "Delay line in shift register speeds M-sequence generation", Electronics, 48 (24), 104-5 (1975).
976. Hermann, A. L., Narashimham, M. S. and Suleman, U., "Isolation procedure for stuck-latch defects in shift registers", IBM Tech. Disclosure Bull., 19 (6), 2029-31 (1976), EEA80-20383.
977. Kasperkovitz, D., "Two-phase plasma-coupled static shift register", IEEE J. Solid-State Circuits, SC-10 (3), 143-51 (1975).
978. Latawiec, K. J., "New method of generation of shifted linear pseudorandom binary sequences", Proc. Inst. Electr. Eng., 121 (8), 905-6 (1974).
979. Mykkeltveit, J., "Generating and counting the double adjacencies in a pure circulating shift register", IEEE Trans. Comput., C-24 (3), 299-304 (1975).
980. Nagami, H., Kitahashi, T., Tanaka, K., Ae, T. and Yoshida, N., "Autonomous behaviors of feedback shift register with single threshold element", Electron. & Commun. Jpn., 58 (8), 37-45 (1975), EEA80-20370.
981. Sastri, K. S., "Generation of delayed replicas of linear p-level m-sequences", Int. J. Syst. Sci., 6 (5), 409-16 (1975).
982. Sastri, K. S., "Specified sequence linear feedback shift registers", Int. J. Syst. Sci., 6 (11), 1009-19 (1975).
983. Sastri, K. S., "Sequential behaviour of linear product feedback shift registers", J. Inst. Electron. & Telecommun. Eng., 22 (10), 653-60 (1976), CCA12-8837.
984. Simonov, V. S. and Shapoval'yants, V. G., "High-speed ring shift register", Instrum. Exp. Tech., 17 (5), 1334-6 (1974).
985. Svoboda, J., Cheng, R. M. H. and Kwok, C. C. K., "Pneumatic queueing controller", Instrum. Technol., 22 (4), 51-4 (1975).
986. Warner, B. H., Jr., "Very long shift registers can be built with RAMs", Electron. Des., 25 (1), 128 (1977), EEA80-28374.

10. Pulse Generators

987. Anon., "Driving lessons (digital logic driving pulses)", Elektor, 2 (2), 238-40 (1976), CCA11-14962.

988. Anon., "A multi-phase clock pulse generator using CMOS", Mundo Electron., no. 58, 81 (1976), Spanish, CCA12-10614.

989. Appleby, V. L., "Digitally controlled pulse generator for FM demodulation", Comput. Des., $\underline{13}$ (7), 83-7 (1974).

990. Aver'yanov, G. A., Voronov, A. S. and Khaitun, F. I., "Generator of arbitrary signal waveforms", Sov. J. Opt. Technol., $\underline{43}$ (5), 330-1 (1976), EEA80-2967.

991. Baker, P. W., "Suggestion for a fast binary sine/cosine generator", IEEE Trans. Comput., C-$\underline{25}$ (11), 1134-6 (1976), CCA11-31575.

992. Basiladze, S. G., Sun, L. W. and Parfenov, A. N., "A 24-channel controlled pulse generator for light-emitting diodes, which has a speed of response of 80 MHz", Instrum. & Exp. Tech., $\underline{18}$ (6), pt. 1, 1757-60 (1975), EEA80-254.

993. Beneking, H. and Filensky, W., "The GaAs MESFET as a pulse regenerator in the gigabit per second range", IEEE Trans. Microwave Theory & Tech., MTT-$\underline{24}$ (6), 385-6 (1976), EEA79-27970.

994. Berezin, A. A., Raikhtsaum, G. A., Ryabichev, Y. A., Smirnov, A. N. and Togatov, V. V., "A generator producing high-voltage trapezoidal voltage pulses with a controllable leading-edge slope", Instrum. & Exp. Tech., $\underline{19}$ (2), pt. 1, 413-15 (1976), EEA80-16762.

995. Berezin, A. A., Raikhtsaum, G. A. and Smirnov, A. N., "A high-power thyristor rectangular-pulse generator", Instrum. & Exp. Tech., $\underline{19}$ (2), pt. 1, 421-3 (1976), EEA80-893.

996. Blomeyer-Bartenstein, P., "Simple pulse generators in TTL-technique", Int. Elektron. Rundsch., $\underline{29}$ (9), 195-7 (1975), German, CCA11-4707.

997. Bronson, B. and Chan, A. Y., "A multifunction, multifamily logic pulser", Hewlett-Packard J., $\underline{28}$ (4), 12-17 (1976), CCA12-8812.

998. Buhler, O. R., "Pulser", IBM Tech. Disclosure Bull., $\underline{14}$ (2), 383 (1971), CCA6-24400.

999. Dick, I. R., "A pulse programmer for high-power nuclear resonance", J. Phys. E., $\underline{9}$ (12), 1054-6 (1976), EEA80-6229.

1000. Divnov, I. I., Gus'kov, Y. A., Zotov, N. I. and Karpov, O. P., "Explosion magnetic generator with an inductive load", Fiz. Goremya & Vzryva, $\underline{12}$ (6), 959-62 (1976), Russian, EEA80-16760.

1001. Dolkart, V. M., Novik, G. K. and Pats, V. B., "A pulse shaper continuously adjustable pulse duration based on transistor-transistor logic integrated units", Telecommun. & Radio Eng. Pt. 2, $\underline{28}$ (8), 102-5 (1973), CCA9-23195.

1002. Dunn, E. C. and Rayburn, K. C., "Pulse former", IBM Tech. Disclosure Bull., $\underline{19}$ (7), 2464-5 (1976), EEA80-24399.

1003. German, V. R., "A push-pull transistor magnetic sweep generator", Telecommun. & Radio Eng. Pt. 2, $\underline{31}$ (3), 124-9 (1976), EEA80-16770.

1004. Gimber, G., "Pulse generator with programmable pulse width", Electron. Eng., $\underline{48}$ (575), 19 (1976).

1005. Godfrey, T. J., Cripps, R. M. and Smith, G. D. W., "A high-
      voltage mercury-wetted reed pulse generator with secondary
      pulse suppression", J. Phys. E., $\underline{10}$ (4), 329-30 (1977),
      EEA80-16768.
1006. Gopinath, B. and Kurshan, R. P., "Periodic sequence generators
      using ordinary arithmetic", IEEE Trans. Circuits & Syst.,
      CAS-$\underline{23}$ (11), 677-83 (1976), CCA11-31515.
1007. Hubbard, J. H. and LeClere, R. J., "Noise rejection timing
      pulse generation", IBM Tech. Disclosure Bull., $\underline{19}$ (5),
      1610-11 (1976), EEA80-16819.
1008. Izadpanah, S. H., "Pulse generation and processing with GaAs
      bistable switches", Solid-State Electron., $\underline{19}$ (2), 129-32 (1976).
1009. Key, E. L., "An analysis of the structure and complexity of
      nonlinear binary sequence generators", IEEE Trans. Inf.
      Theory, IT-$\underline{22}$ (6), 732-45 (1976), EEA80-6226.
1010. Koshcheev, L. G., "A generator based on semiconductor gates
      which produces current pulses having a rectangular shape",
      Instrum. & Exp. Tech., $\underline{19}$ (1), 132-4 (1976), EEA80-6627.
1011. Krisch, P. and Sauberlich, W. D., "Universal pulse-source
      generator for clock frequencies up to 40 MHz", Fernmelde-
      technik, $\underline{13}$ (4), 202-4 (1973), German, CCA9-18316.
1012. Kurochkina, N. M. and Bykova, L. E., "Circuits for excluding
      or isolating one pulse from a continuous pulse series",
      Instrum. & Exp. Tech., $\underline{19}$ (3), pt. 2, 782-4 (1976), EEA80-24401.
1013. Lal, M. and Parshad, R., "Double pulse generator for determi-
      nation of recovery times of binary and other counters", IEEE
      Trans. Ind. Electron. & Control Instrum., IECI-$\underline{23}$ (4), 466-8
      (1976), CCA12-3818.
1014. Lu, J. H., Lo, H. Y. and Yang, C. M., "Simple circuit generates
      selected number of pulses", Electron. Eng., $\underline{48}$ (578), 23 (1976).
1015. Mazumdar, H. S. and Bandyopadhyay, K. K., "Low-cost data
      generator uses only two IC's", Electron. Eng., $\underline{47}$ (573), 17
      (1975).
1016. Nasyrov, M. S., "Digital control of pulse generators", Izv.
      Akad. Nauk. UzSSR Ser. Tekh. Nauk., no. 1, 18-20 (1977),
      Russian, EEA80-16767.
1017. Pettijohn, D., "Two-channel random pulse generator", Behav.
      Res. Methods & Instrum., $\underline{8}$ (3), 287-9 (1976), EEA80-6224.
1018. Posiello, J., "Generator of pulses from 0.1 Hz to 10 MHz",
      Rev. Esp. Electron., $\underline{23}$ (263), 46-7, 51 (1976), Spanish,
      EEA80-2966.
1019. Radkevich, A. V. and Yakovlev, Y. A., "Infra-low-frequency
      pulse generators", Instrum. & Exp. Tech., $\underline{19}$ (1), 137-8
      (1976), EEA80-6228.
1020. Razorenov, G. V., Vas'kivskii, F. A., Grechishkin, R. M. and
      Lukin, A. A., "A compact powerful-pulse generator", Instrum.
      & Exp. Tech., $\underline{19}$ (2), pt. 1, 424-5 (1976), EEA80-16765.

1021. Rysakov, V. M. and Aristov, Y. V., "A two-channel rectangular-
      pulse generator based on tubes having secondary emission",
      Instrum. & Exp. Tech., $\underline{19}$ (2), pt. 1, 432-4 (1976), EEA80-16766.
1022. Smith, D. T. and Ward, J. C., "A pseudo-random pulse train
      generator", Electron. Eng., $\underline{49}$ (591), 21, 23 (1977),
      EEA80-20332.
1023. Tan, Z. C. and Lo, S. N., "Generation of periodic sequences
      using shift register and ROM", Monitor, $\underline{37}$ (7), 231-2 (1976),
      CCA12-10615.
1024. Torrens, H., "Dual PRBSG", New Electron., $\underline{9}$ (1), 30 (1976),
      CCA11-12264.
1025. van Dee, E. and Alberts, B., "Pulse generator", Elektro, $\underline{2}$
      (11), 1138-40 (1976), CCA12-8856.
1026. Van Pelt, R. W., "Digital pulse width modulation generator",
      IBM Tech. Disclosure Bull., $\underline{14}$ (1), 61-2 (1971), CCA6-22188.
1027. Vandre, R. H. and Molen, G. M., "High-voltage pulse generator
      offers variable delay and only 3-ns jitter", Electron. Des.,
      $\underline{24}$ (9), 86 (1976).
1028. Wakerly, J. F., "Pulse generator produces programmable burst",
      Electronics, $\underline{49}$ (9), 99, 101 (1976), CCA11-17335.
1029. Zotov, E. V. and Krasovskii, G. B., "A multichannel high-
      voltage nanosecond-pulse generator", Instrum. & Exp. Tech.,
      $\underline{19}$ (2), pt. 1., 416-18 (1976), EEA80-16763.

11. Adders

1030. Agrawal, D. P., "Circuit adds BCD numbers faster with less
      hardware", Electronics, $\underline{48}$ (23), 120 (1975).
1031. Anker-Werke AG., "Computer command address modifier", Patent
      UK 1186521, Publ. April 1970, CCA6-3736.
1032. Carmichael, P., Coekin, J. A., Dow, R. and Wicking, J. R.,
      "New method of ultra-high-speed mod-2 addition", Proc. IEEE,
      $\underline{63}$ (4), 721-2 (1975).
1033. Chinal, J. P., "The logic of modulo $2^k + 1$ adders", 3rd
      Symposium on Computer Arithmetic, 1975, p. 126-36, CCA11-31537.
1034. Fadrhons, J., "Four-bit adders", Sdelovaci Tech., $\underline{24}$ (9),
      319-35 (1976), Czech, CCA12-3830.
1035. Freitag, H., "Binary parallel adders. II.", Radio Fernsehen
      Elektron., $\underline{21}$ (6), 199-302 (1972), German, CCA7-14655.
1036. Friedman, N., Salama, C. A. T., Holmes, F. E. and Thompson,
      P. M., "Multivalued integrated-injection-logic (MI^2L) full
      adder", Electron. Lett., $\underline{13}$ (5), 135-6 (1977), CCA12-8807.
1037. Friedman, N., Salama, C. A. T., Holmes, F. E. and Thompson,
      P. M., "Realization of a multivalued integrated injection
      logic (MI^2L) full adder", IEEE J. Solid-State Circuits, SC-$\underline{12}$
      (5), 532-4 (1977).
1038. Lim, I. C., "A full adder using Schottky-barrier diodes and a
      tunnel diode", J. Korean Inst. Electron. Eng., $\underline{9}$ (3), 22-8
      (1972), Korean, CCA8-13612.

1039. Magerlein, J. H. and Dunkleberger, L. N., "Direct-coupled
      Josephson full adder", IEEE Trans. Magn., MAG-13 (1), 585-8
      (1977), CCA12-10612.
1040. Mine, H., Hasegawa, T. and Shimada, R., "Self-timing asyn-
      chronous logic network with 3-valued NAND circuits and its
      use in binary adder", Syst. Comput. Contr., 1 (5), 40-7
      (1970), CCA7-14658.
1041. Moraga, C., "Hybrid logic (a fast ternary adder)", Proceedings
      of the 1975 International Symposium on Multi-Valued Logic,
      p. 344-58, CCA11-9624.
1042. Morisue, M., Sumi, T. and Imada, H., "Full adder using tunnel
      diode", Syst. Comput. Contr., 1 (5), 51-2 (1970), CCA7-14659.
1043. Murugesan, S., "Complement adders give unambiguous zero out-
      put", Electron. Engineering, 48(586), 25 (1976), CCA12-3809.
1044. Nordquist, W. R., "Full adder employing exclusive-NOR cir-
      cuitry", Patent USA 3590230, Publ. June 1971, CCA6-20083.
1045. Plaksin, I. I., "Analysis of the static operating conditions
      of a single-digit parallel adder using a tunnel diode and
      transistors", Telecommun. Radio Eng. Pt. 2, 25 (12), 129-31
      (1970), CCA6-24418.
1046. Rivero, R. and Alcaide, L., "Multi-adder/comparator with
      hysteresis", Mundo Electron., no. 57, 35-6, 40 (1976),
      Spanish, CCA12-3822.
1047. Shedletsky, J. J., "Comment on the sequential and indeterminate
      behaviour of an end-around-carry adder", IEEE Trans. Comput.,
      C-26 (3), 271-2 (1977), CCA12-8833.
1048. Stewart, Z., "Full adders simplify design of majority-vote
      logic", Electronics, 50 (12), 134, 136 (1977), EEA80-28368.
1049. Svoboda, A., "Self-checking adder for large scale integration",
      3rd Symposium on Computer Arithmetic, 1975, p. 108-12,
      CCA11-31536.
1050. Taylor, H. F., "Electro-optic technique for adding binary
      numbers", Electron. Lett., 11 (15), 313-14 (1975).
1051. Vekhtev, N. N., Kozlovskii, V. N. and Podsevakina, L. V.,
      "State diagnostics of adders in an electronic computer",
      Mekh. & Avtom. Proiz., no. 12, 34-7 (1973), Russian, CCA9-14284.
1052. Weinberger, A., "A 32-bit adder using partitioned memory
      arrays as universal logic elements", IBM Tech. Disclosure
      Bull., 14 (1), 205-7 (1971), CCA6-22193.
1053. Wojcik, A. S. and Metze, G., "On the cost of base N adders",
      IEEE Trans. Comput., C-20 (10), 1196-203 (1971), CCA6-22186.
1054. Zurla, F. A., "Decimal addition with two passes through binary
      adder", IBM Tech. Disclosure Bull., 14 (2), 621-2 (1971),
      CCA6-24408.

12. Multipliers/Dividers

1055. Arnit, M. A., "A frequency divider with variable division ratio", Latv. PSR Zinat. Akad. Vestis, no. 11, 128 (1976), Russian, EEA80-9083.

1056. BAC Ltd., "Binary multiplication circuit", Patent UK 1256221, Publ. December 1971, CCA7-14665.

1057. Baker, P. W. and McCrea, P. G., "A high-speed serial tree multiplier", Digital Processes, $\underline{1}$ (4), 343-9 (1975), CCA11-14955.

1058. Bezuglyi, V. V. and Zhukov, V. P., "Signal and noise at the output of a frequency multiplier", Izv. VUZ Radioelektron., $\underline{19}$ (12), 45-9 (1976), Russian, EEA80-20468.

1059. Bogdanov, V. S., Bortnikov, V. T. and Tsukanova, N. I., "Syn thesis of optimal circuits for low-sensitivity devices", Izv. VUZ Radioelektron., $\underline{19}$ (11), 118-22 (1976), Russian, EEA80-16904.

1060. Bruk, Y. M., "Period synthesizer with rational approximation to the period grid", Telecommun. & Radio Eng. Pt. 2, $\underline{31}$ (2), 91-4 (1977), EEA80-6347.

1061. Dadda, L., "On parallel digital multipliers", Alta Freq., $\underline{45}$ (10), 574-80 (1976), CCA12-15448.

1062. Dauby, A. and Roger-Rodes, D., "Multiplier circuit", IBM Tech. Disclosure Bull., $\underline{19}$ (5), 1722-4 (1976), EEA80-20381.

1063. Del Campo, J., "Universal frequency divider", Rev. Esp. Electron., $\underline{20}$ (219), 42-4 (1973), Spanish, CCA8-13616.

1064. Durrant, C. J. E., "Cheap logic trace multiplier", Pract. Electron., $\underline{12}$ (10), 830 (1976), CCA11-31498.

1065. Fisher, D. E., "Frequency doubler from clock-pulse recon-structer", Electron. Eng., $\underline{47}$ (570), 7 (1975).

1066. Foster, C., Riseman, E., Stockton, F. and Wogrin, C., "A novel multiply-by-three circuit", 3rd Symposium on Computer Arithmetic, 1975, p. 185-7, CCA11-31500.

1067. Gavalda, J., "1 Hz to control digital logic circuits", Rev. Esp. Electron., $\underline{20}$ (218), 38-41 (1973), Spanish, CCA8-11580.

1068. Geist, D. J., "MOS processor picks up speed with bipolar multiplier", Electronics, $\underline{50}$ (14), 113-15 (1977).

1069. Gimmel, B. A., "An improved type of digital frequency multi-plier", Proc. IEEE, $\underline{65}$(5), 815 (1977), EEA80-20475.

1070. Gusev, V. I. and Petrov, B. E., "Steady-state operational stability of a frequency multiplier with a large multiplication factor", Poluprovodn. Prib. & Tekh. Elektrosvyazi, no. 16, 95-105 (1975), Russian, EEA80-16913.

1071. Hampel, D., McGuire, K. E. and Prost, K. J., "CMOS/SOS serial-parallel multiplier", IEEE J. Solid-State Circuits, SC-$\underline{10}$ (5), 307-14 (1975).

1072. Herrell, D. J., "Experimental multiplier circuit based on super-conducting Josephson devices", IEEE J. Solid-State Circuits, SC-$\underline{10}$ (5), 360-8 (1975).

1073. IBM Corp., "Multiplier", Patent UK 1136522, Publ. December
      1968, CCA5-2698.
1074. Iizuka, T., "Frequency dividers by Schottky coupled I^2L gates",
      IEEE J. Solid-State Circuits, SC-12 (5), 530-2 (1977).
1075. Kane, J., "A low-power, bipolar, two's complement serial pipe-
      line multiplier chip", IEEE J. Solid-State Circuits, SC-11 (5),
      669-78 (1976), CCA12-1465.
1076. Kartalopoulos, S. V., "Hex inverter and or gates for frequency
      doubler", Electron. Engineering, 48 (586), 26-7 (1976),
      EEA80-6246.
1077. Knight, R. B. D. and Stockton, J. R., "Development of the Chung
      and Bedrosian iterative digital cellular multiplier", Electron.
      Lett., 12 (8), 182-3 (1976).
1078. Korneichuk, V. I., Romankevich, A. M. and Tarasenko, V. P.,
      "On the multiplication of digits represented by decade PCM
      pulses", Izv. VUZ Elektromekh., no. 11, 1285-6 (1976),
      Russian, EEA80-9070.
1079. Mick, J., Springer, J. and Ghest, C., "A very rapid series-
      parallel multiplier", Elektron. Ind., 7 (9), 196-7 (1976),
      German, CCA12-3815.
1080. Myasoedov, V. E., Filatov, V. M., Lyubimov, S. I. and Sharkov,
      V. I., "Study of ferromagnetic frequency dividers operating
      in a single-polarity source of periodic current", Izv. VUZ
      Elektromekh, no. 2, 206-8 (1977), Russian, EEA80-20467.
1081. Ninke, W. H. and Ritchie, G. R., "Shift register binary rate
      multipliers", IEEE Trans. Comput., C-26 (3), 276-8 (1977),
      CCA12-8834.
1082. Novozhilov, O. P., "The use of an alloyed semiconductor diode
      with charge storage in a frequency multiplier circuit", Tele-
      commun. & Radio Eng. Pt. 2, 31 (2), 85-90 (1977), EEA80-6346.
1083. Pezaris, S. D., "A 40-ns 17-bit by 17-bit array multiplier",
      IEEE Trans. Comput., C-20 (4), 442-7 (1971), CCA6-9290.
1084. Pipe, C. W., "Variable divider logic systems using integrated
      circuits", Colloquium on Digital Frequency Synthesis in
      Communication Systems (Digest), 1972, Paper 10/1, 3 pp.,
      CCA7-14663.
1085. Ramasubba Rao, N. P., Cleetus, G. M., Johari, V. M. and Mad-
      havan, R., "An ultrafast digital multiplier", Electro-Technol.,
      20 (3), 67-72 (1976), EEA80-20365.
1086. Rao, D. N., Glen, J. L. and Taylor, D., "A square root func-
      tion generator using a binary rate divider", Nucl. Instrum.
      & Meth., 100(3), 381-5 (1972), CCA7-14651.
1087. Reiner, R., "Static MOS frequency divider with three outputs",
      Funkschau, 48 (24), 1055-7 (1976), EEA80-6260.
1088. Sapozhnikov, B. A. and Nikitin, G. A., "On factor transforma-
      tion in fast multipliers", Izv. VUZ Priborostr., 19 (6), 62-6
      (1976), Russian, CCA11-31576.

1089. Shkalikov, V. N., "Phase-shift keyers employing frequency
       multipliers with oppositely connected varactors", Telecommun.
       & Radio Eng. Pt. 2, 31 (1), 109-10 (1976), EEA80-2977.
1090. Springer, J. and Alfke, P., "Parallel multiplier gets boost
       from IC iterative logic", Electronics, 43 (21), 89-93 (1970),
       CCA6-3721.
1091. van Holten, C., "Divide frequencies by nonintegers", Electron.
       Des., 24 (18), 104-7 (1976), CCA12-1404.
1092. Vedernikov, V. M., Kir'yanov, V. P. and Shcherbachenko, A. M.,
       "Method of decreasing multiplication errors in digital-pulse
       multipliers used in laser measurements of displacement",
       Avtometriya, no. 5, 87-92 (1976), Russian, CCA12-6573.
1093. Verjans, J. R., "A serial-parallel multiplier using the NENDEP
       technology", IEEE J. Solid-State Circuits, SC-12 (3), 323-5
       (1977), EEA80-24396.
1094. Zimin, V. V., "A frequency doubler", Telecommun. & Radio Eng.
       Pt. 2, 31 (3), 92-5 (1976), EEA80-16911.

13. Counters

1095. Agrawal, D. P., "Negabinary parallel counters", Digital Pro-
       cesses, 1 (1), 75-85 (1975).
1096. Alimov, V. A., "A high-speed decade counter with integrated
       circuits and a digital display", Prib. & Sist. Upr., no. 3,
       39 (1977), Russian, EEA80-28400.
1097. Anon., "MOSTEK MK 50395, a six decade counter and display
       module", Elektron. Ind., 7 (12), 357-8 (1976), German,
       EEA80-24390.
1098. Anon., "A new programmable counter modulo 10/11", Electron.
       & Appl. Ind., no. 231, 63 (1977), French, EEA80-28379.
1099. Anon., "A bidirectional counter used as a cyclic pseudo-
       random number generator", Electron. & Appl. Ind., no. 231,
       61-2 (1977), French, EEA80-28378.
1100. Artyukh, Y. N., Bespal'ko, V. A. and Zagurskii, V. Y., "A
       high-speed reversing counter with asynchronous counter",
       Instrum. & Exp. Tech., 16 (5), pt. 1, 1402-4 (1973),
       CCA9-25005.
1101. Artyukh, Y. N., Gotlib, G. I. and Zagurskii, V. Y., "Nano-
       second decade counter", Instrum. Exp. Tech., 18 (1), pt. 1,
       105-8 (1975).
1102. Barrett, R. P., "Electronic counter-timers for $6.00 per digit
       [student laboratory apparatus]", Phys. Teach., 15 (1), 54-6
       (1977), EEA80-9073.
1103. Behnke, H., "Electronic counter", Radio Fernsehen Elektron.,
       25 (15), 481-5 (1976), EEA80-288.
1104. Blaho, D., "Application of digital integrated circuits in a
       decade counter", Sdelovaci Tech., 19 (7), 201-4 (1971),
       Czech, CCA6-22201.

1105. Blatt, V., "An I^2L 4-decade programmable counter", 1st European
      Solid State Circuits conference - ESSCIRC, Extended Abstracts,
      1975, p. 22-3, EEA79-28149.
1106. Briggs, A. G., "Improvements to 100 MHz digital frequency
      counter", Electron. Aust., 38 (6), 75 (1976), EEA80-16812.
1107. Carruthers, J., Evans, J. H., Kinsler, J. and Williams, P.,
      "An introduction to digital counters - to accompany series 14
      of Circards", Wireless World, 80 (1461), 130-2 (1974),
      CCA9-23196.
1108. Cherkashin, F. A., Shinkarenko, V. L., Zinchenko, V. M. and
      Burenko, A. S., "A directional pulse counter", Prib. & Sist.
      Upr., no. 12, 40-1 (1976), Russian, EEA80-20396.
1109. Cowking, J., "'Preset-to-one' counter", Pract. Electron., 12
      (12), 989 (1976), EEA80-20398.
1110. Czech, Z., Hajduk, Z., Turek, L. and Kamela, A., "Decimal
      ring counter with digital readout", Pomiary Autom. Kontr.,
      18 (3), 119-21 (1972), Polish, CCA7-14653.
1111. Davio, M. and Bioul, G., "Interconnection structure of in-
      jective counters composed entirely of JK flip-flops", Inf. &
      Control, 33(4), 304-32 (1977), EEA80-28380.
1112. DeLaune, J., "50 MHz programmable counter designed with ECL
      integrated circuits", Contr. & Instrum., 3 (2), 39-42 (1971),
      CCA6-10714.
1113. Diaz, J., "Calculation pre-selection with an operational
      amplifier. [counter presetting]", Mundo Electron., no. 49,
      89 (1976), Spanish, CCA11-14966.
1114. Doak, T., Ville, J., Zuckswert, S. and Cuccia, C. L., "Giga-
      hertz rate counter logic and clock generation using high Ft
      transistors", 1971 IEEE-GMTT International Microwave Symposium,
      Digest of Technical Papers, p. 184-5, CCA6-24420.
1115. Dorr, R. C., "Self-checking combinational logic binary
      counters", IEEE Trans. Comput., C-21 (12), 1426-30 (1972),
      CCA8-2512.
1116. Dunn, E. C. and Rayburn, K. C., "Ring counter checking cir-
      cuits", IBM Tech. Disclosure Bull., 19 (7), 2460-1 (1976),
      EEA80-28384.
1117. Durgavich, T. and Abrams, D., "TTL decade counter divides
      pulse train by any integer", Electronics, 49 (14), 90 (1976),
      EEA79-37131.
1118. Engler, P., "Counting and memory decades with U 106 D", Radio
      Fernsehen Elektron., 25 (13), 420-1 (1976), German,
      CCA11-31622.
1119. Fadrhons, J., "The two-way counters", Sdelovaci Tech., 22
      (8), 291-7 (1974), Czech, CCA9-25013.
1120. Farr, T. M., Jr., "Read-only memory controls universal counter",
      EDN, 21 (9), 114 (1976).
1121. Hanisko, J., "Timer/counter functions as phase-locked loop
      component", EDN, 21 (6), 98 (1976).

1122. Hlavaty, J. and Kolesar, M. L., "Several synchronous reversing counters of impulses with integrated circuits", Sdelovaci Tech., 22 (8), 301-5 (1974), Slovak, CCA9-25014.
1123. Holub, I., "A simple and reliable decade counter with digitron display", Sdelovaci Tech., 19 (7), 205-7 (1971), Czech, CCA6-22202.
1124. Arutyunyan, G. G. and Ter-Israelov, G. S., "Decade pulse counter", Mekh. & Avtom. Upr., no. 3, 53-4 (1971), Russian, CCA6-24414.
1125. Kaestner, O., "Implementing branch instructions with polynomial counters", Comput. Des., 14 (1), 69-75 (1975).
1126. Kammerer, J., "The application of digital networks. IV.", Elektromeister & Disch. Elektrohandwek, 51 (20), 1322-3 (1976), German, CCA12-3814.
1127. Klein, H. J., "Fast automatic counter with highly variable cycle length", Nachrichtentech. Elektron., 27 (2), 61-3 (1977), German, EEA80-20390.
1128. Knallinger, G., "Enabling of prescaler frequency counters without additional gates", Elektronik, 26 (2), 65 (1977), German, EEA80-16804.
1129. Kumar, P., Arora, V. K. and Parshad, R., "Synchronous up/down decade counter", IEEE Trans. Ind. Electron. & Control Instrum., IECI-23 (3), 342-3 (1976), CCA11-31529.
1130. Laidman, H. S., "Upgrading inexpensive counters", CQ Radio Amat. J., 31(8), 16-22 (1975), CCA11-1007.
1131. Levy, D., Alexandrovitz, A. and Ben Uri, J., "A new high-speed up/down counter used as velocity-servo error register (digital phase comparator)", Proc. Inst. Radio Electron. Eng. Aust., 32 (2), 55-8 (1971), CCA6-15137.
1132. McNatt, M. S., "Use digital counting logic", Electron. Des., 23 (2), 62-6 (1975).
1133. Manning, F. B. and Fenichel, R. R., "Synchronous counters constructed entirely of J-K flip-flops", IEEE Trans. Comput., C-25 (3), 300-6 (1976).
1134. Marcinkowska, W., "Dynamic method of checking parallel counter", Prace Przemysl. Inst. Telekoman, 20 (68), 51-3 (1970), Polish, CCA6-20070.
1135. Masuyama, H. and Yoshida, N., "A simplification of counter using R-S flip-flops", Mem. Fac. Eng. Hiroshima Univ., 6 (1), 9-17 (1975), CCA11-22568.
1136. Medvedev, V. T., "A binary reversing counter with pulse logic which is based on integrated circuits", Instrum. & Exp. Tech., 18 (4), 1082-3 (1975), CCA11-17349.
1137. Pehoucek, M., "Tolerance analysis of a simple binary counter", Sdelovaci Tech., 24 (9), 325-6 (1976), Czech, CCA12-3829.
1138. Regalado, A., "Magnetic core counting circuit", Patent USA 3438014, Publ. April 1969, CCA5-2688.
1139. Runyon, S., "Focus on electronic counters", Electron. Des., 25 (1), 54-63 (1977).

1140. Sandell, J., "Simplified design of counters", Aust. Electron.
      Eng., 9 (9), 19-22 (1976), CCA12-6574.
1141. Schreiber, H., "30 MHz TTL counting decade using a biquinary
      numicator", Toute Electron., no. 356, 23-7 (1971), French,
      CCA6-17583.
1142. Scrupski, S. E., "All classes of electronic counters reap
      performance benefits from custom-designed LSI chips", Elec-
      tronics, 50 (9), 130-3 (1977).
1143. Sequin, C. H., "Logic drive circuit for 3-phase charge-coupled
      devices", Electron. Lett., 11 (16), 371-2 (1975).
1144. Shurygin, I. T. and Novikov, L. G., "Decimal counters on
      hybrid integrated micro-circuits", Prib. & Sist. Upr., no. 11,
      44-5 (1976), Russian, EEA80-20395.
1145. Skarzhepa, V. A. and Morozov, A. A., "Reversible pulse counters
      using blocking thyristors with a cathode load", Izv. VUZ
      Radioelektron., 19 (12), 80-2 (1976), Russian, EEA80-28381.
1146. Thomas, H., "Digital difference counter", Elektronik, 26 (2),
      72 (1977), German, EEA80-16806.
1147. Wagner, F., "Codes of counter registers", Arch. Autom. &
      Telemech., 21 (2), 225-32 (1976), Polish, CCA11-31487.
1148. Westowski, K., "Accumulating adder works with reversible
      counter", Electron. Engineering, 49 (588 , 18 (1977),
      CCA12-10604.
1149. Yuen, C. K., "Note on base-2 arithmetic logic", IEEE Trans.
      Comput., C-24 (3), 325-9 (1975).

14. Clocks

1150. Bentley, J. H., "Build a multiphase clock that never loses its
      sequence", Electron. Des., 23 (6), 96, 98 (1975).
1151. Birnie, P. A., "Digital event timer. II. Construction",
      Wireless World, 83 (1493), 65-8 (1977), EEA80-9034.
1152. Bolon, C., "Decoders drive flip-flops for clean multiphase
      clock", Electronics, 50 (8), 107-9 (1977), EEA80-16799.
1153. Bose, S., "Retriggerable one shot using a 555 timer", Electron.
      Engineering, 48 (585), 23, 25 (1976), EEA80-253.
1154. Chandrasekhar, M. G., Manoharan, L. C. and Neelakantan, S.,
      "Programmable start/stop timers using digital integrated
      circuits", J. Inst. Eng. (India) Electron Telecommun. Eng.
      Div., 55 (1), 15-16 (1974).
1155. Chawla, B. R., Gummel, H. K. and Kozak, P., "MOTIS an MOS
      timing simulator", IEEE Trans. Circuits Syst., CAS-22 (12),
      901-10 (1975).
1156. Clarke, E. P., DeOrazio, W. R. and Lilie, M. I., "Computer
      accessible digital clock", IBM Tech. Disclosure Bull., 15 (8),
      2524-6 (1973), CCA8-13621.
1157. Cross, A. F., "Time-code receiver clock-2. control logic and
      display", Wireless World, 82 (1483), 63-6 (1976).

1158. Deshmuth, J., "Clock derived programmable time delays", Electron. Engineering, 48 (582), 20-1 (1976), CCA11-28015.

1159. Deutchmann, G., "Operational behavior of TTL-compatible timing elements", Radio Fernsehen Elektron., 25 (22), 736-40 (1976), German, EEA80-20399.

1160. George, G. A., "Serial data clock control", IBM Tech. Disclosure Bull., 16 (10), 3338-9 (1974), CCA9-18340.

1161. Gergek, F., "Potentiometer and timer control up/down counter", Electronics, 49 (10), 94-5 (1976).

1162. Gindi, A. M. and Moore, J. G., "Synchronized clocking system", IBM Tech. Disclosure Bull., 19 (5), 1900-4 (1976), CCA12-12795.

1163. Hahs, W. R., "Delay line clock-data separator", IBM Tech. Disclosure Bull., 14 (1), 116-17 (1971), CCA6-22191.

1164. Hammel, R. A., "Dual frequency dual duty cycle gated clock generator", Electronics, 50 (2), 99 (1977), CCA12-8805.

1165. Heeren, R. H., "Clocked dynamic inverter", Patent USA 3903431, Publ. September 1975, CCA11-9590.

1166. Helm, W., "Adjustable TTL clock maintains 50% duty cycle", Electronics, 50 (3), 107, 109 (1977), CCA12-10607.

1167. Huertas, J. L., "Programmable delay synchronised with a clock signal", Electron. Eng., 47 (565), 21 (1975).

1168. Jung, W. G., "Take a fresh look at new IC timer applications", EDN, 22 (6), 127-35 (1977).

1169. Kozak, P., Gummel, H. K. and Chawla, B. R., "Operational features of an MOS timing simulator", Design Automation Conference, 12th, Proceedings, p. 95-101.

1170. Lamdan, T., "Asynchronous timing in logic systems", Digital Processes, 2 (2), 157-62 (1976).

1171. Macfarlane, I. P., "A digital phaseable clock and time code generator", Proc. Inst. Radio Electron. Eng. Aust., 32 (2), 38-50 (1971), CCA6-15136.

1172. Marion, A. and Gouyon, A., "Application of integrated logic circuits: construction of a digital clock. I.", Toute Electron., no. 353, 33-6 (1971), French, CCA6-15143.

1173. Meggitt, G. C., "TTL system programmer for cyclic activation analysis", Nucl. Instrum. Methods, 124 (2), 455-9 (1975).

1174. Millman, B., "A modification to the AX-08 to enable the crystal clock to function as a one millisecond clock", Decuscope, 9 (5), 6-7 (1970), CCA6-12643.

1175. Mitchell, C. J., "Timing circuit", Patent UK 1256638, Publ. December 1971, CCA7-14666.

1176. Mizuno, M., "Two-phase dynamic logic circuit", Patent USA 3909627, Publ. September 1975, CCA11-6348.

1177. Muller, R., "I^2L timing circuit for the 1ms-10s range", IEEE J. Solid-State Circuits, SC-12 (2), 139-43 (1977), EEA80-9017.

1178. Muller, R., "I^2L timing circuit without external components", 1977 International Solid-State Circuits Conference, Digest of Technical Papers, p. 112-13.

1179. Nye, D. D., "Running time clock", IBM Tech. Disclosure Bull.,
      18 (8), 2428-9 (1976), CCA11-12272.
1180. Peltz, G., "Long duration timing circuit", Toute Electron.,
      no. 409, 78-83 (1976), CCA11-17357.
1181. Pfleiderer, H. J. and Knauer, K., "Generating overlapped
      clock phases for CCD array", Electronics, 48 (3), 97, 99 (1975).
1182. Reid, R., "Pushbutton gated clock pulse", New Electron., 10
      (1), 40 (1977), EEA80-16794.
1183. Roberts, J. and Chan, C., "A universal bipolar/MOS clock
      driver", Comput. Des., 13 (4), 152 (1974), CCA9-18307.
1184. Ruchli, A. E., Wolff, P. K., Sr. and Goertzel, G., "Power
      and timing optimization of large digital systems", Proceedings
      of the 1976 IEEE International Symposium on Circuits and Sys-
      tems, p. 402-5, CCA12-3886.
1185. Tenny, R., "Build a programmable sequencer with a broad opera-
      ting range", EDN, 21 (10), 90-2 (1976).
1186. Thompson, P. M. and Belanger, A. J., "Digital arithmetic units
      for a high data rate", Radio Electron. Eng., 45 (3), 116-20
      (1975).
1187. Veto, K., "Phase-locked loop generates clock from nonreturn-
      to-zero data", Electron. Des., 23 (9), 90 (1975).
1188. Wheelock, N. G., "Micropower comparators generate 2-phase
      clock", Electronics, 49 (17), 94-5 (1976), CCA11-28129.
1189. Wilkinson, A. B., "Clock for teletype serial data", Electron.
      Eng., 46 (561), 25 (1974).
1190. Williams, M. D., "Up/down latching sequencer keeps order",
      Electronics, 49 (25), 99 (1976), CCA12-3811.

15. Converters
a. A/D and D/A

1191. Abdullaev, I. M. and Akhmedov, R. M., "A functional analog-
      digital converter based on integrated microcircuits", Instrum.
      & Exp. Tech., 19 (2), pt. 1, 87-8 (1976), EEA80-20472.
1192. Aitchison, C. S., "Fast A.D. converter using a digital dis-
      criminator and a V.C.O.", Electron. Lett., 13 (10), 281-2 (1977).
1193. Akella, B. S. and Srinivasa, Y. G., "A new A/D converter for
      non-linear input-output relationships", Int. J. Electron.,
      42 (3), 295-8 (1977), CCA12-8851.
1194. Albarran, J. F., "A charge-transfer multiplying digital-to-
      analog converter", IEEE J. Solid-State Circuits, SC-11 (6),
      772-9 (1976), EEA80-6337.
1195. Anon., "General purpose analogue to digital converter",
      Electron. Engineering, 48 (584), 31-3 (1976), CCA12-1423.
1196. Anon., "Principle of an A/D circuit with log-output", Radio
      Elektron. Schau, 52 (11), 75 (1976), German, EEA80-20479.
1197. Antonio, J., "Digital analogue converter with a 'priority'
      encoder", Mundo Electron., no. 57, 98 (1976), Spanish,
      CCA12-3859.

1198. Belomestnykh, V. A., V'yukhin, V. N. and Kasperovich, A. N.,
"An experimental method of determining the dynamic properties
of a fast-acting A/D convertor", Avtometriya, no. 5, 83-7
(1976), Russian, CCA12-6586.

1199. Blazhkevich, B. I. and Pogribnoi, V. A., "Follow-up electromechanical A/D converters using multistable elements", Izv.
VUZ Elektromekh., no. 2, 187-90 (1977), Russian, EEA80-16902.

1200. Brugger, R., "From analogue to digital", Micomp, $\underline{1}$ (1), 58-60
(1976), German, CCA12-3862.

1201. Candy, J. C., Ching, Y. C. and Alexander, D. S., "Using tri-
angularly weighted interpolation to get 13-bit PCM from a
sigma-delta modulator", IEEE Trans. Commun., COM-$\underline{24}$ (11),
1268-75 (1976).

1202. Cecil, J. B. and Solomon, J. E., "Present and future trends
in monolithic A/D and D/A converter art", 13th IEEE Computer
Society International Conference, COMPCON 76, Digest of
Papers, p. 59-62, CCA12-1428.

1203. Clark, V. R., "D/A conversion with single supply circuit",
Analog Diagloue, $\underline{10}$ (1), 15 (1976), CCA12-1421.

1204. Collins, F. A. and Sicking, C. J., "Properties of low precision
analogue-to-digital converters", IEEE Trans. Aerosp. & Electron.
Syst., AES-$\underline{12}$ (5), 643-6 (1976), CCA12-3855.

1205. Comer, D. T., "A monolithic 12-bit D/A converter", 1977 IEEE
Solid-State Circuits Conference, Digest of Technical Papers,
p. 104-5.

1206. Daleiden, G., "An economically priced DA converter", Elektronik, $\underline{26}$ (1), 44 (1977), German, EEA80-16894.

1207. Davies, C., "A high-speed multiplying ADC [A/D convertor]",
Electron., no. 107, 59 (1976), EEA80-16899.

1208. Dendinger, S., "Try the sampling comparator in your next A/D
interface design", EDN, $\underline{21}$ (17), 91-5 (1976), CCA12-6589.

1209. Dobkin, R. C., "Don't forget reference stability when designing
A-to-D converters", EDN, $\underline{22}$ (12), 105-8 (1977).

1210. Eichelberger, C. W. and Butler, W. J., "An analog-to-digital
converter using charge-transfer technology", 1977 IEEE International Solid-State Circuits Conference, Digest of Technical
Papers, p. 94-5.

1211. Evans, L., "Building blocks take the problem out of A/D converter designs", EDN, $\underline{21}$ (14), 68-70 (1976), CCA12-3853.

1212. Fadem, R. J., "Function synthesizer avoids D/A discontinuities",
EDN, $\underline{21}$ (11), 118 (1976).

1213. Fang, F. F. and Herrell, D. J., "Analog-to-digital converter
for Josephson tunneling memory", IBM Tech. Disclosure Bull.,
$\underline{18}$ (8), 2668-70 (1976), CCA11-12328.

1214. Gobbur, S. G., Landis, D. A. and Goulding, F. S., "Fast suc-
cessive approximation analog-to-digital converter", Nucl.
Instrum. & Methods, $\underline{140}$ (2), 405-6 (1977), CCA12-10629.

1215. Gordon, B. M., "Analog-to-digital conversion, a tutorial", Tutorial Proceedings of the 22nd International ISA Instrumentation Symposium, 1976, p. 1-16, EEA80-20482.

1216. Hareyama, K., Yoshimura, K. and Murase, K., "A bipolar monolithic analog subsystem for a 4-1/2 digit A/D converter", 1977 IEEE International Solid-State Circuits Conference, Digest of Technical Papers, p. 100-1.

1217. Hartley, R. W., "Special purpose analogue to digital converter", Electron. Engineering, $\underline{48}$ (585), 28 (1976), CCA12-1424.

1218. Hell, R., "Digital sample and hold", Radio Elektron. Schau, $\underline{52}$ (10), 55 (1976), German, CCA12-6593.

1219. Jenkins, A., "The ZN425E – a new device for analogue/digital conversion", SERT J., $\underline{10}$ (9), 206-7 (1976), EEA80-16910.

1220. Klein, M., "Analog digital converter using Josephson junctions", 1977 International Solid-State Circuits Conference, Digest of Technical Papers, p. 202-3.

1221. Korotkova, N. A. and Chistyakov, N. P., "A comparison circuit for an analogue/digital convertor of digit-by-digit balancing", Prib. & Sist. Upr., no. 8, 44-5 (1976), Russian, CCA12-1427.

1222. Kosinskii, A. V., "Analogue-digital converter of displacements in a multichannel phase shifter", Meas. Tech., $\underline{18}$ (11), 1602-6 (1975), CCA12-3858.

1223. Krendelev, V. A. and Rychenkov, V. N., "A multichannel analog-digital converter", Instrum. & Exp. Tech., $\underline{18}$ (6), pt. 1, 1752-6 (1975), CCA12-1425.

1224. Landsburg, G. F., "A charge-balancing monolithic A/D converter", 1977 IEEE International Solid-State Circuits Conference, Digest of Technical Papers, p. 98.

1225. McCharles, R. H., Saletore, V. A., Black, W. C. and Hodges, D. A., "An algorithm analog-to-digital converter", 1977 IEEE International Solid-State Circuits Conference, Digest of Technical Papers, p. 96-7.

1226. Markov, B. F. and Sotirov, G. R., "Estimation of noise stability in digit by digit balancing convertors", Izv. VUZ Priborostr., $\underline{19}$ (11), 58-63 (1976), Russian, CCA12-6590.

1227. Mattera, L., "Monolithic converters gain strength", Electronics, $\underline{50}$ (12), 78-9 (1977).

1228. Memishian, J. and Marshall, W., "Really-good low-power modular converters call for more than just a few off-the-shelf IC's", Analog Dialogue, $\underline{10}$ (2), 12-13 (1976), CCA12-10627.

1229. Nagaikin, A. S., Torchin, A. L., Eremina, G. I. and Gordienko, A. M., "A converter for converting an alternating-sign code into a voltage", Instrum. & Exp. Tech., $\underline{19}$ (1), 101-3 (1976), CCA12-8852.

1230. Ortiz, M. T., "Design for an analogue-digital convertor. I. Analogue section", Control Cibern. & Autom., $\underline{10}$ (1), 36-47 (1976), Spanish, CCA12-12801.

1231. Ortiz, M. T. and Karoly, V., "Design for an analogue-digital convertor. II. Digital section", Control Cibern. & Autom., 10 (1), 48-53 (1976), Spanish, EEA80-16893.

1232. Plassche, R. J. v. d. and Grift, R. E. J. v.d., "A five-digit A/D converter", 1977 IEEE International Solid-State Circuits Conference, Digest of Technical Papers, p. 102-3.

1233. Popov, V. P., "Automatic error correction for the results of analogue-digital conversion", Avtometriya, no. 5, 62-79 (1976), Russian, CCA12-6585.

1234. Post, H. U. and Waldschmidt, K., "Quasi-digital filtering for indirect DACS in MOS-technology", Proceedings of the 1976 IEEE International Symposium on Circuits and Systems, p. 303-6, CCA12-3863.

1235. Pretzl, G., "High speed analog/digital converters", Elektronik, 25 (12), 36-42 (1976), German, CCA12-3852.

1236. Pretzl, G., "Monolithic D/A and A/D converters", Elektronik, 26 (3), 38-46 (1977), German, EEA80-20463.

1237. Price, J. J., "A passive laser-trimming technique to improve the linearity of a 10-bit D/A converter", IEEE J. Solid-State Circuits, SC-11 (6), 789-94 (1976), EEA80-6338.

1238. Raphael, H. A., "Low cost A-to-D conversion during microcomputer idle time", 16 (3), 112, 114, 116 (1977), CCA12-12800.

1239. Ritmanich, W., Langley, D. and Wold, I., "13-bit monolithic CMOS A/D converter", Analog Dialogue, 10 (1), 3-5 (1976), CCA12-1420.

1240. Schoeff, J. A., "A monolithic companding D/A converter", 1977 IEEE International Solid-State Circuits Conference, Digest of Technical Papers, p. 58.

1241. Tarnay, K., "Remarks on the analysis of a stabilized analog-digital converter", Probl. Control & Inf. Theory, 4 (3), 231-9 (1975), EEA80-20477.

1242. Trofimov, A. S. and Chelnokov, L. P., "A pulsed analog-digital converter in the CAMAC standard", Instrum. & Exp. Tech., 19 (2), pt. 1, 367-8 (1976), EEA80-20470.

1243. Tseng, S. C. C., "Analog-to-digital converter", IBM Tech. Disclosure Bull., 19 (5), 1940-3 (1976), EEA80-16905.

1244. Van De Plassche, R. V., "Dynamic element matching for high-accuracy monolithic D/A converters", IEEE J. Solid-State Circuits, SC-11 (6), 795-800 (1976), CCA12-3854.

1245. Watson, J. D., "A low cost precision A/D system", Electron. Ind., 2 (11), 21-3 (1976), CCA12-6588.

1246. Wheelock, N., "A/D converter chip sets do more than just cut DVM parts count", EDN, 22 (3), 88-91 (1977).

1247. Whitelaw, D., "Digital-analogue converters solving the price/ performance problem", Electron. Power, 20 (9), 892-3 (1974).

1248. Whitelaw, D., "D/A and A/D converters: making the right choice", Electron, no. 107, 60 (1976), EEA80-16900.

1249. Woodward, C. E., Konkle, K. H. and Naiman, M. L., "Monolithic
      voltage-comparator array for A/D converters", IEEE J. Solid-
      State Circuits, SC-10 (6), 392-9 (1975).
1250. Yamaguchi, T. and Sato, S., "Monolithic clockless-A/D-converter
      integrated circuits", IEEE Trans. Electron. Devices, ED-22
      (5), 295-7 (1975).
1251. Zuch, E. L., "Converters match bit machines to the real world",
      Can. Controls & Instrum., 15 (1), 20-2 (1976), CCA12-8850.
1252. Zuch, G., "The wonderful world of A/D/A. II. Converters match
      bit machines to the real world", Can. Controls & Instrum.,
      15 (10), 32-3 (1976), CCA12-6587.

                           b.   Code

1253. Anon., "BCD-binary converter", New Electron., 9 (1), 29 (1976),
      CCA11-12263.
1254. Anon., "Seven-segment to BCD convertor", Elektro, 2 (12), 1235
      (1976), EEA80-16901.
1255. Benedek, M., "Developing large binary to BCD conversion struc-
      tures", 3rd Symposium on Computer Arithmetic, 1975, p. 188-96,
      CCA11-31581.
1256. Beougher, L. C., "A method for high speed BCD-to-binary con-
      version", Comput. Des., 12 (3), 53-9 (1973), CCA8-13600.
1257. Blandford, D., "BCD to binary converter", New Electron., 9
      (2), 14 (1976), CCA11-14968.
1258. Brockman, D. M., "Special PROM mode effects binary-to-BCD
      converter", Electronics, 50 (7), 105, 107 (1977), EEA80-20362.
1259. Burgel, E., "Fast code converters with MSI-TTL circuits. II.
      [Binary to BCD]", Elektronik, 24 (12), 77-80 (1975), German,
      CCA11-9565.
1260. Chandraschkar, V. T. and Ramadurai, R., "Seven segment to BCD
      converter uses standard TTL", Electron. Engineering, 49
      (587), 19 (1977), CCA12-6575.
1261. Chetty, P. R. K., "Economical code convertor for digital
      systems", J. Inst. Electron. & Telecommun. Eng., 22 (8),
      523-5 (1976), CCA12-1415.
1262. Jurecka, J., "The TESLA MH 7442 integrated circuit for BCD
      to decimal conversion", Sdelovaci Tech., 24 (8), 283-90
      (1976), Czech, EEA80-3028.
1263. Lanning, W. C., "General algorithms for direct radix conver-
      sion", Comput. Des., 14 (6), 61-73 (1975).
1264. Murugasan, S. and Kameswara, Rao, C., "Binary/b.c.d. conver-
      sion made simple", Electron. Engineering, 48 (585), 25 (1976),
      CCA12-1401.
1265. Peshkov, A. T. and Vishnyakov, V. A., "A decimal-to-binary
      converter for computers", Izv. VUZ Priborostr., 18 (9), 71-4
      (1975), Russian, CCA11-12258.

1266. Singh, H., Manohar, M. and Gupta, A. C., "Sixteen bit fractional binary to digital medium speed converter with 4-digit display", J. Inst. Electron. & Telecommun. Eng., 21 (8), 421-4 (1975), CCA11-12261.
1267. Shouler, P., "Improved circuit for seven segment converter", Electron. Eng., 49 (592), 21 (1977), EEA80-24384.
1268. Southway, J., "IC trio converts 7-segment code to decimal", Electronics, 47 (24), 113 (1974).
1269. Valuev, Y. M., Grebenyuk, V. M. and Zinov, V. G., "A time interval to digital code converter based on integrated circuits", Instrum. & Exp. Tech., 19 (2), pt. 1, 371-4 (1976), EEA80-20471.
1270. Wydro, K. and Klobukowski, W., "Conversion of integer codes", Pomiary Autom. Kontr., 17 (6), 255-8 (1971), Polish, CCA6-20068.

c. Voltage/Frequency

1271. Bosnjakovic, P. and Kon, J., "New voltage-to-frequency convertor with nonlinearity correction", Electron. Lett., 13 (1), 5-6 (1977), EEA80-6335.
1272. Cate, T., "IC V/F converters readily handle other functions such as F/V, A/D", EDN, 22 (1), 82-6 (1977), EEA80-20466.
1273. Czarnul, Z. and Bialko, M., "Utilisation of a single inductorless neuristor line section as a voltage-to-frequency convertor", Electron. Lett., 13 (9), 251-6 (1977), EEA80-20465.
1274. Gilbert, B., "A versatile monolithic voltage-to-frequency converter", IEEE J. Solid-State Circuits, SC-11 (6), 852-64 (1976), EEA80-6340.
1275. Gorka, W., "A current/frequency convertor linear over five decades", Elektronik, 26 (3), 64 (1977), German, EEA80-20464.
1276. Harjung, A., "A voltage-frequency convertor with high linearity", Elektronik, 26 (1), 73 (1977), German, EEA80-16896.
1277. Klynin, V. A., Tikhonov, Y. N. and Korneev, V. N., "A linear voltage-frequency converter", Instrum. & Exp. Tech., 19 (2), pt. 1, 397-9 (1976), EEA80-16906.
1278. Knight, M. B., "Voltage-to-current converters", RCA Tech. Not., no. TN 1167, 1-5 (1976), EEA80-20480.
1279. Kokorev, E. N. and Chernov, V. G., "Integrating voltage-to-frequency converter with high input impedance", Izv. VUZ Priborostr., 20(1), 58-62 (1977), Russian, EEA80-16908.
1280. Kraus, K., "Simple voltage-frequency converters using 74121 monoflops", Nachr. Elektron., 31 (2), 34 (1977), German, EEA80-20474.
1281. Kress, D. and Gilbert, B., "Versatile monolithic V/f or I/f converter", Analog Dialogue, 10 (2), 6-8 (1976), CCA12-10626.
1282. Mel'nikov, O. N., "A logarithmic frequency-voltage convertor", Meas. Tech., 19 (1), 129-33 (1976), CCA12-6592.

1283. Siluyanov, B. P., "Transistorized DC voltage-to-frequency
      convertor", Poluprovodn. Prib. & Tekh. Elektrosvyazi, no. 16,
      175-82 (1975), Russian, EEA80-16914.
1284. Stoenescu, A., "Wobbulated function generator", Electroteh.
      Electron. & Autom. Autom. & Electron., 20 (4), 163-6 (1976),
      Rumanian, CCA12-10622.
1285. Taniguchi, K. and Sakai, T., "A method for improving the
      linearity of reset-type voltage-to-frequency converters",
      Syst.-Comput.-Control, 6 (5), 1-7 (1975), EEA80-20481.

                    d.   Others

1286. Asatryan, T. P. and Ambartsumyan, G. G., "A sine-cosine func-
      tion convertor", Izv. Akad. Nauk Arm. SSR, Ser. Tekh. Nauk,
      29 (2), 10-16 (1976), Russian, EEA80-20469.
1287. Bessant, M. J., "Pulse train/sine-wave converter", New
      Electron., 9 (23), 39 (1976), EEA80-6344.
1288. Bliss, B. E. and Dervan, J. T., "Character code to dot video
      converter", IBM Tech. Disclosure Bull., 14 (1), 14-15 (1971),
      CCA6-22187.
1289. Bogdanovskaya, M. V., Zharkov, A. P., Nikitin, A. M., Petro-
      pavlovskii, V. P. and Sinitsyn, N. V., "Digital angle con-
      verters with synchronous elements", Meas. Tech., 1596-8
      (1975), CCA12-3857.
1290. Damljanovic, D. D., "[Low] current to frequency converter
      without insulation", Electron. Engineering, 48 (586), 21
      (1976), EEA80-6332.
1291. Dmitriev, V. I., "Resistance-to-PRF (pulse repetition fre-
      quency) converter", Prib. & Sist. Upr., no. 11, 41-3 (1976),
      Russian, EEA80-16909.
1292. Domrachev, V. G. and Meiko, B. S., "Criteria for evaluating
      the precision of digital angle converters", Meas. Tech.,
      18 (11), 1590-5 (1975), CCA12-3856.
1293. Dooley, D. J., "Companding data conversion", Advances in
      Instrumentation, Vol. 31, pt. 1, 1976, p. 565, EEA80-20483.
1294. Fedorovskii, A. E. and Ivanyushkin, E. M., "A synchronised
      transistorised current converter", Instrum. & Exp. Tech.,
      19 (1), pt. 2, 168-70 (1976), EEA80-9081.
1295. Il'nitskii, L. Y. and Golego, A. G., "Two-circuit functional
      convertor with one non-linear element", Mat. Model & Teor.
      Elektr. Tsepei, no. 13, 84-8 (1975), Russian, CCA11-9586.
1296. Ivanov, B. N. and Dubenets, A. L., "Angle-of-rotation con-
      version errors of circular magnetoelectric convertors", Meas.
      Teach., 19 (1), 86-91 (1976), CCA12-6591.
1297. Kirenskiy, I. G., "Simplification of the equivalent circuit
      of a transistor frequency converter", Telecommun. & Radio
      Eng. Pt. 2, 31 (3), 100-3 (1976), EEA80-16912.

1298. Konishi, Y. and Hoshino, N., "12 GHz low-noise converter", Proceedings of the IEEE 1972 International Conference on Communications, 26-19/4 pp., EEA76-4227.
1299. Lazarev, V. B., "A high-speed phase-voltage converter", Instrum. & Exp. Tech., $\underline{19}$ (2), pt. 1, 407-9 (1976), EEA80-16907.
1300. Lukasiewicz, Z. and Matejkowski, W., "Angular-digital convertors", Pr. Przem. Inst. Telekomun., $\underline{26}$ (85), 35-40 (1976), Polish, EEA80-6345.
1301. McDonnell, D., "Application of velocity pin on SDC 1700 converters", Electron. Engineering, $\underline{49}$ (590), 17, 19 (1977), EEA80-20462.
1302. Pauly, A., "Input/output signal converter of URTL circuit system U1", Siemens Rev., $\underline{42}$ (4), 154-7 (1975).
1303. Sancholuz, A. G., "Current inversion impedance converter using operational amplifiers", Rev. Sci. Instrum., $\underline{47}$ (11), 1407-9 (1976), EEA80-9085.
1304. Sander, T. P. Y., "Temperature to pulse-length converter", Wireless World, $\underline{83}$ (1493), 76 (1977), EEA80-9087.
1305. Schwarz, J., "Analysis of behaviour of supply-triggered static converters. II.", Z. Elektr. Inf.- & Energietech., $\underline{6}$ (6), 481-93 (1976), German, EEA80-9088.
1306. Stepanov, A. V., "Logic converter for angular-displacement sensor", Autom. Control. Comput. Sci., $\underline{8}$ (1), 79-80 (1974).
1307. Urakseev, M. A., "Flux distribution of electromechanical computing converters", Izv. VUZ Elektromekh., no. 12, 1377-81 (1976), Russian, CCA12-10623.
1308. Vasin, N. N., Ioffe, V. G., Boltyanskii, A. A. and Pshenichnikov, Y. V., "A multichannel measuring DC-to-DC convertor for low mV ranges", Prib. & Sist. Upr., no. 10, 39-40 (1976), Russian, EEA80-20476.

16. Multiplexers

1309. Anon., "A three-input, four-bit digital multiplexer IC array", Comput. Des., $\underline{9}$ (12), 76 (1970), CCA6-9287.
1310. Crabbe, E. P., Jr., "Multiplexer chip forms majority-vote circuit", Electronics, $\underline{50}$ (1), 106-7 (1977), CCA12-3813.
1311. Hicks, W. T., "Generate any 6-input function with two multiplexers", EDN, $\underline{20}$ (8), 74, 76 (1975).
1312. Le Van, T. and Van Houttl, N., "Delayed universal logic modules and sequential machine synthesis", IEEE Trans. Comput., C-$\underline{24}$ (8), 853-5 (1975).
1313. Lesea, A. and Urkumyan, N., "Multiplexer system reduces cost of terminal interfacing", Comput. Des., $\underline{16}$ (8), 109-13 (1977).
1314. Mause, K., "Multiplexing and demultiplexing techniques with Gunn devices in the gigabit-per-second range", IEEE Trans. Microwave Theory Tech., MTT-$\underline{24}$ (12), 926-9 (1976).

1315. Mitarai, H. and Kuo, H., "ROM micro-reduction techniques",
      USA-Jpn. Computer Conference, 2nd, Proceedings, 1975, p. 126-30.
1316. Murugesan, S., "Programmable universal logic module", Int. J.
      Electron., 40 (5), 509-12 (1976), CCA11-22561.
1317. Siebert, J. E., "Digital multiplexers reduce chip count in
      logic design", Electronics, 50 (9), 120-1 (1977), CCA12-15418.
1318. Szajnowski, W. J., "A universal sequential element provides
      flexibility", Electron. Engineering, 49 (587), 21, 23 (1977),
      CCA12-6576.
1319. Tabloski, T. F. and Mowle, F. J., "Numerical expansion tech-
      nique and its application to minimal multiplexer logic cir-
      cuits", IEEE Trans. Comput., C-25 (7), 684-702 (1976).
1320. Whitehead, D. G., "Algorithm for logic-circuit synthesis by
      using multiplexers", Electron. Lett., 13 (12), 355-6 (1977).

17.   Interfaces

1321. Bird, P., "A novel shaper circuit for d.t.l. and t.t.l. input
      interfacing", New Electron., 7 (13), 28, 32 (1974), CCA9-23194.
1322. Brokaw, P., "Interface adapter multiplexes many DA converters",
      Electronics, 50 (12), 151-2 (1977), EEA80-28370.
1323. Bunge, P. J., "Common-gate, common-base circuits shift voltage
      levels", Electronics, 49 (4), 115 (1976), CCA11-14973.
1324. Cole, A., "T.t.l.-c.m.o.s. interface", New Electron., 9 (6),
      17 (1976), CCA11-17352.
1325. Craney, P. M., "Relays and logic ICs can be working partners",
      Electronics, 49 (2), 107-10 (1976), CCA11-12249.
1326. Dingwall, A. G. F., "TTL-to-CMOS buffer circuit", Report
      TN-1114, RCA, Princeton, N. J., 1975, 3 pp.
1327. Hellwarth, G. A. and Harrison, T. J., "Noise and ground prob-
      lems in digital signal interfaces", Autom. & Control, 6 (2),
      12-13 (1976), CCA12-15395.
1328. Hilsher, R. W., "Universal interface: TTL to diode array",
      EDN, 20 (5), 74, 76 (1975).
1329. Lscelius, R. H., "Simple circuit interfaces TTL to CMOS with
      use of only a single 12-V supply", Electron. Des., 24 (25),
      112 (1976), CCA12-10609.
1330. Morris, R. L., "C-MOS Schmitt trigger can be more than an
      interface", Electronics, 47 (16), 124-5 (1974), CCA9-21951.
1331. Netzer, Y., "Interface between bipolar or CMOS logic and JFET",
      Electron. Engineering, 48 (578), 29 (1976), EEA79-24526.
1332. O'Brien, T. E., "Monolithic level shifter lets MOS, TTL share
      same network", Electronics, 50 (1), 115 (1977), EEA80-2996.
1333. Payne, A. J., "Linear and interface circuits", Microelectron.
      & Reliab., 13 (5), 373-8 (1974), EEA78-5364.
1334. Sahakian, A., "Computer/cassette interface takes tone from
      clock", Electronics, 50 (1), 115 (1977), EEA80-2996.

1335. Sarpangal, S., "Interface CMOS to TTL with diodes and save the cost of expensive buffers", Electron. Des., $\underline{24}$ (23), 74 (1976), EEA80-16814.

1336. Turner, P., "Faster optically-coupled isolator is compatible with TTL interfaces", Electron. Engineering, $\underline{46}$ (556), 46-8 (1974), CCA9-18308.

18.   Others

1337. Ae, T. and Yoshida, N., "Multistable circuits and their coding", Syst. Comput. Control, $\underline{6}$ (2), 22-8 (1975), CCA11-28117.

1338. Aharon, M. and Lamdan, T., "Carrier acceleration circuit in FET logic", 9th Convention of Electrical and Electronic Engineers in Israel, 1975, p. B-1-5/1-8, Hebrew, CCA11-31502.

1339. Anon., "Mains synchronising circuit", New Electron., $\underline{9}$ (15), 14 (1976), CCA11-28131.

1340. Arnit, M. A., Artyukh, Y. N., Baums, A. K. and Zagurskii, V. Y., "Fast logic circuits using hybrid components with nonreactive coupling", Conference on Tunnel Diodes in Computing and Measurement Technology, 1970, p. 79-84, Russian, Publ. 1972, CCA8-2531.

1341. Arzubi, L. M., "In-phase driver circuit", IBM Tech. Disclosure Bull., $\underline{19}$ (1), 31-2 (1976), CCA11-22566.

1342. Bloom, G. E., "A magnetic approach to upset prediction of logic latches", IEEE Trans. Nucl. Sci., NS-$\underline{23}$ (6), 1743-8 (1976), EEA80-9021.

1343. Bond, G. L., "Dual-redundant logic system", IBM Tech. Disclosure Bull., $\underline{15}$ (4), 1145-6 (1972), CCA8-5222.

1344. Brown, D., Greenwood, A., Pratt, J. C. and Spencer, J. E., "Multichannel remotely programmable logic delay system", Nucl. Instrum. & Methods, $\underline{138}$ (4), 695-713 (1976).

1345. Cavalli Sforza, M., Goggi, G. and Rossini, B., "A programmable NIM-standard logic unit", Nucl. Instrum. & Methods, $\underline{137}$ (3), 609-11 (1976), CCA12-3826.

1346. Chaney, T. J., "Beware the synchronizer", Digest of Papers, Sixth Annual IEEE Computer Society International Conference, 1972, p. 317-19, CCA8-2533.

1347. Dabrowski, G., "Digital integrated circuits – features and concept. II.", Elektron. Int., no. 23-24, 415-16 (1974), German, EEA78-8607.

1348. Damaye, R., "Introduction to the use of numerical integrated circuits. V: Sequential circuits", Toute Electron., no. 358, 45-50 (1971). French, CCA6-24417.

1349. Dansky, A. H., "Multiple threshold IGFET ternary circuits", IBM Tech. Disclosure Bull., $\underline{27}$, 1356-7 (1974), EEA78-8186.

1350. Davidson, E. E. and Lane, R. D., "Diodes damp line reflections without overloading logic", Electronics, $\underline{49}$ (4), 123-7 (1976), CCA11-14956.

1351. De Mong, R. D., "Power out warning interrupt circuit", IBM Tech. Disclosure Bull., $\underline{18}$ (12), 4147-9 (1976), CCA11-28130.

1352. de Rivas, J., "Waveforms-in-quadrature generator", Wireless World, 82 (1492), 43 (1976), EEA80-2979.
1353. Dennison, R. T., Ludlow, P. J. and Park, S. J., "Harper cell read reference circuit", IBM Tech. Disclosure Bull., 18 (9), 2902-4 (1976), CCA11-12330.
1354. Dvorak, T. J., Lowdermil, D. J. and Plant, J. W., "Hardware assist for microcode execution of storage-to-storage move instructions", IBM Tech. Disclosure Bull., 19 (1), 67-70 (1976), CCA11-22624.
1355. Epstein, G., "Multiple-valued signal processing with limiting", In: Computer Science and multiple-valued logic. Theory and applications, Rine, D. C., (Ed.), North-Holland, Amsterdam, Netherlands, p. 71-80 (1977), CCA12-10641.
1356. Freytag, D., "A compact time digitizer in CAMAC format", Nucl. Instrum. & Methods, 138(4), 685-9 (1976), CCA12-3860.
1357. Gersbach, J. E., "ΔI suppressor", IBM Tech. Disclosure Bull., 19 (1), 30 (1976), CCA11-31499.
1358. Hartley, C. J., "Digital command inverts signal", Electronics, 48 (5), 85 (1975).
1359. Hauser, M. W., "Programmable sine-wave oscillator uses only one CMOS IC", Electron. Des., 22 (24), 200 (1974).
1360. Herald, R. F. and Shidler, K. A., "Index selection logic", IBM Tech. Disclosure Bull., 18 (7), 2260-2 (1975), CCA11-9575.
1361. Herzog, O., "Universal cellular array", International Symposium of Unif. Struct. Autom. and Logic, Proceedings, 1975, p. 7-12.
1362. Higuchi, T. and Koyama, K., "Ternary Φ type fail-safe logic circuits using ferrite cores", Trans. Soc. Instrum. & Control Eng., 12 (5), 593-9 (1976), Japanese, EEA80-20400.
1363. Hlavaty, J. and Kolesar, M. L., "A parallel comparator with a fast transfer channel", Elektrotech. Cas., 27 (10), 766-8 (1976), Slovak, CCA12-10608.
1364. Holzgrebe, H. G., "Hardware concepts in commercial data processing systems. IV.", Fernmelde-Praxis, 53 (7), 317-40 (1976), German, CCA11-22619.
1365. Homan, M. E., "FET depletion load push-pull logical circuit", IBM Tech. Disclosure Bull., 18 (3), 910-11 (1975), CCA11-9572.
1366. Horvath, P., "Digitally controlled fast logic modules", IEEE Trans. Nucl. Sci., NS-20 (1), 193-8 (1973), CCA8-13606.
1367. Huertas, J. L. and Civit, A., "Universal logic modules with function selection by means of an autonomous sequential machine", Electron. & Fis. Apl., 18 (2), 73-7 (1975), Spanish, CCA11-17338.
1368. Kalisz, J., "LUBIN-unified notation system for digital circuits with positive and negative logic", Elektronika, 15 (3), 125-8 (1974), Polish, CCA9-21956.
1369. Kambayashi, Y. and Yajima, S., "Applications of two-way logic circuits", Inst. Electronics Commun. Engrs. Japan. ABC, 52 (11), 216-23 (1969), Japanese, CCA6-5597.

1370. Kashtanov, Y. V. and Ugryumov, E. P., "A noise-resistant long-interval integrator", Izv. VUZ Priborostr., 19 (8), 54-8 (1976), Russian, CCA11-31491.

1371. Khizhnichenko, V. I., "The transmission of random signals through m-circuits", Telecommun. & Radio Eng. Pt. 2, 30 (2), 62-8 (1975), CCA11-14970.

1372. Kinniment, D. J. and Woods, J. V., "Synchronisation and arbitration circuits in digital systems", Proc. Inst. Electr. Eng., 123(10), 961-6 (1976), EEA80-20393.

1373. Kondrat'ev, R. M., "Synthesis of logic circuits with provision for resetting starting conditions", Izv. VUZ Radioelektron, 14 (7), 725-8 (1971), Russian, CCA6-24399.

1374. Kopperschmidt, G., "Majority logic in practical circuitry", Elektronik, 25 (10), 71-8 (1976), German, CCA12-1403.

1375. Kornev, N. A., "Time locking circuits", Instrum. Exp. Tech., 18 (2), 445-7 (1975).

1376. Kostuch, D. J., "Time delay for MOSFET integrated logic", IBM Tech. Disclosure Bull., 13 (2), 519-20 (1970), EEA74-6943.

1377. Kotter, E. and Svoboda, B., "Control device for integrated logic circuits", Sdelovaci Tech., 19 (6), 168-9 (1971), Czech, CCA6-22198.

1378. Krausener, J. M., "Integrated logic circuits: anticipated retention", Electronique, no. 140, 31-4 (1971), French, CCA6-9288.

1379. Kroupa, V., "Pulse subtractor for frequency synthesis", Electron. Engineering, 49 (587), 25 (1977), EEA80-9002.

1380. Krumrein, G., "Delay circuits and astable circuits with TTL integrated circuits", Polytech. Tijdschr. Elektrotech. Electron., 26(7), 269-70 (1971), CCA6-12768.

1381. Lamdan, T. and Aharon, M., "A circuit for high-speed carry propagation in LSI-FET technology", Radio & Electron. Eng., 46 (7), 337-42 (1976), CCA11-28115.

1382. Leininger, J. C., "Universal logic module", IBM Tech. Disclosure Bull., 13 (5), 1294-5 (1970), CCA6-3727.

1383. Leung, W. C., "Digital phase-locked loop circuit", IBM Tech. Disclosure Bull., 18 (10), 3334-7 (1976), CCA11-17348.

1384. Levy, A., "Keep your power flowing with an uninterruptible supply", Electron. Des., 24 (5), 62-4 (1976), CCA11-17356.

1385. McCann, M., "Worse case TTL line driving", Microelectronics, 3 (10), 45-55 (1971), CCA6-20067.

1386. Maley, G. A. and Walsh, J. L., "Interchanger I circuit", IBM Tech. Disclosure Bull., 14 (2), 629 (1971), CCA6-24409.

1387. Manukyan, Y. S., "Digital time interval generator with automatic error correction", Meas. Tech., 15 (9), 1292-4 (1972), CCA8-13614.

1388. Maruyama, K., "Circuit for conditional substring function", IBM Tech. Disclosure Bull., 18 (9), 3099-104 (1976), CCA11-14964.

1389. Mead, C. A., Pashley, R. D., Britton, L. D., Daimon, Y. T. and
      Sando, S. F., Jr., "128-bit multicomparator", IEEE J. Solid-
      State Cicuits, SC-11 (5), 692-5 (1976), EEA80-3011.
1390. Murugesan, S., "Negabinary arithmetic circuits using binary
      arithmetic", IEEE J. Electron. Circuits & Syst., 1 (2), 77-8
      (1977), EEA80-20386.
1391. Nevzlin, B. I., "A current integrator with frequency output",
      Izv. VUZ Priborostr., 19 (10), 64-8 (1976), Russian, CCA12-10625.
1392. Ngai, C. H. and Wassel, E. R., "Destructive overlap detection
      hardware", IBM Tech. Disclosure Bull., 19 (1), 61-4 (1976),
      CCA11-22623.
1393. Okumura, I. and Suzuki, S., "Proposal for a ternary logic sys-
      tem", Syst. Comput. Controls, 5 (1), 59-67 (1974).
1394. O'Leary, P. M., "Simple sequence delay for digital data",
      Electron. Engineering, 49 (588), 24 (1977), CCA12-10605.
1395. Plant, J. W. and Wassel, E. R., "Page boundary crossing detec-
      tion hardware", IBM Tech. Disclosure Bull., 19 (1), 57-60
      (1976), CCA11-22622.
1396. Pomerleau, A., Fournier, M. and Buijs, H. L., "Realization of
      a discrete Fourier transform (DFT) module for incorporation in
      FET processors", Proc. IEEE, 65 (1), 173-4 (1977), EEA80-6266.
1397. Pratapa Reddy, V. C. V., "Threshold-logic instrumentation to
      identify the signal-arrival order", IEEE Trans. Instrum. &
      Meas., IM-25 (3), 268-70 (1976), CCA11-28110.
1398. Priebe, D., "Multifamily logic clip shows all pin states simul-
      taneously", Hewlett-Packard J., 28 (4), 18-20 (1976), CCA12-8813.
1399. Quick, P., "Logic recorders-indispensible in digital technique",
      Mess. & Pruef., no. 11, 658-66 (1976), German, CCA12-8901.
1400. Reinet, J. R. and Glazer, M. A., "A 32x9 ECL dual address regi-
      ster using an interleaving cell technique", 1977 International
      Solid-State Circuits Conference, Digest of Technical Papers,
      p. 72-3.
1401. Schiller, D., "Power supply unit for medium voltages and cur-
      rents", Radio Fernsehen Elektron., 25 (14), 456-8, 463-4
      (1976), German, CCA11-31506.
1402. Schwarz, S. and Erdmeier, U., "Reducing the interference
      susceptibility of logic circuitry", Elektronik, 19 (12), 407-9
      (1970), German, CCA6-5594.
1403. Segawa, S., Kusaka, T. and Yoshitake, T., "An equivalent uni-
      junction transistor - a new trigger device", NEC Rec. & Dev.,
      no. 44, 41-8 (1977), EEA80-17265.
1404. Srinivasan, R., "Anomalous behaviour of transistor-transistor
      logic circuits with totem-pole output", Thesis, Washington
      University, St. Louis, Mo., 1974, 215 pp., Order No. 75-14937,
      CCA11-9587.
1405. Stockle, H., "Indicator systems", Regelungstech. Prax, 18 (7),
      M35-8 (1976), German, CCA11-31507.
1406. Stuckert, P. E., "Digital directional-coupling circuit", IBM
      Tech. Disclosure Bull., 19 (5), 1923-6 (1976), EEA80-20382.

1407. Swanson, R. M., "Complementary MOS transistors in micropower
      circuits", Thesis, Stanford University, California, 1975,
      216 pp., Order No. 75-13609, CCA11-4692.
1408. Tenny, R., "Build a programmable sequencer with a broad opera-
      ting range", EDN, 21 (10), 90-2 (1976).
1409. Tlaczala, W. and Basiladze, S. G., "Digitally controlled nano-
      second delay box in CAMAC standard", Nukleonika, 19 (9), 785-9
      (1974), Russian, CCA11-17414.
1410. Unger, S. H., "Tree realization of iterative circuits", IEEE
      Trans. Comput., C-26 (4), 365-83 (1977), CCA12-10636.
1411. Val'skis, C. Y., "Compensation of pulse distortions using
      ferrite rings", Meas. Tech., 18 (10), 1515-17 (1975),
      CCA11-31533.
1412. Vingron, P., "Combinational circuits and stores - a mapping-
      theoretical representation of their synthesis", MSR, 20 (3),
      140-5 (1977), German, EEA80-28392.
1413. Voith, R. P., "ULM implicants for minimization of universal
      logic module circuits", IEEE Trans. Comput., C-26 (5), 417-24
      (1977).
1414. Wachter, R. and Schubert, W., "Possibilities of the pneumo-
      electrical conversion of fluidic signals", Mess. Steuern
      Regeln Mit Automatisierungsprax, 19 (9), 321-4 (1976), German,
      CCA11-31486.
1415. Wallace, L. J., "High-resolution document stopping and clutch
      stop correction system", IBM Tech. Disclosure Bull., 19 (6),
      2025-7 (1976), CCA12-15393.
1416. Waxman, J. and Rootenberg, J., "Logic circuit for cycle detec-
      tion in a state diagram", IEEE Trans. Comput., C-26 (3),
      303-5 (1977), CCA12-8835.
1417. Widlar, R. J., "Local IC regulator for logic circuits", Comput.
      Des., 10 (5), 115-26 (1971), CCA6-17561.
1418. Williams, M. D., "Up/down latching sequencer keeps order",
      Electronics, 49 (25), 99 (1976), EEA80-2990.
1419. Wilson, M. J. D. and Aleksander, I., "Ring-structured adaptable
      logic circuit for the classification of binary vectors", Elec-
      tron. Lett., 19 (16), 337-9 (1974), CCA9-21954.
1420. Woodward, F. I., "A device for translating chart recorder in-
      formation into digital form", J. Phys. E, 10 (3), 213-16
      (1977), CCA12-10628.
1421. Wooley, B. A. and Baugh, C. R., "Integrated m-out-of-n detec-
      tion circuit using threshold logic", IEEE J. Solid-State Cir-
      cuits, SC-9 (5), 297-306 (1974).
1422. Wormald, E. G., "A note on synchronizer or interlock malopera-
      tion", IEEE Trans. Comput., C-26 (3), 371-8 (1977), CCA12-8861.
1423. Wright, M. J., "A bidirectional pulse stretcher has many
      uses", Electron. Des., 24 (22), 180-3 (1976), CCA12-8808.
1424. Zubov, V. G. and Yurkovskii, D. A., "FET nanovolt modulator
      with temperature compensated zero drift", Otbor & Peredacha
      Inf., no. 48, 89-92 (1976), Russian, EEA80-28354.

B.   MEMORY

I.   BOOKS

1. Dolotta, T. A., et al, "Data Processing in 1980-1985: A study of potential limitations to progress", Wiley-Interscience, New York, 1976.
2. Eimbinder, J., (Ed.), "Semiconductor memories", Wiley-Interscience, New York, 1971.
3. Gray, H. J., "High-speed digital memories and circuits", Addison-Wesley Publishing Co., Reading, Mass., 1976.
4. Hnatek, E. R., "A user's handbook of semiconductor memories", Wiley-Interscience, New York, 1977.
5. Hodges, D. A., (Ed.), "Semiconductor memories", IEEE Press, New York, 1972.
6. Luecke, G., Mize, J. P. and Carr, W. N., "Semiconductor memory design and applications", McGraw-Hill Book Co., New York, 1973.
7. Middlhoek, S., George, P. and Dekker, P., (Eds.), "Physics of computer memory devices", Academic Press, Inc., New York, 1976.
8. Miller, S. W. and Gagliardi, U. O., "Symposium on advanced memory concepts", Report AD-A029631/9SL, SRI, Menlo Park, CA, 1976.
9. Riley, W. B., (Ed.), "Electronic computer memory technology", McGraw-Hill Book Co., New York, 1971.
10. Soma, J. T., "The computer industry: An economic-legal analysis of its technology and growth", Lexington Books, D. C. Heat and Co., Lexington, Mass., 1976.
11. Turn, R., "Computers in the 1980's", Columbia University Press, New York, 1974.

II.   REVIEW ARTICLES

12. Altman, L., "Memory types multiply, microprocessor families grow", Electronics, $\underline{49}$ (22), 76-8, 81-2 (1976), EEA80-684.
13. Altman, L., "Memories - It's a user paradise: cheaper RAM's, reprogrammable ROM's, CCD's and bubbles coming along", Electronics, $\underline{50}$ (2), 81-96 (1977).
14. Anon., "Memories in the making", Data Syst., no. 4, 10, 12, 14 (1976), CCA11-31585.
15. Backler, J., "Read/write memory modules", Digital Des., $\underline{6}$ (5), 91-4 (1976), CCA11-31597.
16. Ball, L. F., "Computer devices - memories", J. R. Signals Inst., $\underline{11}$ (3), 87-98 (1973), CCA9-14346.
17. Beuter, R., "Introduction to semiconductor memory techniques", Radio Ind., $\underline{42}$ (8), 37-40 (1975), Italian, CCA11-4814.
18. Billeter, E., "Semiconductor memory stores", Radio TV Electron., $\underline{36}$ (3), 74-8 (1976), German, EEA79-24634.

19. Chang, J. J., "Nonvolatile semiconductor memory devices", Proc.
    IEEE, 64 (7), 1039-59 (1976).
20. Devries, H. G., "Memories are made of this", Electron Prod.,
    18 (5), 45-8 (1975).
21. Frankenberg, R. J., "Designer's guide to: Semiconductor
    memories - 1", EDN, 20 (14), 22-9 (1975).
22. Frankenberg, R. J., "Designer's guide to: Semiconductor
    memories - 2", EDN, 20 (15), 58-65 (1975).
23. Frankenberg, R. J., "Designer's guide to: Semiconductor
    memories - 3", EDN, 20 (16), 68-74 (1975).
24. Franson, P., "Semiconductors and IC's - new technologies and
    clever designers promise to solve your every problem", EDN,
    21 (13), 20-4 (1976), CCA12-1405.
25. Franson, P., "Semiconductor memories - technology advances spur
    new products; manufacturers strive for standardization", EDN,
    22 (12), 46-58 (1977).
26. Gardiner, K. M., "Semiconductors for computer memories", SME
    Technical Papers Series MS, 1976, Paper MS76-365, 12 pp.
27. Giffard, R. P., Gallop, J. C. and Petlery, B. W., "Applications
    of the Josephson effects", In: Progress in Quantum Electronics,
    Vol. 4 (4), Sanders, J. H. and Stenholm, S., (Eds.), Pergamon
    Press, Fairview Park, Elmsford, New York, 1977.
28. Harloff, H. J., "Semiconductor memories in data processing
    equipment", Phys. Bl., 32 (3), 119-29 (1976), German, CCA11-17411.
29. Hodges, D. A., "Microelectronic memories", Sci. Am., 237 (3),
    130-45 (1977).
30. Irie, F. and Nakamura, A., "The present status of applications of
    the Josephson effect", Proceedings of the 6th International
    Cryogenic Engineering Conference, 1976, p. 11-19.
31. Jetter, J., "Application of semiconductor stores in production
    equipment", Ind.-Anz., 98(95), 1689-90 (1976), German, CCA12-10689.
32. Kirchner, J. H., "Semiconductor storage, technology and applica-
    tions", Fernmelde-Ing., 25 (1), 1-36 (1971), German, CCA6-10778.
33. Kohyama, S. and Ohuchi, K., "Semiconductor memories", J. Inst.
    Telev. Eng. Jap., 29 (9), 680-7 (1975), Japanese, CCA11-17408.
34. Koppel, R. J. and Maltz, I., "Predicting the real costs of semi-
    conductor-memory systems", Electronics, 49 (24), 117-22 (1976),
    EEA80-6251.
35. Koppel, R. L. and Maltz, I. R., "Cost, performance, and relia-
    bility tradeoffs in semiconductor memory systems", 13th IEEE
    Computer Society International Conference, Digest of Papers,
    1976, p. 218-22, CCA12-1520.
36. Luce, R., "Latest development in solid state technology and
    their application", Proceedings of National Electron Conference,
    Vol. 29, 1974, p. 29.
37. Luis-Gonzalez-de-Torre, J., "Memory systems", Rev. Telecomun.,
    28 (111), 8-11 (1973), Spanish, CCA9-16050.
38. Matick, R. E., "Review of current proposed technologies for mass
    storage systems", Proc. IEEE, 60, 266-89 (1972).

39. Mitterer, R. W., "A review on random access MOS memories", 1st
European Solid State Circuits Conference - ESSCIRC, Extended
Abstracts, 1975, p. 125-30, EEA79-33155.
40. Ninomiya, S., "Memories for large scale computers", J. Inst.
Electron. & Commun. Eng. Jap., 54 (4), 587-91 (1971), Japanese,
CCA6-17812.
41. Reiner, H., "Memory storage elements", NTG-Fachber., 54, 121-32
(1975), German, EEA80-286.
42. Reymond, R., "Semiconductor memories", Man Made Memories, Conference Summaries, 1971, 1 pp., CCA6-12746.
43. Rogge, H., "Comparison of different high speed memories", Entwicklungsber Siemens Halske, 32 (3), 9-14 (1969), German,
CCA5-11585.
44. Shenton, G., "Semiconductor memories - which direction?", New
Electron., 8 (4), 44-6, 50, 54 (1975), EEA78-24486.
45. Strehlow, W. H., "Memories are made of this", Technol. Forecast.
& Soc. Change, 6 (1), 65-74 (1974), CCA9-21985.
46. Sutherland, I. E. and Mead, C. A., "Microelectronics and computer
science", Sci. Am., 237 (3), 210-28 (1977).
47. Tarui, Y., Nagai, K. and Hayashi, Y., "Nonvolatile semiconductor
memory", Oyo Buturi, 43 (10), 990-1002 (1974), Japanese,
EEA78-15366.
48. Terman, L. M., "The role of microelectronics in data processing",
Sci. Am., 237 (3), 162-77 (1977).
49. Torrero, E. A., "Focus on semiconductor memories", Electron.
Des., 23 (7), 98-107 (1975), EEA78-24453.
50. Vinsani, M. and Cislaghi, E., "Semiconductor memories", Alta
Freq., 42 (9), 430-43 (1973), Italian, CCA9-9208.
51. Wilcock, J. D., "Semiconductor memories - an electronics revolution", Electron. & Power, 20 (1), 8-11 (1974), CCA9-11962.
52. Wilcock, J. D., "RAMs - a review", New Electron., 7 (4), 37-8,
41-2, 45, 49 (1974), CCA9-14353.
53. Zdrazil, Z., "On semiconductor stores", Automatizace, 19(5),
123-5 (1976), Czech, CCA11-28278.

III. HIERARCHY
    1. General

54. Adkins, G. and Pooch, U. W., "A simulation study of the IBM 370/
168 and AMDAHL 470/V6", 10th Annual Simulation Symposium, 1977,
p. 17-48, CCA12-12945.
55. Anon., "Burroughs reports (B1800 Series)", Burroughs Clearing
House, 61 (3), 20-2 (1976), CCA12-12917.
56. Anon., "Supercomputers: their structure and performance", Elettron Oggi, no. 10, 1295-7 (1976), CCA12-12940.
57. Batcher, K. E., "The multidimensional access memory in STARAN",
IEEE Trans. Comput., C-26 (2), 174-7 (1977), CCA12-10656.

58. Beausoleil, W. F., Clark, W. A., Ho, I. T. and Pricer, W. D.,
    "Hierarchical storage chip", IBM Tech. Disclosure Bull., $\underline{16}$ (5),
    1364-5 (1973), CCA9-9216.
59. Briggs, F. A. and Davidson, E. S., "Organization of semiconductor
    memories for parallel-pipelined processors", IEEE Trans. Comput.,
    C-$\underline{26}$ (2), 162-9 (1977).
60. Chang, D. Y., Kuck, D. J. and Lawrie, D. H., "On the effective
    bandwidth of parallel memories", IEEE Trans. Comput., C-$\underline{26}$ (5),
    480-90 (1977), CCA12-12841.
61. Chiang, M. M., "Semiconductor memories in the Amdahl 470/V6 com-
    puter", 13th IEEE Computer Society International Conference,
    COMPCON 76, Digest of Papers, p. 223-4.
62. Chow, C. K., "Algorithm for determining optimum configurations of
    memory hierarchies", IBM Tech. Disclosure Bull., $\underline{15}$ (7), 2330-3
    (1972), CCA8-11623.
63. Doyle, M. S. and Graham, J. W., "Some parameters affecting the
    performance of paged storage hierarchies", Info. J., $\underline{13}$ (2),
    197-207 (1975).
64. Falk, H., "Reaching for a gigaflop", IEEE Spectrum, $\underline{13}$ (10),
    65-9 (1976), CCA12-4057.
65. Finch, T. R., "Semiconductor memory", 1971 IEEE International
    Convention Digest, p. 272-3, CCA6-15262.
66. Gecsei, J., "Determining hit ratios for multilevel hierarchies",
    IBM J. Res. & Dev., $\underline{18}$ (4), 316-27 (1974), CCA9-25087.
67. Glock, H. and Mitterer, R., "High speed Read-Write memories in
    bipolar technology", Siemens Forsch Entwicklungsber Res. Dev.
    Rep., $\underline{4}$ (4), 250-4 (1975).
68. Kaufmann, H., "Technical information stores", Ber. Bunsenges.
    Phys. Chem., $\underline{80}$ (11), 1160-8 (1976), CCA12-10681.
69. Langdon, G. G., Jr., "Some equations to aid the preliminary
    design of a two-level memory hierarchy", 2nd USA-Japan Computer
    Conference Proceedings, 1975, p. 135-8, CCA11-15047.
70. Makino, T. and Ohno, N., "Characteristics of not found probabil-
    ity (NFP) in memory hierarchy system", NEC Res. Dev., no. 43,
    51-8 (1976).
71. Malyutin, V. I. and Shatashvili, N. G., "Some design questions
    of hierarchical shift-register memories using flat magnetic
    domains", Autom. & Remote Control, $\underline{37}$ (4), Pt. 2, 604-15 (1976),
    Russian, CCA12-3988.
72. Mattson, R. L. and Traiger, I. L., "Storage Hierarchy design",
    Sixth Annual IEEE Computer Society International Conference,
    Digest of Papers, 1972, p. 145-8, CCA8-2627.
73. Rege, S. L., "Cost performance, and size tradeoffs for different
    levels of a memory hierarchy", Computer, $\underline{9}$ (4), 43-51 (1976).
74. Schmitt, R., "Semiconductor memories for medium-to-small data
    systems", Siemens Rev., $\underline{43}$ (11), 482-6 (1976), EEA80-28912.
75. Semkin, V. A. and Shuvikov, V. I., "Determination of the number
    of inter-level transfers in a hierarchical memory when solving a
    class of problems", Avtom. & Vychist. Tekh., no. 6, 83 (1976),
    Russian, CCA12-12909.

76. Singh, Y. and Goldfeder, M. E., "Design of optimal storage hier-
    archies for multiprogrammed computer systems", 7th Asilomar
    Conference on Circuits, Systems and Computers, 1973, p. 428-3,
    Publ. 1974, CCA9-20750.
77. Tan, W. P. H., "Closed queueing analysis of multiprogrammed
    demand-paging computer systems", 6th Australian Computer Con-
    ference, Proceedings, 1974, p. 406-23.
78. Tasso, J. and Salle, F., "Memory hierarchy and technology", Onde
    Electr., 54 (6), 261-72 (1976), French, CCA-20745.

                    2. Main

79. Anon., "Main memory unit", Report NASA-CR-120733, IBM Corp.,
    Huntsville, Ala., 1975, 31 pp., CCA11-2219.
80. Asano, L., "Main memory construction of U-1110", Inf. Process.
    Soc. Jap., 16 (4), 353-8 (1975), Japanese, CCA10-22906.
81. Ayling, J. K., "Monolithic main memory is taking off", 1971 IEEE
    International Convention Digest, p. 70-1, CCA6-15259.
82. Ayling, J. K. and Moore, R. D., "Monolithic main memory", 1971
    International Solid State Circuits Conference, Digest of Technical
    Papers, p. 76-7, CCA6-15271.
83. Frankenberg, R. J., "Minicomputer designer chooses semiconductor
    over core memory", Comput. Des., 14 (3), 59-68 (1975).
84. Gates, H. R., McKinney, J. D. and North, W. D., "Bipolar LSI for
    a main memory", 1971 International Solid State Circuits Con-
    ference, Digest of Technical Papers, p. 78-9, CCA6-15272.
85. Horikoshi, H. and Momose, T., "Main memory architecture of multi-
    processor", Inf. Process. Soc. Jap., 16 (4), 314-21 (1975),
    Japanese, CCA10-25977.
86. Ishii, O., "A survey of main frame memories", Inf. Process. Soc.
    Jap., 16 (4), 258-74 (1975), Japanese, CCA10-22901.
87. Kokubu, A., "The logical structure of main memory in the SYMBOL
    computer", Inf. Process. Soc. Jap., 16 (4), 377-81 (1975),
    Japanese, CCA10-22909.
88. Lauffer, D. and Lim, P., "A user's look at MOS-RAMS for main
    frame memory", 1971 IEEE International Convention Digest,
    p. 72-3, CCA6-15260.
89. Lerner, E. M. and Someriu, M. D., "Application of a semiconductor
    main memory in a fault-tolerant real-time system", Proceedings of
    the IEEE Milwaukee Symposium on Automatic Computation and Control,
    1976, p. 413-19, CCA11-28306.
90. Levine, L. and Myers, W., "Timing: A crucial factor in LSI-MOS
    main-memory design", Electronics, 48 (14), 107-11 (1975).
91. Miwa, O., "Main memory construction of FACOM 230-75 system",
    Inf. Process. Soc. Jap., 16 (4), 331-4 (1975), Japanese,
    CCA10-22903.
92. Nakagawa, I., "The configuration of main memory for OUK-90 sys-
    tems", Inf. Process. Soc. Jap., 16 (4), 344-8 (1975), Japanese,
    CCA10-22904.

93. Neuhauser, C., "Functional description of the EMMY main memory
    system", Report AD-A021148/2SL, Stanford Univ., Calif., Digital
    Systems Lab., 1975, 16 pp.
94. Raisanen, W., "Motorola MOS-bipolar main-frame memory", 1970
    IEEE International Convention Digest, p. 132-3, CCA5-22458.
95. Suzuki, I., "Logical and physical structure of B7700 main
    memory", Inf. Process. Soc. Jap., 16 (4), 359-63 (1975),
    Japanese, CCA10-22907.
96. Suzuki, S., Sato, K. and Kobayashi, H., "Composition of the
    ACOS series 77 main memory", Inf. Process. Soc. Jap., 16 (4),
    335-9 (1975), Japanese, CCA10-25830.
97. Takei, K., "The logical structure of main memory systems", Inf.
    Process. Soc. Jap., 16 (4), 305-13 (1975), Japanese, CCA10-22902.
98. Thomas, R. T., "Main memory for user microprogram residence -
    an analysis", 6th Annual Workshop on Microprogram, Conference
    Proceedings, p. 13-20.
99. Wijnhoven, R. M. G., "Main memories and control memories",
    Informatie, 15 (7-8), 359-64 (1973), Dutch, CCA9-3386.

3.   Cache/Buffer

100. Ackland, B. D. and Pucknell, D. A., "Cache store concepts in
     real time engineering minicomputer systems", Symposium on
     Computers in Engineering, 1976, p. 120-4, CCA12-13129.
101. Agrawal, O. P. and Pohm, A. V., "Cache memory systems for
     multiprocessor architecture", Proceedings of the AFIPS National
     Computer Conference, 1977, p. 955-64.
102. Atterbury, G. F. and Adams, G. F., "Building today's technol-
     ogies into a large-scale time-sharing system", Comput. Des.,
     14 (9), 79-85 (1975).
103. Barsamian, H. and DeCegama, A. L., "System design considerations
     of cache memories", Sixth Annual IEEE Computer Society Inter-
     national Conference, Digest of Papers, 1972, p. 107-10,
     CCA8-2625.
104. Bennett, M., Berard, P., Boksenbaum, C. and Veran, M., "Effic-
     iency of a cache memory", Rev. Fr. Autom. Inf. Rech. Oper.,
     10 (5), 1-15 (1976).
105. Brown, D., "A scratch pad memory for buffering multiwire pro-
     portional chamber data", Nucl. Instrum. & Methods, 117 (2),
     561-7 (1974), CCA9-20774.
106. Chow, C. K., "Determination of cache's capacity and its matching
     storage hierarchy", IEEE Trans. Comput., C-25 (2), 157-64
     (1976).
107. Edrington, J., "Buffer speeds response time of first-in, first-
     out memory", Electronics, 49 (20), 89-90 (1976).
108. Fegan, D. J., "Fast inexpensive 1024-bit buffer memory data
     acquisition system for cosmic ray data using TTL integrated
     circuits and a paper tape punch", Nucl. Instrum. Methods, 131
     (3), 541-7 (1975).

109. Glushchenko, B. G., "Comparison of two methods of organization
     of the buffer memory of a digital computer", Eng. Cybern, $\underline{12}$
     (1), 151-4 (1974).
110. Halatsis, C., Philokyprou, G. and Maritsas, D., "Designing
     block-structured bubble-up buffers", Electron. Engineering,
     $\underline{48}$ (578), 45-8 (1976), CCA11-17402.
111. Hessel, G. and Seddig, H. P., "64-channel 1-bit buffer store
     for fast time analysis", Radio Fernsehen Elektron., $\underline{24}$ (20),
     659-60 (1975), German, CCA11-6384.
112. Hunter, G. M., "Memory system including buffer memories", Patent
     USA 3699533, Publ. October 1972, CCA8-7356.
113. Iizuka, H., "Cache memory simulation", Bull. Electrotech. Lab.,
     $\underline{37}$ (8), 751-69 (1973), Japanese, CCA9-16067.
114. Jones, J. D., Junod, D. M., Partridge, R. L. and Shawley, B. L.,
     "Updating cache data array's with data stored by other CPU's",
     IBM Tech. Disclosure Bull., $\underline{19}$ (2), 594-6 (1976), CCA12-1452.
115. Kosky, F. G., Millham, E. H., Scaccia, R. J. and Villante, F.
     J., "High-speed programmable buffer", IBM Tech. Disclosure
     Bull., $\underline{18}$ (12), 3993-6 (1976), CCA11-22620.
116. Kunz, A., "Acquisition system of data in a MOS dynamic register
     used as a buffer memory", Z. Angew. Math. & Phys., $\underline{24}$ (3),
     460-2 (1973), French, CCA9-8284.
117. Marsh, A. H., "Digital sample/hold simplifies communications
     timing", EDN, $\underline{20}$ (1), 59 (1975).
118. Merrill, B., "System 370/168 cache memory performance", Confer-
     ence on Computer Measurement and Evaluation (CME) - Selected
     Papers from the Share Project, 1975, p. 576-9.
119. Nessett, D. M., "Effectiveness of cache memories in a multipro-
     cessor environment", Aust. Comput. J., $\underline{7}$ (1), 33-8 (1975).
120. Philokyprou, G. and Halatsis, C. C., "Electronic buffer mem-
     ories", Int. J. Electr. Eng. Educ., $\underline{12}$ (1), 25-36 (1975).
121. Pohm, A. V., Agrawal, O. P. and Monroe, R. N., "Cost and per-
     formance tradeoffs of buffered memories", Proc. IEEE, $\underline{63}$ (8),
     1129-35 (1975).
122. Rajaraman, V. and Vikas, O., "Buffered stack memory organiza-
     tion", Electron. Lett., $\underline{11}$ (14), 305-7 (1975).
123. Ramchendani, C. B., "An analytical model for memory contention
     in cache memory multiprocessor systems", Proceedings of the 6th
     Hawaii International Conference on Systems Sciences, 1973,
     p. 87-9, CCA8-11551.
124. Strecker, W. D., "Cache memories for PDP-11 family computers",
     Symp. on Comput. Archit., 3rd Annu., Conf. Proc., 1976, p. 155-8.
125. White, G., "MOS buffer memories after economies", Electron,
     no. 2, 12-14 (1972), EEA75-24355.
126. Wiecek, H., "Buffer registers", Informatyka, $\underline{12}$ (1), 9-11 (1976),
     Polish, CCA11-17413.

IV.   RANDOM-ACCESS MEMORIES
      1.   General

127. Anon., "Density, power, clocking tradeoffs favor 16-pin 4k
     RAMs in small systems", Digital Des., $\underline{6}$ (12), 60-2 (1976),
     EEA80-24965.
128. Brown, J. R., Jr., "Timing pecularities of multiplexed RAM's",
     Comput. Des., $\underline{16}$ (7), 85-92 (1977).
129. Coker, D., "16-k RAM eases memory design for mainframe and
     microcomputers", Electronics, $\underline{50}$ (9), 115-19 (1977).
130. Coker, D. and Davis, K., "16K RAM - from micros to mainframes",
     13th IEEE Computer Society International Conference, Digest of
     Papers, 1976, p. 225-7, CCA12-1522.
131. Cunningham, J. and Jaffe, J., "Insight into RAM costs aids
     memory-system design", Electronics, $\underline{48}$ (25), 101-3 (1975),
     CCA11-6378.
132. Dobree Wilson, M. J. and Aleksander, I., "Pattern recognition
     properties of RAM/ROM arrays", Electron. Lett., $\underline{13}$ (9), 253-4
     (1977).
133. Ernst, H., Glock, H., Rathbone, R., Schwabe, U. and Burker, U.,
     "A fast bipolar 1024 bit RAM for high-performance memory
     systems", Siemens Forsch.- & Entwicklungsber., $\underline{4}$ (2), 86-91
     (1977), EEA80-21012.
134. Foss, R. C., Harland, R. and Roberts, J. A., "Check list for
     4096-bit RAMs flags potential problems in memory design",
     Electronics, $\underline{49}$ (18), 103-7 (1976), CCA11-28281.
135. Hackmeister, D., "Focus on semiconductor RAM's", Electron.
     Des., $\underline{25}$ (17), 56-62 (1977).
136. Haseloff, E., "Operation of the TMS4030/60 RAMs", New Electron.,
     $\underline{9}$ (22), 47-8, 53, 57, 60, 62 (1976), EEA80-6265.
137. Koch, R., Herbst, H. and Jespers, P., "Three-terminal CID as
     random access memory cell", IEEE J. Solid-State Circuits,
     SC-$\underline{12}$ (5), 534-6 (1977).
138. Lockhart, J., "Predict as 4-k RAM's average I_{DD}", Electron.
     Des., $\underline{25}$ (17), 78-9 (1977).
139. Mukai, H., Kawarada, K., Kundo, K. and Toyoda, K., "Ultra high-
     speed 1K-bit RAM with 7.5ns access time", 1977 International
     Solid-State Circuits Conference, Digest of the Technical
     Papers, p. 78-9.
140. Schlenther, M., "Focus on: bipolar and MOS RAMs", Elektronik,
     $\underline{25}$ (10), 57-68 (1976), German, EEA80-273.
141. Schneider, B., "LSI components are pattern sensitive", EC-Nyt,
     no. 49, 10-12 (1976), Danish, CCA11-22666.
142. Texas Instruments Inc., "Random access memory array", Patent
     UK 1401203, Publ. July 1975, CCA11-9655.
143. Vadasz, L. L., Chua, H. T. and Grove, A. S., "Semiconductor
     random-access memories", IEEE Spectrum, $\underline{8}$ (5), 40-8 (1971),
     CCA6-15243.

144. Wilcock, J. D., "Semiconductor r.a.m.s an expanding activity",
     New Electron., 6 (4), 39-40, 3, 4, 6 (1973), CCA8-11648.
145. Woelkers, R. C., "Random access semiconductor memory system",
     1971 IEEE International Convention Digest, p. 78-9, CCA6-15261.

        2.   Dynamic

146. Abbott, R. A., Regitz, W. M. and Karp, J. A., "A 4K MOS dynamic
     random-access memory", IEEE J. Solid-State Circuits, SC-8,
     292-8 (1973), EEA76-35352.
147. Ahlquist, C. N., Breivolgel, J. R., Koo, J. T., McCollum, J. L.,
     Oldham, W. G. and Renninger, A. L., "A 16 384-bit dynamic RAM",
     IEEE J. Solid-State Circuits, SC-11 (5), 570-4 (1976),
     EEA79-46416.
148. Allen, C. A. and Lund, D. F., "Dynamic m.o.s. memory array
     chip", Patent USA 3685027, Publ. August 1972, EEA76-821.
149. Anon., "Data input/output to the MCM 6604", New Electron., 9
     (7), 42 (1976), CCA11-17410.
150. Anon., "Dynamic RAMs pick up speed, quadruple in density",
     Electronics, 50 (2), 84-9 (1977), CCA-8891.
151. Bapat, D. and Mrazek, D., "System considerations in dynamic
     MOSRAM's", Electron. Equip. News, 13 (10), 30-2, 5 (1972),
     EEA75-13757.
152. Barnes, J. J., Shabte, G. N. and Jenne, F. B., "The buried-
     source VMOS dynamic RAM device", 1977 International Electron
     Devices Meeting, Technical Digest, p. 272-6.
153. Bernacchi, J. R. and Palfi, T. L., "Why not the 8K RAM?", 1975
     Semiconductor Test Symposium, Digest of Papers, p. 19-22.
154. Bizjak, J. F., "A new MOS random access memory", Aust. Electron.
     Eng., 5 (12), 10-14 (1972), EEA76-11952.
155. Boll, H. J., Fuls, E. N., Nelson, J. T. and Yau, L. D., "Auto-
     matic refresh dynamic memory", 1976 International Solid-State
     Circuits Conference, Digest of Technical Papers, p. 132-3,
     237, EEA79-46467.
156. Cahuvel, G., "Cyclic/refresh and control circuits for dynamic
     memories", Electron. & Microelectron. Ind., no. 225, 55-8
     (1976), French, EEA80-3001.
157. Clemens, J. T., Doklan, R. H. and Nolen, J. J., "An n-channel
     Si-gate integrated circuit technology (RAM applications)",
     1975 International Electron Devices Meeting, Technical Digest,
     p. 299-302, EEA79-46531.
158. Coleman, D. W. and Temkin, B. M., "Steady state gamma testing
     of a 4K NMOS dynamic RAM", IEEE Trans. Nucl. Sci., NS-23 (3),
     1301-3 (1976), CCA11-22642.
159. Dames, U., "SAB 1103 - a Siemens semiconductor memory", Com-
     ponents Rep., 10 (2), 33-4 (1975), CCA11-9673.
160. Dockerty, R. C., Abbas, S. A. and Barile, C. A., "Low-leakage
     N- and P-channel silicon-gate FET's with an SiO_2-Si_3N_4-gate
     insulator", IEEE Trans. Electron. Devices, ED-22 (2), 33-9 (1975).

161. Feryszka, R. and Yung, A. K., "Two transistor dynamic memory cell", Report TN-1007, RCA, Princeton, N. J., 1975, 2 pp., EEA78-41704.

162. Foss, R. C. and Harland, R., "Peripheral circuits for one-transistor cell MOS RAM's", IEEE J. Solid-State Circuits, SC-10 (5), 255-61 (1975), CCA10-25844.

163. Foss, R. C. and Harland, R., "Standards for dynamic MOS RAM's", Electron. Des., 25 (17), 66-70 (1977).

164. Gerardin, C. J. and Levis, J. P., "Transparent refreshing for semiconductor dynamic random-access memory", IBM Tech. Disclosure Bull., 16 (3), 934-6 (1973), CCA9-3424.

165. Hoff, M. E., Jr., "Silicon-gate dynamic MOS crams 1024 bits on a chip", Electronics, 43 (16), 68-73 (1970), CCA5-22440.

166. Honeywell Info., "Addressable dynamic memory system", Patent UK 1406117, Publ. September 1975, CCA11-9690.

167. Itoh, K., Shimohigashi, K., Chiba, K., Taniguchi, K. and Kawamoto, H., "A high-speed 16-kbit n-MOS random-access memory", IEEE J. Solid-State Circuits, SC-11 (5), 585-90 (1976), EEA79-46418.

168. Karp, J. A., "Dynamic refresh memories", 1971 IEEE International Convention Digest, p. 36-7, EEA74-22715.

169. Kinniment, D. J., "Low-power semiconductor memory cell", Proc. Inst. Electr. Eng., 120 (10), 1212-5 (1973), CCA9-3425.

170. Kluge, W., "Cellular dynamic memory array with reduced data-access time", Electron. Lett., 9 (19), 458-60 (1973), CCA9-3421.

171. Kluge, W., "Dynamic memory with fast random access and page transfer properties", Digital Processes, 1 (4), 279-93 (1975).

172. Landers, G., "Choosing among 4-k MOS RAMs? A head-to-head comparison of the available package sizes aids in selection of the best memory for your system", Electron. Des., 24 (12), 138-42 (1976).

173. Lenfant, J., "Fast random and sequential access to dynamic memories of any size", IEEE Trans. Comput., C-26 (9), 847-55 (1977).

174. Lund, T., "Semiconductor memory system design", Man Made Memories, Conference Summaries, 1971, 1 pp., CCA6-12748.

175. Masuhara, T., Nagata, M., Hashimoto, N. and Adachi, Y., "Low-voltage dynamic MOS memory", Electron. Commun. Jap., 57 (8), 129-37 (1974), EEA79-1041.

176. McCracken, T. E., "Development of a solid state refresh memory tonal display", Thesis, Univ. Missouri, Columbia, 1974, 323 pp., Order No. 75-5767, EEA78-37657.

177. Mehta, R. J., Geilhufe, M. and Palfi, T. L., "Random access read/write semiconductor memory", Patent USA 3795898, Publ. March 1974, CCA9-23257.

178. Mitterer, R. W., "Dynamic 4-kilobit MOS read-write memories", International Conference on Large Scale Integrated Circuits, 1974, p. 253-63, EEA78-11241.

179. Proebsting, R., "Dynamic MOS RAM's: An economic solution for
     many system designs", EDN, 22 (12), 61-6 (1977).
180. Reed, J. A., "MOS dynamic memory array and refreshing system",
     Patent USA 3858185, Publ. December 1974, EEA78-15420.
181. Sander, W. B. and Early, J. M., "A 4096 x 1 (I^3L) (isoplanar
     integrated injection logic) bipolar dynamic RAM", 1976 Inter-
     national Solid-State Circuits Conference, Digest of Technical
     Papers, p. 182-3, EEA79-46475.
182. Sander, W. B., Shepherd, W. H. and Schinelle, R. D., "Dynamic
     I^2L random-access memory competes with MOS designs", Electronics,
     49 (17), 99-102 (1976), EEA79-41588.
183. Schroeder, P. R. and Proebsting, R. J., "A 16K x 1-bit dynamic
     RAM", 1977 International Solid-State Circuits Conference,
     Digest of Technical Papers, p. 12-13.
184. Shiga, K., Itoh, T., Ikeda, N., Arai, M., Sudoh, M. and Maki,
     M., "Mono-stable 4-transistor CMOS RAM", Natl. Tech. Rep. Mat-
     sushita Electr. Ind., 22 (5), 607-12 (1976), Japanese.
185. Stein, K. U., Sihling, A. and Doering, E., "Storage array and
     sense/refresh circuit for single-transistor memory cells", IEEE
     J. Solid-State Circuits, SC-7, 336-40 (1972), EEA75-36284.
186. Stone, H. S., "Dynamic memories with fast random and sequential
     access", IEEE Trans. Comput., C-24 (12), 1167-74 (1975),
     CCA11-4794.
187. Sun, R. C. and Clemens, J. T., "Characterization of reverse-
     bias leakage currents and their effect on the holding time
     characteristics of MOS dynamic RAMs", 1977 International
     Electron Devices Meeting, Technical Digest, p. 254-7.
188. Tasch, A. F., Jr., Frye, R. C. and Fu, H. S., "Charge-coupled
     RAM cell concept", IEEE J. Solid-State Circuits, SC-11 (1),
     58-63 (1976).
189. Tasch, A. F., Jr., Fu, H. S., Holloway, T. C. and Frye, R. C.,
     "Charge capacity analysis of the charge-coupled RAM cell", IEEE
     J. Solid-State Circuits, SC-11 (5), 575-85 (1976), CCA11-31608.
190. Terman, L. M., "Avoidance of refresh times in dynamic memories",
     IBM Tech. Disclosure Bull., 15 (7), 2056-8 (1972), CCA8-11645.
191. Vittera, J., "Nonvolatile µp memory uses 4k n-channel MOS
     RAM's", EDN, 20 (8), 53-6 (1975).
192. Walther, T. R., "Dynamic N-MOS RAM with simplified refresh",
     Comput. Des., 12 (2), 53-8 (1973), CCA8-13692.
193. Walther, T. R. and McCoy, M. R., "A three transistor MOS memory
     cell with internal refresh", Microelectronics, 4 (3), 41-3
     (1972), EEA76-26102.

         3.   Static

194. Anon., "A static 4096-bit RAM with a particularly high packing
     density", Elektronik, 23 (4), 116 (1974), German, EEA77-23039.
195. Anon., "Static MOS RAMs gain ground", Electron. & Microelectron.
     Ind., no. 223, 41-3 (1976), French, EEA79-33109.

196. Bardell, P. H., "Static MOS memory", 1971 IEEE International
     Convention Digest, p. 32-3, CCA6-15256.
197. Budinsky, J., "The MOS RAM static memory of the 1101 type with
     the capacity of 256 bites (256 x 1)", Sdelovaci Tech., 24 (2),
     43-8 (1976), Czech, EEA79-28693.
198. Budinsky, J., "The use of the MOS RAM static memory of the 1101
     type", Sdelovaci Tech., 24 (3), 87-91 (1976), Czech,
     EEA79-28603.
199. Chauvel, G., "Static type memories for storing information",
     Electron. & Microelectron. Ind., no. 211, 45-9 (1975), French,
     CCA11-9675.
200. Dingwall, A. G., "Compact C.O.S./M.O.S. 256-bit random-access
     memory", International Electron Devices Meeting (abstracts),
     1970, p. 16, CCA6-5694.
201. Dingwall, A. G. F. and Stricker, R. E., "High density COS/MOS
     1024-bit static random access memory", IEEE J. Solid-State
     Circuits, SC-10, 197-200 (1975), EEA78-33208.
202. Herndon, W. H., Ho, W. and Ramirez, R., "A static 4096-bit
     bipolar random-access memory", IEEE J. Solid-State Circuits,
     SC-12 (5), 524-6 (1977).
203. Hnatek, E. R., Graves, W. and Schmitt, R. G., "How static is
     the static 4K RAM?", 1976 Semiconductor Test Symposium, Digest
     of Papers, p. 3-8, EEA80-3042.
204. Hume, S., "Consider 1024-bit C-MOS RAMs for small static-memory
     systems", Electronics, 48 (15), 102-6 (1975), EEA79-1034.
205. Kaneko, K., Kurobe, T. and Ogawa, Y., "Four-transistor static
     memory cell using CMOS", Electron. & Commun. Jpn., 58 (10),
     110-18 (1975), EEA80-24992.
206. Kroeger, J. H., "Operation of the Am9130/40 4k static r.a.m.s",
     New Electron., 9 (19), 38, 41, 43 (1976), CCA11-31621.
207. McKenny, V. G., "A 5-V only 4k static RAM", 1977 International
     Solid-State Circuits Conference, Digest of Technical Papers,
     p. 16-17.
208. Ochii, K., Suzuki, Y., Ueno, M., Sato, K. and Asahi, K., "C^{2}MOS
     4k static RAM", 1977 International Solid-State Circuits Con-
     ference, Digest of Technical Papers, p. 18-19.
209. O'Connell, T. R., Hartman, J. M., Errett, E. B. and Leach, G.
     S., "Two static 4K clocked and nonclocked RAM designs", IEEE
     J. Solid-State Circuits, SC-12 (5), 497-501 (1977).
210. O'Connell, T. R., Hartman, J. M., Errett, E. D., Leach, G. and
     Dunn, W. C., "A 4k static clocked and nonclocked RAM design",
     1977 International Solid-State Circuits Conference, Digest of
     Technical Papers, p. 14-15.
211. Ohuchi, K., Tanaka, S., Ara, K., Tanuma, Y., Kubota, N. and
     Horiuchi, S., "A 1kb N-silicon-gate static random-access
     memory with two different depletion-load thresholds", Oyo
     Buturi, 44 (Suppl.), 211-16 (1975), EEA78-41683.
212. Pashley, R., Owen, W., Kokkonen, K. and Ebel, A., "Speedy RAM
     runs cool with power-down circuitry", Electronics, 50 (16),
     103-7 (1977).

213. Pashley, R. D., Owen, W. H., III, Kokkonen, K. R., Jechmen, R.
     M., Ebel, A. V., Ahlquist, C. N. and Schoen, P., "A high per-
     formance 4k static RAM fabricated with advanced MOS technology",
     1977 International Solid-State Circuits Conference, Digest of
     Technical Papers, p. 22-4.
214. Saito, S., Endo, N., Uchida, Y., Tanaka, T., Nishi, Y. and
     Tamaru, K., "A 256 bit nonvolatile static random access memory
     with MNOS memory transistors", Jap. J. Appl. Phys., $\underline{15}$ (Suppl.
     15-1), 185-90 (1976), EEA80-9522.
215. Schlageter, J. M., Jayakumar, N., Kroeger, J. H. and Sarkissian,
     V., "Two 4K static 5-V RAM's", IEEE J. Solid-State Circuits,
     SC-$\underline{11}$ (5), 602-9 (1976), CCA11-31611.
216. Suzuki, S., Nagahashi, Y., Tanaka, T., Yamada, T., Muta, H.,
     Okabayashi, H. and Yamada, K., "A static random-access memory
     with normally-off-type Schottky barrier FET's", IEEE J. Solid-
     State Circuits, SC-$\underline{8}$, 326-31 (1973), EEA76-35357.
217. Young, S., "Uncompromising 4k static RAM runs fast on little
     power", Electronics, $\underline{50}$ (10), 99-103 (1977), EEA80-28906.

V.   READ ONLY MEMORIES
     1.   General

218. Anon., "Which type of memories supply the words chosen by ap-
     plication of the control system?", Elektron. & Elektrotech.,
     $\underline{30}$ (759), 78-80, 89 (1975), Dutch, CCA10-17626.
219. Anon., "Freely programmable process control equipment", Tech.
     Rundsch., $\underline{68}$ (47), 27, 29 (1976), CCA12-6673.
220. Bishop, A., "High speed memory application using field program-
     mable ROMs and ECL RAMs", Man Made Memories, Conference Sum-
     maries, 1971, 1 pp., CCA6-12745.
221. Haynes, G., "Optimising fixed content memories", New Electron.,
     $\underline{9}$ (19), 26, 9 (1976), CCA11-31619.
222. Hilberg, W., "On the structure and capacity of electronic digital
     memories", Elektron. Rechenanlagen, $\underline{18}$ (3), 114-22 (1976),
     German, CCA11-22635.
223. Kahng, D., "Developments in dual dielectric charge-storage (DDC)
     REPROM (reprogrammable read often memories)", 14th Annual Pro-
     ceedings Reliability Physics Conference, 1976, p. 192,
     EEA80-17275.
224. Lau, S., "Recent developments in the design and application of
     a bipolar control store sequencer and other MPU associated LSI
     components", Microelectron. & Reliability, $\underline{15}$ (4), 329-33 (1976).
225. McDowell, J., "Large bipolar ROMs and p/ROMs revolutionize logic
     and system design", Comput. Des., $\underline{13}$ (6), 100, 102-4 (1974),
     CCA9-21973.
226. Sweeney, M., "Large bipolar r.o.m.s and p.r.o.m.s in logic and
     system design", New Electron., $\underline{7}$ (15), 29, 32, 39 (1974),
     CCA9-23256.

227. Timm, V., "Focus on: ROMs, PROMs and PLAs", Elektronik, $\underline{25}$ (5), 38-47 (1976), German, CCA11-22662.
228. Westcott, D. W., "Replacing portions of read-only store modules with random-access memory bits", IBM Tech. Disclosure Bull., $\underline{17}$ (10), 2915-16 (1975), CCA10-15154.

2.   ROM

229. Alberts, G. S., Farrar, P. A. and Hallen, R. L., "Constructing a read only memory for redundancy and other applications", Patent USA 3959047, Publ. May 1976, CA85-55539.
230. Armstrong, L., "Big ROM's begin to make waves", Electronics, $\underline{50}$ (19), 73-4 (1977).
231. Askin, H. O. and Sonoda, G., "Double-bit line ROS array", IBM Tech. Disclosure Bull., $\underline{19}$ (5), 1683-5 (1976), EEA80-20379.
232. Baltinger, U., Frantz, H., Haug, W. and Remshardt, R., "IGFET read-only memory", IBM Tech. Disclosure Bull., $\underline{14}$, 1429-30 (1971), EEA75-7541.
233. Baker, P. W., "High-speed BCD multiplier using ROMs", Digital Processes, $\underline{2}$ (1), 89-93 (1976), CCA11-22553.
234. Finn, C. L., "All-semiconductor memory system includes read-only and read/write chips", Hewlett-Packard J., $\underline{24}$ (4), 22-4 (1972), CCA8-11643.
235. Gaskin, J., "Measuring the access time of bipolar r.o.m.s", New Electron., $\underline{7}$ (11), 27 (1974), CCA9-20768.
236. Geisselhardt, W., "A multistable transfluxor memory for purposes in telephony", IEEE Trans. Magn., MAG-$\underline{8}$ (3), 358-61 (1972), CCA8-13667.
237. Greene, R. M., "A 32K ROM using differential ramp techniques", 1976 International Solid-State Circuits Conference, Digest of Technical Papers, p. 186-7, CCA11-28302.
238. Grochowski, E. G., "Rewritable IGFET read-only store", IBM Tech. Disclosure Bull., $\underline{14}$, 263 (1971), EEA74-34089.
239. Gunn, J. F., Lynes, D. H. and Pritchett, R. L., "A bipolar 16K ROM utilizing Schottky diode cell", Computer, $\underline{10}$ (7), 14-17 (1977).
240. Gunn, J. F., Pritchett, R. L. and Lynes, D. J., "A bipolar 16K ROM utilizing Schottky diode cells", 1977 International Solid-State Circuits Conference, Digest of Technical Papers, p. 118-19.
241. Hardin, D. K., "Read-only storage control for high-speed instruction execution", IBM Tech. Disclosure Bull., $\underline{16}$ (4), 1106-8 (1973), CCA9-3359.
242. Herrell, D. J. and Park, K. C., "Write once read-only store", IBM Tech. Disclosure Bull., $\underline{15}$ (3), 949-50 (1972), CCA8-2694.
243. Herzenberg, L. and Andres, K., "A large, static MOS/bipolar ROM with combinatorial addressing for keyboard encoding", 1971 International Solid State Circuits Conference, Digest of Technical Papers, p. 64-5, CCA6-15268.

244. Iizuka, H., Masuoka, F., Sato, T. and Ishikawa, M., "Electrically alterable avalanche-injection-type MOS read-only memory with stacked-gate structure (SAMOS)", IEEE Trans. Electron Devices, ED-23 (4), 379-87 (1976), CCA11-15067.

245. Katholing, G., "MOS read only memory (MOS-ROM) GDR 101", Siemens Electron. Components Bull., 7 (2), 33-6 (1972), EEA75-36311.

246. Kawagoe, H. and Tsuji, N., "Minimum size ROM structure compatible with silicon-gate E/D MOS LSI", IEEE J. Solid-State Circuits, SC-11 (3), 360-4 (1976), CCA11-22668.

247. Kobylar, A., Lindsay, R. L. and Pitroda, S. G., "ROMs cut cost, response time of m-out-of-N detectors", Electronics, 46 (4), 112-14 (1973), CCA8-9533.

248. Leininger, J. C. and Parchinski, K. J., "Read only and writable control storage", IBM Tech. Disclosure Bull., 13 (5), 1288-9 (1970), CCA6-3726.

249. Mazda, F. F., "The components of computers. V. The read only memory", Electron. Compon., 13 (20), 999-1001, 4, 5, 6 (1972), CCA8-2662.

250. McDonald, B. A., "The M.N.O.S. bipolar transistor", International Electron Devices Meeting (abstracts), 1970, p. 18, CCA6-5696.

251. Mukhopad, Y. F., Fedchenko, A. I. and Popkov, V. K., "Method of enhancing the homogeniety of read-only memories", Autom. Control Comput. Sci., 9 (5), 79-82 (1975).

252. Muller, R. G., Nietsch, H., Rossler, B. and Wolter, E., "An 8192-bit electrically alterable ROM employing a one-transistor cell with floating gate", IEEE J. Solid-State Circuits, SC-12 (5), 507-14 (1977).

253. Muoio, A. W., "Character/symbol generation by MOS read only memory", Proc. Soc. Inform. Display, 11, 6-15 (1970), EEA73-25697.

254. Nat. Cash Register Co., "Read-only memory", Patent UK 1289938, Publ. September 1972, CCA8-5263.

255. Norberg, R., "The symbol generator: a link between man and computer", Eltek. Aktuell Elektron., 16 (7), 38-40 (1973), Swedish, CCA9-3422.

256. North, J. C. and Weick, W. W., "Laser coding of bipolar read-only memories", IEEE J. Solid-State Circuits, SC-11 (4), 500-5 (1976), EEA79-37685.

257. Peatman, J. B., Dack, D. G. and Warren, D. A., "ROMs in microprocessors can test themselves", Electronics, 47 (23), 153, 155 (1974).

258. Quinones, M., "An associative-capacitive ROM for reprogrammable logic applications", Comput. Des., 13 (1), 98-101 (1974), CCA9-9213.

259. Rodgers, T. J., Hiltpold, R., Zimmer, J. W., Marr, G. and Trotter, J. D., "VMOS ROM", IEEE J. Solid-State Circuits, SC-11 (5), 614-22 (1976), EEA79-46422.

260. Schmidt, S. A. and Woessner, R. J., "Read-only store using
     large scale integration", IBM Tech. Disclosure Bull., 15 (5),
     1574-5 (1972), CCA8-9538.
261. Selleck, J. E., "Positive polarity bistable resistor read-
     mostly memory", IBM Tech. Disclosure Bull., 15 (7), 2119-20
     (1972), CCA8-11646.
262. Shah, A., "Application of read-only memories (ROMs) for combi-
     nation logic and for arithmetic problems: memories with custom-
     designed address decoders", Mitt. AGEN, no. 19, 37-44 (1975),
     German, CCA11-12333.
263. Shum, N. Y., "On-line tester for read only memory", Electron.
     Engineering, 48 (582), 15 (1976), EEA79-41587.
264. Siemens AG, "Read-only memory", Patent UK 1288994, Publ.
     September 1972, CCA8-7339.
265. Siemens AG, "Read-only memory", Patent UK 1288995, Publ.
     September 1972, CCA8-7340.
266. Sonoda, G., "Charge distribution read-only array", IBM Tech.
     Disclosure Bull., 19 (5), 1690-2 (1976), CCA12-12906.
267. Spence, W., Shen, M. K., Lockwood, G. C. and Trudel, M. L., "A
     4096-bit word-alterable ROM", IEEE Trans. Electron Devices,
     ED-24 (5), 610-13 (1977), CCA12-15512.
268. Stewart, R. G., "High-density CMOS ROM arrays", IEEE J. Solid-
     State Circuits, SC-12 (5), 502-6 (1977).
269. Sutherlin, K. K., Lauer, J. P. and Olenick, R. W., "Holoscan:
     a commercial holographic ROM", Appl. Opt., 13 (6), 1345-54
     (1974), CCA9-20770.
270. Tektronix Inc., "Read-only memory", Patent UK 1288440, Publ.
     September 1972, CCA8-2658.
271. Travis, T., "Patching a program into a ROM", Electron. Des.,
     24 (18), 98-101 (1976), CCA12-1450.
272. Trostyanetskii, D. S., "Method of increasing the signal to
     noise ratio of the storage element of a transformer read only
     memory", Avtom. & Vychisl. Tekh., no. 2, 79-84 (1975),
     Russian, CCA10-15166.
273. Trostyanetskii, D. S., "Structural partitioning of a matrix
     read-only memory", Autom. Control Comput. Sci., 9 (4), 63-7
     (1975).
274. Wada, T., Nakagiri, M., Iwase, K. and Nakanuma, S., "MAS-ROM-
     the fabrication and memory characteristics", Microelectronics,
     4 (2), 8-19 (1972), EEA75-36301.
275. Western Electric Co., Inc., "Read-only memory", Patent UK
     1294933, Publ. November 1972, CCA8-9551.

3.   PROM/EAROM

276. Ammann, R. W. and Donnelly, J. M., "Semiconductor read-only
     memory applications", GTE Autom. Electr. Tech. J., 14 (8),
     370-81 (1975), CCA11-6380.

277. Anon., "Programmable read-only memories", New Electron., $\underline{9}$ (14), 24-6 (1976), EEA79-37692.
278. Anon., "PROM power supply strobe", New Electron., $\underline{9}$ (1), 30 (1976), CCA11-12334.
279. Anon., "Quartz window permits memory to be erased by ultra-violet light", Insul./Circuits, $\underline{22}$ (3), 17-18 (1976), CCA11-17406.
280. Anon., "Alterable ROMs make design more flexible", Electronics, $\underline{50}$ (2), 89-94 (1977), CCA12-8892.
281. Barnes, J., Linden, J. and Edwards, J., "Operation and characterization of N-channel EPROM cells", 1976 International Electron Devices Meeting, Technical Digest, p. 173-6.
282. Blacksher, R., "PROM decoder replaces chip-enabling logic", Electronics, $\underline{49}$ (18), 100-2 (1976).
283. Brady, J., "PROM provides linear or logarithmic display", Electronics, $\underline{49}$ (23), 101, 103 (1976), EEA80-272.
284. Brockhoff, W. R., "Electrically shorted semiconductor junctions utilized as programmable read only memory elements", 14th Annual Proceedings Reliability Physics Conference, 1976, p. 202-6, EEA80-17268.
285. Dirks, C., "Programmable ROM's simplify sequential digital networks", Elektronik, $\underline{22}$(2), 68-70 (1973), German, CCA8-11564.
286. Dodson, W. H., Heightly, J. D., Alfisi, V. J. and Lodi, R. J., "Bootstrap-pumped control circuitry for an MNOS EAROM", 1977 International Solid-State Circuits Conference, Digest of the Technical Papers, p. 182-3.
287. Donnelly, T. M., Powell, W. W., Dobson, J. and Devaney, J., "Effects of programming variations on nichrome link PROMS", Proceedings of the 13th Annual Reliability Physics Symposium, 1975, p. 168-73, CCA11-28311.
288. Emberty, R. G., McNeil, J. R. and Collins, G. J., "EPROM high-speed erasing scheme", IEEE Trans. Electron Devices, ED-$\underline{24}$ (2), 159 (1977), EEA80-6679.
289. Farr, T. M., Jr., "Read-only memory controls universal counter", EDN, $\underline{21}$ (9), 114 (1976).
290. Frenzel, L. E., "DIP switches and diodes form programmable ROM", Electronics, $\underline{47}$ (11), 126-7 (1975).
291. Friedberg, H., "Properties and applications of electronically programmable MOS-ROMs", Elektron. Ind., $\underline{4}$ (4), 57-60 (1973), German, EEA76-29310.
292. Frohman-Bentchkowsky, D., "A fully decoded 2048-bit electrically programmable FAMOS read-only memory", IEEE J. Solid-State Circuits, SC-$\underline{6}$ (5), 301-6 (1971), EEA74-34081.
293. Godbole, V. R., "PROM converts binary code to drive 1-1/2-digit display", Electronics, $\underline{48}$ (25), 105, 107 (1975).
294. Gosney, W. M., "DIFMOS - a floating-gate electrically erasable nonvolatile semiconductor memory technology", IEEE Trans. Electron Devices, ED-$\underline{24}$ (5), 594-9 (1977), EEA80-25003.
295. Greene, B., "The Intel 2708 erasable 8k PROM", Elektron. Ind., $\underline{7}$ (12), 359-60 (1976), German, EEA80-24993.

296. Greene, R., Perlegos, G., Salsbury, P. J. and Morgan, W. L.,
     "The erasable PROM yet puts 16384 bits on a chip", Electronics,
     <u>50</u> (5), 108-11 (1977), CCA12-8894.
297. Kelley, J. W. and Millet, D. F., "An electrically alterable
     ROM - and it doesn't use nitride", Electronics, <u>49</u> (25), 101-4
     (1976), EEA80-2991.
298. Kerber, R. J., "Build a PROM programmer and program your own
     devices", Electron. Des., <u>24</u> (22), 172-3, 176 (1976), EEA80-9012.
299. Kikuchi, M., Ohya, S., Kamaya, M., Koike, M. and Yamamoto, H.,
     "A 2048-bit N-channel fully decoded electrically writable/
     erasable nonvolatile read only memory", 1st European Solid State
     Circuits Conference - ESSCIRC, Extended Abstracts, 1975, p. 66-7,
     EEA79-28701.
300. Koo, T. K., "Non-volatility characteristics of MNOS memory FET
     and EAROM arrays", IEEE Proc. Natl. Aerosp. Electron. Conf.,
     1974, p. 37-41.
301. Lockwood, G. C., Naber, C. T. and Koo, T. K., "A 1024 bit MNOS
     electrically-alterable ROM", 1972 WESCON Technical Papers, Vol.
     16, Paper 4/5, 6 pp., EEA76-8752.
302. Masuoka, F., Ishikawa, M., Sato, T. and Iizuka, H., "An elec-
     trically programmable and erasable 2048-bit stacked-gated MOS
     ROM", Presented at 1973 International Electron Devices Meeting
     Supplement, Washington, D. C., December, EEA77-18819.
303. Muller, R. G., Nietsch, H., Rossler, B. and Wolte, E., "Elec-
     trically-alterable 8K/bit N-channel MOS PROM", 1977 International
     Solid-State Circuits Conference, Digest of Technical Papers,
     p. 188-9.
304. Murrmann, H. and Scharbert, J., "High-speed 1024-bit ECL pro-
     grammable read-only memory", Siemens Forsch. Entwicklungsber
     Res. Dev. Rep., <u>4</u> (4), 245-9 (1975).
305. Neugebauer, C. A., Burgess, J. F. and Stein, L., "Electrically
     erasable buried-gate nonvolatile read-only memory", IEEE Trans.
     Electron Devices, ED-<u>24</u> (5), 613-18 (1977), CCA12-15513.
306. Rai, Y., Sasami, T. and Hasegawa, Y., "Mo-gate EAROM process
     cuts write/erase voltage and prices, too", JEE, no. 111, 26-31
     (1976), CCA11-22669.
307. Rawlinson, J. S., "Thyristor switched power supply (PROM pro-
     grammer)", Electron. Engineering, <u>48</u> (585), 17 (1976), CCA12-1507.
308. Roberts, M. J., "PROM function generator", IEEE Trans. Ind.
     Electron. & Control Instrum., IECI-<u>23</u> (3), 312-13 (1976),
     CCA11-31587.
309. Rossler, B., "Electrically erasable and reprogrammable read-
     only memory using the n-channel SIMOS one-transistor cell",
     IEEE Trans. Electron Devices, ED-<u>24</u> (5), 606-10 (1977),
     EEA80-25004.
310. Rossler, B. and Muller, R. G., "Erasable and electrically repro-
     grammable read-only memory using the N-channel SIMOS one-tran-
     sistor cell", Siemens Forsch.-& Entwicklungsber., <u>4</u> (6), 345-51
     (1975), CCA11-4815.

311. Salsbury, P. J., Morgan, W. J., Perlegos, G. and Simko, R. T.,
"High performance, MOS EPROM's using a stacked-gate cell", 1977
International Solid-State Circuits Conference, Digest of
Technical Papers, p. 186-7.

312. Scheibe, A. and Schulte, H., "Technology of a new n-channel
one-transistor EAROM cell called SIMOS", IEEE Trans. Electron
Devices, ED-24 (5), 600-6 (1977), CCA12-15511.

313. Schroeder, J. E. and Goslin, R. L., "A 1024-bit fused link CMOS
PROM", 1977 International Solid-State Circuits Conference,
Digest of Technical Papers, p. 190-1.

314. Schusser, M., "Programmable read-only memories: PROMs with
nickel-chrome fuse gaps", Elektron. Ind., 7 (11), 300-2 (1976),
German, CCA12-8896.

315. Smith, R. L., "How a PROM works", Radio-Electron., 46 (11),
72-4, 119 (1975), CCA11-9685.

316. Timm, V., "TTL-program controller with PROM", Elektronik, 24
(1), 53-8 (1974), German, CCA10-5489.

317. Wada, T., "Electrically programmable ROM using N-channel memory
transistor with floating gate", Solid-State Electron., 20 (7),
623-7 (1977).

318. Wada, T., Onoda, K., Ishiguro, H. and Nakanuma, S., "MAS-ROM -
electrically reprogrammable ROM with decoder", IEEE J. Solid-
State Circuits, SC-7, 375-81 (1972), EEA75-36287.

319. Waser, S., "Computer program reduces fusible-FROM errors",
Electronics, 49 (6), 120-1 (1976), CCA11-15061.

320. Waser, S., "p/ROM card simplifies computer diagnosis", Comput.
Des., 16 (1), 98-101 (1977).

321. Yamaguchi, T. and Sato, S., "Circuit array and system of MAOS
type fully decoded electrically alterable read only memory-IC",
Jap. J. Appl. Phys., 13 (9), 1414-20 (1974), EEA77-38490.

322. Zdrazil, Z., "Utilization of read-only storage for combinatorial
logic circuits", Automatizace, 19 (12), 322-7 (1976), Czech,
CCA12-10631.

VI.   TECHNOLOGIES
1.   General

323. Altman, L., "Five technologies squeezing more performance from
LSI chips", Electronics, 50 (17), 91-3 (1977).

324. Altman, L. and Mattera, L., "Memories score new highs in speed
and density", Electronics, 49 (4), 105-10 (1976), EEA79-20520.

325. Anon., "Data stores for computers", Neue Tech., 18 (11), 729,
731, 733 (1976), German, CCA12-15492.

326. Bomhardt, K., "Semiconductor techniques in international compe-
tition", Elektro-Monteur, 27 (7), 59-61 (1976), German,
EEA80-3369.

327. Chang, J. J., "Nonvolatile semiconductor memory devices", Proc.
IEEE, 64 (7), 1039-59 (1976), CCA11-22672.

328. Coady, J., "High density memory cell fabrication", IBM Tech.
     Disclosure Bull., 19 (6), 2007-9 (1976), CCA12-15516.
329. Dance, M., "Memory devices", Electron. Ind., 3 (4), 29-31, 33,
     35 (1977), EEA80-24394.
330. DeTroye, N. C., "Digital integrated circuits with low dissipa-
     tion", Philips Tech. Rev., 35 (7-8), 212-20 (1975).
331. Goser, K., "Low power memories in modified P/MOS techniques and
     CMOS technique", Nachrichtentech. Z., (NTZ), 26 (1), 9-15 (1973),
     German, EEA76-11973.
332. Henle, R. A., Ho, I. T., Johnson, W. S., Pricer, W. D. and
     Walsh, J. L., "The application of transistor technology to
     computers", IEEE Trans. Comp., C-25 (12), 1289-303 (1976),
     EEA80-279.
333. Hnatek, E., "Chipping away at core: another round", Digital
     Des., 6 (7), 31, 34, 36, 38-42 (1976), CCA12-3979.
334. Ichikawa, T., Sakamura, K. and Aiso, H., "ARES - A memory,
     capable of associating stored information through relevancy
     estimation", Proceedings of the AFIPS National Computer Con-
     ference, 1977, p. 947-54.
335. Katholing, G., "Highly integrated circuits - technique and
     region of application", Elektro-Anz., 29 (23), 557-60 (1976),
     German, EEA90-9515.
336. Leventhal, L. A., "Semiconductor technologies and semiconductor
     memories", Simulation, 27 (2), 65-71 (1976).
337. McGrenera, D., "MOS memories and display systems", Man Made
     Memories, Conference Summaries, 1971, 2 pp., CCA6-12749.
338. Oldham, W. G., "The fabrication of microelectronic circuits",
     Sci. Am., 237(3), 110-28 (1977).
339. Russell, D., "10k e.c.l. - the systematic choice", New Electron.,
     8 (24), 37-8, 42 (1975), EEA79-12498.
340. Sautter, D., "Points of interest of modern semiconductor tech-
     nologies", Elektroniker, 15(12), EL1-3 (1976), German,
     EEA80-17184.
341. Vaughan, W. C. M., "Mini's most important part", Electron.
     Des., 23 (14), 52-5 (1975).
342. Zatz, S., "Fundamentals of integrated circuits: integrated
     circuit component technology", Electron. Electro-optic Infrared
     Countermeas., 2 (10), 22-3, 29-30, 46 (1976), EEA80-24966.

2. Bipolar Transistors
   a. General

343. Alkfe, P., "Bipolar memories", Microelectron. & Reliab., 13
     (5), 339-43 (1974).
344. Anon., "Active bipolar memories", Electron. & Microelectron.
     Ind., no. 227, 49 (1976), French, EEA80-9510.
345. Baker, W. D., Herndon, W. H., Longo, T. A. and Pelzer, D. L.,
     "Oxide isolation brings high density to production bipolar
     memories", Electronics, 46 (7), 65-70 (1973), CCA8-11638.

346. Barber, M. R., "Bipolar memories", 1971 IEEE International
     Convention Digest, p. 30-1, CCA6-15255.
347. Beausoleil, W. F., Ho, I. T., Jen, T. S. and Pricer, W. D.,
     "Two device monolithic bipolar memory array", Patent USA
     3697962, Publ. October 1972, CCA8-5279.
348. Gersbach, J. E., "Current steering simplifies and shrinks 1k
     bipolar RAM", Electronics, 47 (9), 110-14 (1974), CCA9-16068.
349. Greene, F. S., Jr. and Phan, N. U., "Megabit bipolar LSI memory
     systems. IV.", 1970 International Solid-State Circuits Con-
     ference, Digest of Technical Papers, p. 40-1, CCA5-13756.
350. Lynes, D. J. and Hodges, D. A., "A diode-coupled bipolar tran-
     sistor memory cell", 1970 International Solid-State Circuits
     Conference, Digest of Technical Papers, p. 44-5, CCA5-13758.
351. Lynes, D. J. and Hodges, D. A., "Memory using diode-coupled
     bipolar transistor cells", IEEE J. Solid-State Circuits, SC-5
     (5), 186-91 (1970), CCA5-22444.
352. Murrmann, H., "Integrated circuits in bipolar technology",
     NTG-Fachber., 54, 31-42 (1975), German, CCA5-22444.
353. Newman, P. C., "High performance bipolar technology for LSI",
     Proceedings of the Agard Lecture Series, No. 75, Paper 2, 22 pp.
354. Palfi, T. L., "Bipolar memory cell having low power consumption",
     IBM Tech. Disclosure Bull., 13 (6), 1730-1 (1970), CCA6-5687.
355. Panousis, P. T., "A TRIM bipolar charge storage memory", Inter-
     national Electron Devices Meeting (abstracts), 1970, p. 18,
     CCA6-5696.
356. Peltzer, D., "Isoplanar processing (memory circuits)", 1972
     WESCON Technical Papers, Vol. 16, Paper 19/2, 4 pp., CCA8-7362.
357. Peltzer, D. and Herndon, B., "Isolation method shrinks bipolar
     cells for fast, dense memories", Electronics, 41 (5), 52-5
     (1971), CCA6-10775.
358. Phan, N. N., Shier, J. S. and Evans, A. L., "A fast 1024-bit
     bipolar RAM using JFET", 1977 International Solid-State Circuits
     Conference, Digest of Technical Papers, p. 70-1.
359. Raisanen, W., Hillis, D., Marley, J. and Trolsen, G., "Large
     scale integrated circuit bipolar memories (MOS)", Onde Elec.,
     50 (9), 763-7 (1970), French, CCA6-3772.
360. Sanders, T. J. and Morcom, W. R., "Polysilicon-filled notch
     produces flat, well-isolated bipolar memory", Electronics, 46
     (8), 117-20 (1973), CCA8-13693.
361. Sanders, T. J., Morcom, W. R. and Kim, C. S., "An improved
     dielectric-junction combination isolation technique for inte-
     grated circuits", 1973 International Electron Devices Meeting,
     Technical Digest, p. 38-40, CCA9-11967.
362. Scrupski, S. E., "High-density bipolars spur advances in com-
     puter design", Electronics, 48 (22), 81-6 (1975), CCA11-2188.
363. Wiedmann, S. K. and Berger, H., "Small-size, low-power bipolar
     memory cell", 1971 International Solid State Circuits Conference,
     Digest of Technical Papers, p. 18-19, CCA6-15267.

364. Wilcock, J. D., "Bipolar semiconductor technologies for bulk
     memories", Colloquium Digest on Computer Memories, 1973,
     p. 2, CCA8-11650.

                    b.   ECL

365. Buerker, U. and Rydval, P., "High-speed bipolar 128-bit random-
     access memory", Siemens Rev., 42 (9), 402-6 (1975).
366. Ebel, M. S., Gionis, J. and Regitz, W. M., "A 4096-bit high-
     speed emitter-coupled-logic (ECL) compatible random-access
     memory", IEEE J. Solid-State Circuits, SC-10 (5), 262-7 (1975),
     CCA10-25845.
367. Miles, T. E. and Phillips, C. D., "The evolution of ECL memories
     for high-speed applications", New Electron., 7 (7), 43-4, 47
     (1974), CCA9-18411.
368. Rathbone, R., Ernst, H., Glock, H. and Schwabe, U., "A 1024-bit
     ECL RAM with 15-ns access time", 1976 International Solid-
     State Circuits Conference, Digest of Technical Papers, p. 188-9,
     CCA11-28303.

                    c.   TTL

369. Anon., "1969 panorama of integrated circuits: The TTL family",
     Electronique Industr., no. 127, 613-18 (1969), French, CCA5-8296.
370. Baker, W. D., Herndon, W. H., Longo, T. A. and Pelzer, D. L.,
     "Isoplanar technique of higher density, faster memories", New
     Electron., 7 (4), 58-9, 61 (1974), CCA9-14354.
371. MacLeod, A. M., "An elementary introduction to TTL", Phys.
     Educ., 10 (6), 440-5 (1975), CCA11-9583.
372. Mayumi, H., Nokubo, J., Okada, K. and Shiba, H., "A 25-ns read
     access bipolar 1 kbit TTL RAM", IEEE J. Solid-State Circuits,
     SC-9 (5), 283-4 (1974), CCA9-23253.
373. Moussie, M. and Surtouc, D., "High speed dead memories in bi-
     polar technology", Onde Elec., 51 (4), 292-7 (1971), French,
     CCA6-15245.
374. Priel, U., "Take a look inside the TTL IC", Electron. Des.,
     19 (8), 68-71 (1971), CCA6-15127.

                    d. I^2L

375. Anon., "First I^2L processor in four-bit design surpasses N-
     MOS", Electronics, 48 (2), 29-32 (1975).
376. Bishop, A., "I^2L explained and compared with c.m.o.s.", New
     Electron., 9 (10), 36, 38, 40 (1976), CCA11-22570.
377. Blossfeld, L., "I^2L and standard bipolar combined", Elektronik,
     26 (4), 57-60 (1977), EEA80-28903.
378. Hewlett, F. W., Jr., "The Schottky I^2L technology and its appli-
     cation in a 24 x 9 sequential access memory", IEEE J. Solid
     State Circuits, SC-12 (2), 119-23 (1977), CCA12-8897.

379. Horton, R. L., Englade, J. and McGee, G., "I^{2}L takes bipolar integration a significant step forward", Electronics, <u>48</u> (3), 83-90 (1975).
380. Pederson, R. A., "Integrated injection logic: a bipolar LSI technique", Computer, <u>9</u> (2), 24-9 (1976), CCA11-22552.
381. Ragonese, L. J. and Yang, N. T., "Enhanced integrated injection logic performance using novel symmetrical cell topography", 1977 International Electron Devices Meeting, Technical Digest, p. 166-9.
382. Sloan, B. J. and Stehlin, R. A., "Advances in I^{2}L development", 13th IEEE Computer Society International Conference, Digest of Papers, 1976, p. 56-8, EEA80-291.

3.   Field-Effect Transistors
      a. MOS

383. Altman, L., "New MOS processes set speed, density record", Electronics, <u>50</u> (22), 92-4 (1977).
384. Anon., "An in-depth look at the MK 4027 (semiconductor memories)", New Electron., <u>9</u> (12), 58, 60, 64 (1976), CCA11-22671.
385. Baranyai, A., "MOS-LSI memory systems", Hiradstech. Ipari. Kut. Intez. Kozl., <u>12</u> (2), 33-52 (1972), Hungarian, EEA75-39937.
386. Bayliss, J., Johnston, R. and Freeman, J., "New 4,096-bit MOS chip is heart of fast, compact computer memory", Electronics, <u>45</u> (26), 97-100, 2-3 (1972), CCA8-7347.
387. Beer, A., "LOCMOS lends itself to full scale integration", New Electron., <u>10</u> (3), 28-32 (1977), EEA80-28910.
388. Boleky, E. J., Burns, J. R., Meyer, J. E. and Scott, J. H., "MOS memory travels in fast bipolar crowd", Electronics, <u>43</u> (15), 82-5 (1970), CCA5-22439.
389. Broeker, B., "System applications of 4K RAMs", 1973 WESCON Technical Papers, Vol. 17, Paper 16/2, 1 pp., CCA9-9220.
390. Busch, J. E., Henrikson, R. L. and Oltman, B. A., "No. 1 EAX semiconductor memory systems", GTE Autom. Electr. Tech. J., <u>14</u> (7), 320-33 (1975).
391. Chapple, I. K., "The application of MOS memories to replace core storage in large computers", Microelectron. & Reliab., <u>13</u>, 415 (1974), EEA78-5368.
392. Clemens, J. T., Mowery, G. L., Doklan, R. H., Nolen, J. J. and Drobek, J., "A p-channel Si-gate integrated circuit technology (RAM applications)", 1975 International Electron Devices Meeting, Technical Digest, p. 291-4, EEA79-46529.
393. Cook, H., "Multiplexing MOS RAMs for high speed", Electron. Eng., <u>47</u> (567), 27 (1975).
394. Critchlow, D. L., "High speed MOSFET circuits using advanced lithography", Computer, <u>9</u> (2), 31-7 (1976), EEA79-33104.
395. Dailey, J. R. and Knutzleman, H. C., "IGFET functional memory cells", IBM Tech. Disclosure Bull., <u>13</u>, 3122-4 (1971), EEA74-18799.

396. Dubois, R., "M.O.S. and their applications in memories. I.",
     Automatisme, 18 (2), 78-84 (1973), French, EEA76-26085.
397. Dubois, R., "MOS and their applications in memories. II.",
     Automatisme, 18 (4), 165-74 (1973), French, EEA76-26086.
398. Dubois, R., "M.O.S. devices and their applications in memories.
     III.", Automatisme, 18 (5), 206-17 (1973), EEA76-29307.
399. Ebel, M. S., Gionis, J. and Regitz, W. M., "4096-bit high-speed
     emitter-coupled-logic (ECL) compatible random-access memory",
     IEEE J. Solid-State Circuits, SC-10 (5), 262-7 (1975).
400. Fink, E., "An asynchronous memory system using the MK 4008-9
     r.a.m.", New Electron., 7 (12), 33-4 (1974), CCA9-23255.
401. Fontenay, R., "Development of integrated live (RAM) memories",
     Electron. & Microelectron. Ind., no. 162, 41-3 (1972), French,
     CCA8-5268.
402. Fontenay, R., "A 4096-bit random access memory", Electron. &
     Microelectron. Ind., no. 168, 32-3 (1973), French, EEA76-23021.
403. Forbes, L., "N-channel ion-implanted enhancement/depletion
     MOSFET's", IEEE J. Solid-State Circuits, SC-8 (2), 184-5 (1973).
404. Foss, R. C. and Harland, R., "Should MOS RAMs be 'TTL compa-
     tible'?", Electron. Des., 24 (12), 144-6 (1976), EEA79-37146.
405. Foss, R. C. and Monico, J., "MOS memory cells and their engi-
     neering implications", International Conference on Large Scale
     Integrated Circuits, 1974, p. 23-41, EEA78-11691.
406. Frankenberg, R. J., "Minicomputer designer chooses semiconductor
     over core memory", Comput. Des., 14 (3), 59-68 (1975).
407. Friedberg, H., "An n-channel MOS memory for 5 V and its appli-
     cations", Elektron. Anz., 5 (2), 39-41 (1973), German,
     CCA8-11641.
408. Frohman-Bentchkowsky, D., "Silicon gate MOS technology and its
     memory applications", 1970 WESCON Technical Papers, Vol. 14,
     Paper 7/1, 7 pp., EEA74-7791.
409. Ginovker, A. S., Gusev, A. A., Konkov, V. P., Kuryshev, G. L.,
     Mishin, A. I., Sinitsa, S. P. and Tepman, B. G., "Integrated
     circuit with a programming structure employing MOS storage
     transistors", Radio Eng. & Electron. Phys., 17, 1225-6 (1972),
     EEA76-26151.
410. Gladu, R. G. and Olson, L. T., "Multigated 'dribble' powered
     MOS FET storage cell", IBM Tech. Disclosure Bull., 17, 3336
     (1975), EEA78-28300.
411. Goser, K., "Low power memories in modified P/MOS techniques
     and CMOS technique", Nachrichtentech. Z. (NTZ), 26 (1), 9-15
     (1973), German, CCA8-9539.
412. Harding, J., "MOS developments - memories and microprocessors",
     Electron, no. 78, 21, 24-5 (1975), EEA79-1042.
413. Hausele, H. and Penzel, H. J., "Application of 4096-bit MOS
     memory devices in a 524-k byte memory", Siemens Forsch. - &
     Entwicklungsrer., 4 (4), 203-6 (1975), CCA10-25833.

414. Heissing, H. and Mitterer, R., "4096-bit MOS memory device with single-transistor cells", Siemens Forsch. Entwicklungsber, Res. Dev. Rep., $\underline{4}$ (4), 197-202 (1975).

415. Henderson, R. C., Pease, R. F. W., Voshchenkov, A. M., Mallery, P. and Wadsack, R. L., "A high-speed p-channel random access 1024-bit memory made with electron lithography", 1973 International Electron Devices Meeting, Technical Digest, p. 138-40, CCA9-11972.

416. Hoff, M. E., Jr., "MOS memory and its application", Comput. Design, $\underline{9}$ (6), 83-7 (1970), CCA5-686.

417. Hoffman, W. K. and Kalter, H. L., "An 8K b random-access memory chip using the one-device FET cell", IEEE J. Solid-State Circuits, SC-$\underline{8}$, 298-305 (1973), EEA76-35353.

418. Hollingsworth, R. J., "Memory cell", Report TN-1125, RCA, Princeton, N. J., 1975, 2 pp., CCA11-4818.

419. Huang, J. S. T., "MOS technology tradeoffs", Proceedings of the National Electronics Conference, Vol. 29, 1974, p. 92-7.

420. Huber, R. J., Smith, K. F., Hill, D. R., Fordemwalt, J. N., Hanson, J. W. and Dobelle, W. H., "Simplified n-channel process achieves high performance", Electronics, $\underline{47}$ (5), 117-22 (1974), CCA9-11960.

421. Ieda, N., Yoshimura, H. and Asaoka, T., "High speed MOS memory", Rev. Electr. Commun. Lab., $\underline{21}$ (7-8), 473-81 (1973), CCA9-11966.

422. Itoh, K., Shimohigashi, K., Chiba, K., Taniguchi, K. and Kawamoto, H., "A high-speed 16-kbit n-MOS random-access memory", IEEE J. Solid-State Circuits, SC-$\underline{11}$ (5), 585-90 (1976).

423. Jakits, D., "Survey of the present state of the art of the MOS technology", NTG-Fachber., $\underline{54}$, 43-72 (1975), German, EEA80-666.

424. James, R. P., "Three-device MOSFET charge-storage cell", IBM Tech. Disclosure Bull., $\underline{14}$, 1435 (1971), EEA75-7542.

425. Kolev, K., Smirnov, N. D. and Peikov, P. H., "Memory effect in MOS structures", C. R. Acad. Bulg. Sci., $\underline{27}$, 461-3 (1974), EEA78-11534.

426. Kriegler, R. J., "Recent advances in MOS technology", Telesis, $\underline{4}$ (6), 175-80 (1976), EEA79-46315.

427. Krolikowski, W., Brown, W., Dries, R., Foote, R., Lund, D., Plimley, R., Reuter, J., Sandhu, J., Scow, K. and Tuttle, J., "A 1024 bit n-channel M.O.S. read-write memory chip", International Electron Devices Metering (abstracts), 1970, p. 16, CCA5-5693.

428. Kvamme, E. F., "MOS memory systems. I.", Elektroni, $\underline{20}$ (3), 75-8 (1971), German, CCA6-10777.

429. Lambrechtse, C. W., Salters, R. H. W. and Boonstra, L., "A 4096-bit one-transistor per-bit RAM with internal timing and low dissipation", 1973 International Solid State Circuits Conference, Digest of Technical Papers, p. 26-7, EEA76-23033.

430. Landers, G., "Choosing among 4-k MOS RAMs?", Electron. Des., $\underline{24}$ (12), 138-42 (1976), EEA79-37145.

431. Laughton, W. J., "Molybdenum gates open the door to faster MOS memories", Electronics, $\underline{44}$ (8), 68-73 (1971), CCA6-10776.

432. LeBlanc, A. R., "Field-effect transistor memory circuit", IBM Tech. Disclosure Bull., $\underline{15}$, 1292-3 (1972), EEA76-8691.

433. Lechouchu, J., "Power supply of m.o.s. RAM in pulse rating", Onde Elec., $\underline{51}$ (4), 305-10 (1971), French, CCA6-12677.

434. Lilen, H., "The 16K RAM and its advantages: in what way is it different?", Electron. & Microelectron. Ind., no. 224, 47-9 (1976), French, CCA12-1512.

435. Linford, J., "MOS memories for digital systems", Electron. Equip. News, $\underline{13}$ (9), 42-5 (1972), EEA75-10265.

436. McDonald, A., "Choosing between MOS processes", New Electron, $\underline{6}$ (4), 52-3 (1973), EEA76-16472.

437. Maine, S. G. T., "Will m.o.s. stimulate a new computer generation?", Electron. Eng., $\underline{43}$ (525), 27-30 (1971), EEA75-385.

438. Markkula, A. C., Jr., "n-channel RAMS", 1973 WESCON Technical Papers, Vol. 17, Paper 16/1, 5 pp., CCA9-9219.

439. Marko, S., "Data store with MOSFET memory", Meres & Autom., $\underline{23}$ (12), 441-6 (1975), Hungarian, CCA11-17409.

440. Marvin, C. E., "MOS/LSI throughout", 1971 IEEE International Convention Digest, p. 74-5, CCA6-15183.

441. Masuoka, F., "Avalanche injection-type MOS memory", Patent USA 3868187, Publ. February 1975.

442. Matukura, Y., "MOS memory", J. Inst. Electron. & Commun. Eng. Jap., $\underline{55}$, 544-9 (1972), Japanese, EEA75-39945.

443. McKenny, V., "Depletion-mode transistors boost speed of MOS random access memory", Electronics, $\underline{44}$ (4), 80-5 (1971), EEA74-13270.

444. Mundy, J. L., "High density four-transistor MOS content addressed memory", Patent USA 3701980, Publ. October 1972, CCA8-9544.

445. Myers, D. K., "Radiation effects on commercial 4-bit NMOS memories", IEEE Trans. Nucl. Sci., NS-$\underline{23}$ (6), 1732-7 (1976), CCA12-8899.

446. Nishiyama, A., Yoshino, H., Yoshida, T. and Yamaguchi, T., "Learning machine using MOS IC weighting elements", Electron. Commun. Jap., $\underline{57}$ (1), 112-18 (1974).

447. Oikawa, H., Wada, T. and Amazawa, T., "Molybdenum metallization system for large scale integrated circuits", Rev. Electr. Commun. Lab., $\underline{24}$ (5-6), 407-17 (1976).

448. Owczarek, A. and Guryn, R., "MOS read-write memory using pumping devices", Elektronika, $\underline{17}$ (1), 35-9 (1976), Polish, CCA11-15064.

449. Pashley, R. D. and McCormick, G. A., "A 70-ns 1K MOS RAM", 1976 International Solid-State Circuits Conference, Digest of Technical Papers, p. 138-9, 238, EEA79-46470.

450. Proebsting, R. and Green, R., "A TTL compatible 4096-bit n-channel RAM", 1973 International Solid State Circuits Conference, Digest of Technical Papers, p. 28-9, EEA76-23034.

451. Regitz, W. M. and Karp, J., "Three-transistor-cell 1024-bit 500-ns MOS RAM", IEEE J. Solid State Circuits, SC-5 (5), 181-6 (1970), CCA5-22443.

452. Remshardt, R. and Baitinger, U., "A high performance low power 2048-bit memory chip in MOSFET technology and its application", IEEE J. Solid-State Circuits, SC-11 (3), 352-9 (1976), CCA11-22667.

453. Repchick, D. P., "3-device cell MOSFET memory system", IBM Tech. Disclosure Bull., 14, 285-6 (1971), EEA74-34090.

454. Rideout, V. L., Walker, J. J. and Cramer, A., "A one device memory cell using single layer of polysilicon and a self-registering metal-to-polysilicon contact", 1977 International Electron Devices Meeting, Technical Digest, p. 258-60.

455. Rodgers, T. J., Hiltpold, W. R., Frederick, B., Barnes, J. J., Jenne, F. B. and Trotter, J. D., "VMOS memory technology", IEEE J. Solid-State Circuits, SC-12 (5), 515-23 (1977).

456. Shah, R., Lo, T. C. and Linden, J., "Improve memory systems with 4-K RAMs", Electron. Des., 22 (22), 100-3 (1974).

457. Shenton, G., "The 4K r.a.m. - a change of direction?", New Electron., 8 (10), 47, 50 (1975), EEA78-28867.

458. Shimotori, K., Anami, K., Nagayama, Y., Okhura, I., Ohmori, M. and Nakano, T., "Fully ion implanted 4096-bit high-speed DSA MOS RAM", 1977 International Solid-State Circuit Conference, Digest of Technical Papers, p. 76-7.

459. Smith, G. E., "MOS technology offers the MOST for the least", Bell. Lab. Rec., 54 (10), 281-5 (1976), CCA12-4009.

460. Sonoda, G. Y., "I.g.f.e.t. address powered Schmidt storage cell", IBM Tech. Disclosure Bull., 17, 3337 (1975), EEA78-28301.

461. Strath, A. B., "M.o.s. developments", New Electron, 9 (6), 38 (1976), EEA79-28684.

462. Talamonti, L., "Two-level capacitor boosts MOS memory performance", Electronics, 46 (4), 115-17 (1973), EEA76-12635.

463. Terman, L. M., "MOSFET memory circuits", Proc. IEEE, 59 (7), 1044-58 (1971), CCA6-15246.

464. Tingley, R. G. and Johnson, D. W., "Automation improves MOS Process characterization", Circuits Manuf., 17 (4), 30-4 (1977).

465. Werner, W., "New MOS memories", Elektron. Anz., 5 (2), 34-5 (1973), German, CCA8-11640.

466. Wotruba, G., "A large-scale integrated memory circuit using single-transistor cells with a density of 1600 bit/mm^2 in N-silicon gate technology", Siemens Forsch. - & Entwicklungsber., 4 (4), 207-12 (1975), EEA78-37650.

467. Wynhoven, R. M., "MOST stores provide more capacity", Electron. Equip. News, 14 (4), 67 (1972), EEA75-32677.

468. Yoshida, K., Yamazaki, I., Doi, K., Horiuchi, S. and Shibata, T., "A 16-bit LSI minicomputer", IEEE J. Solid-State Circuits, SC-11 (5), 696-702 (1976), EEA80-3012.

469. Yu, H. N., Dennard, R. H., Chang, T. H. P., Osburn, C. M., Di-
     lonardo, V. and Luhn, H. E., "Fabrication of a miniature 8K-bit
     memory chip using electron-beam exposure", J. Vac. Sci. &
     Technol., 12 (6), 1297-300 (1976), CCA11-12332.
470. Yu, R. T., "MOS memory system", Patent USA 3906463, Publ.
     September 1975, CCA11-4817.
471. Zdravkovic, S. and Sanja, S., "Memory realisations in MOS
     techniques", Tehnika, 28, 744-744e (1973), Croatian, EEA76-26105.

                    b.   MNOS

472. Augusta, B. and Landler, F., "MNOS device characteristics and
     applications in memories", International Conference on Large
     Scale Integrated Circuits, 1974, p. 226-32, EEA78-11690.
473. Brewer, J. E. and Hadden, D. R., Jr., "Block-oriented random
     access MNOS memory", AFIPS National Computer Conference, Pro-
     ceedings, Vol. 43, 1974, p. 837-40.
474. Bostock, D., "MNOS technology and non-volatile logic", Micro-
     electron. & Reliab., 14 (4), 359 (1975), CCA11-9680.
475. Brewer, J. E., "MNOS density parameters", IEEE Trans. Electron
     Devices, ED-24 (5), 618-25 (1977), CCA12-15514.
476. Card. H. C. and Worrall, A. G., "Electrically alterable ava-
     lanche-injection memory", Electron. Lett., 9 (1), 14-15 (1973),
     CCA8-7349.
477. Carlstedt, G. L. and Svensson, C. M., "MNOS memory transistors
     in simple memory arrays", IEEE J. Solid-State Circuits, SC-7,
     382-8 (1972), EEA75-36288.
478. Chawala, A. S. and Lin, H. C., "A cross-point MNOS capacitor
     memory", 1976 International Electron Devices Meeting, Technical
     Digest, p. 181-4.
479. Cricchi, J. R., "MNOS nonvolatile semiconductor storage", 1974
     Digests of the Intermag. Conference, 6-5/1 pp., EEA78-8204.
480. Cricchi, J. R., Blaha, F. C. and Fitzpatrick, M. D., "The drain-
     source protected MNOS memory device and memory endurance", 1973
     International Electron Devices Meeting, Technical Digest,
     p. 126-9, EEA77-14545.
481. Conti, M. and Vanin, A., "Conduction and biasing mechanisms
     into MNOS structure for memory", Ind. Ital. Elettrotec. &
     Elettron., 25 (10), 893-9 (1972), Italian, CCA8-5270.
482. Endo, N. and Nishi, Y., "A double-diffused MNOS transistor as
     a new non-volatile memory", 1973 International Electron Devices
     Meeting, Technical Digest, p. 78-81, EEA77-14543.
483. Endo, N. and Nishi, Y., "Non-volatile semiconductor memory
     device", Patent UK 1429604, Publ. March 1976, CA85-27854.
484. Fagan, J. L., White, M. H. and Lampe, D. L., "A high-density,
     read/write, nonvolatile charge-addressed memory", 1976 Inter-
     national Solid-State Circuits Conference, Digest of Technical
     Papers, p. 184-5, CCA11-28301.

485. Ferris-Prabhu, A. V., "Charge transfer by direct tunneling in thin-oxide memory transistors", IEEE Trans. Electron Devices, ED-24 (5), 524-30 (1977), CCA12-15505.

486. Gentil, P., Le Goascoz, V. and Boel, J., "MNOS memory devices", Onde Electr., 52 (10), 427-34 (1972), French, CCA8-5275.

487. Goodwin, N. W., "Metal-nitride-oxide-semiconductor memory device", Patent USA 3803705, Publ. April 1974, CA80-150264.

488. Gregory, P. J., "MNOS - a solution for non-volatile semiconductor memory", Colloquium Digest on Computer Memories, 1973, p. 2, CCA8-11651.

489. Horninger, K., "New semiconductor storage elements. Electrically re-programmable storage matrix with MNOS transistors", Elektron. Ind., 4 (5/6), 94-5 (1972), German, EEA75-32681.

490. Horninger, K. H., "A fully decoded MNOS store in single-channel technology, using novel storage transistors", Bull. Assoc. Suisse Electr., 64 (20), 1258-63 (1973), German, CCA9-6239.

491. Horninger, K., "Storage arrays with MNOS transistors", Siemens Forsch. & Entwicklungsber., 4, 213-19 (1975), EEA78-37651.

492. Hsia, Y., "Memory applications of the MNOS", 1972 WESCON Technical Papers, Vol. 16, Paper 4/3, 5 pp., CCA8-7360.

493. Hsia, Y., "MNOS LSI memory device data retention measurements and projections", IEEE Trans. Electron Devices, ED-24 (5), 568-77 (1977), CCA12-15507.

494. Huang, J. S., "Switching properties of m.n.o.s. memory transistors", Electron. Lett., 8, 469-70 (1972), EEA75-39932.

495. Jeppson, K., "MNOS - foremost semiconductor memories", Elteknik, 15 (10), 30-3 (1972), Swedish, CCA8-5267.

496. Katsube, T., Adachi, Y. and Ikoma, T., "Trap centers in MNOS memory devices measured by thermally stimulated currents", Jap. J. Appl. Phys., 12 (10), 1633-4 (1973), EEA77-379.

497. Katsube, T., Adachi, Y. and Ikoma, T., "Memory traps in MNOS diodes measured by thermally stimulated current", Solid-State Electron., 19 (1), 11-16 (1976).

498. Katesube, T., Adach, Y. and Ikoma, T., "Trap distribution and memory characteristics of MNOS diodes", Electron. Commun. Jpn., 59 (2), 131-9 (1976).

499. Kendall, E. J. M. and Haslett, J. W., "On the removal of the memory properties of MNOS devices", IEEE Trans. Electron Devices, ED-19, 287-8 (1972), EEA75-10267.

500. Kendall, J. T., "Recent progress in MNOS memory technology", Microelectron. & Reliab., 13 (5), 413-14 (1974), EEA78-5367.

501. Kim, M. J., "Charge storage MISFET memory devices", Jap. J. Appl. Phys., 13 (11), 1847-59 (1974), EEA78-5359.

502. Kim, M. J., "Metal insulator silicon field effect transistor memory", Patent USA 3853496, Publ. December 1974, CA82-67308.

503. Koike, S. and Kambara, G., "Electrically erasable nonvolatile optical MNOS memory device", IEEE J. Solid-State Circuits, SC-11 (2), 303-7 (1976), CCA11-17404.

504. Koike, S., Kambara, G., Matsuda, T. and Tsuru, S., "Nonvolatile
     MNOS memory", Natl. Tech. Rep., 20 (2), 217-26 (1974),
     Japanese, CCA9-25102.
505. Koike, S., Kano, G., Kashiwakura, A. and Teramoto, I., "New
     two-terminal C-MNOS memory cells", IEEE Trans. Electron Devices,
     ED-23 (9), 1036-41 (1976), EEA79-41556.
506. Koo, T. K., "The non-volatility characteristics of MNOS memory
     FET and EAROM arrays", Proceedings of the IEEE 1974 National
     Aerospace and Electronics Conference, p. 37-41, EEA78-8623.
507. Korobov, I. V., Plotnikov, A. F., Popov, Y. M. and Seleznev,
     V. N., "Reversible recording of optical information in metal-
     dielectric-semiconductor structures", Sov. J. Quantum Electron.,
     5 (9), 1092-5 (1975), CCA11-12335.
508. Krick, P. J., "Implanted stepped-oxide MNOSFET", IEEE Trans.
     Electron Devices, ED-22 (2), 62-3 (1975).
509. Lehovec, K., "Charge distribution in the nitride of MNOS memory
     devices", J. Electron. Mater., 6 (2), 77-93 (1977).
510. Lodi, R. J., Wegener, H. A. R., Borovikca, M. B., Kosicki, B.
     B., Pogemiller, T. A. and Eklund, M. W., "MNOS-BORAM memory
     characteristics", IEEE J. Solid-State Circuits, SC-11 (5),
     622-30 (1976), EEA79-46423.
511. Mavor, J., "Metal-nitride-oxide-silicon avalanche-injection
     memory", Electron. Lett., 9 (5), 111-12 (1973), CCA8-11639.
512. Mavor, J., "Reply to comment on metal-nitride-oxide-silicon
     avalanche-injection memory", Electron. Lett., 9 (8), 349 (1973),
     EEA77-6025.
513. Mellor, P. J. T. and Dunn, P. J., "Comment on metal-nitride-
     oxide-silicon avalanche-injection memory", Electron. Lett., 9
     (11), 252-3 (1973), CCA9-1233.
514. Moschwitzer, A. and Pasch, L., "MNOS storage FET's for integrated
     semiconductor memories", Z. Elek. Inf. - & Energietech., 1 (1),
     37-50 (1971), German, EEA75-10300.
515. Naber, C. T. and Lockwood, G. C., "Processing of MNOS nonvola-
     tile memories", Semiconductor Silicon 1973, 2nd Int. Symp. on
     Silicon Mater. Sci. and Technol., p. 401-13.
516. National Cash Register Co., "Bistable memory cell", Patent UK
     1281148, Publ. July 1972, CCA8-2670.
517. Ning, T. H. and Yu, H. N., "Erasable nonvolatile memory device
     using hole trapping in SiO_2", IBM Tech. Disclosure Bull., 18
     (8), 2740-2 (1976), CCA11-12329.
518. Oakley, R. E., "Memories in MNOS", Man Made Memories, Conference
     Summaries, 1971, 2 pp., CCA6-12756.
519. Oakley, R. E. and Browne, V. A., "The effect of repeated
     switching cycles on the MNOS memory transistor", 1972 European
     Solid State Devices Research Conference, p. 233, Publ. 1973,
     EEA76-32282.
520. Pasch, L., "The MNOS-LAD (metal-nitride-oxide-silicon avalanche-
     diode) - a new semiconductor storage element", Nachrichtentech.
     Elektron., 23 (2), 73-6 (1973), German, EEA76-19906.

521. Penner, L. and Kalet, M., "An electronic tuning system for TV receivers utilising a nonvolatile MNOS memory", 1976 International Solid-State Circuits Conference, Digest of Technical Papers, p. 70-1, EEA79-48632.
522. Pepper, M., "A non-volatile solid state memory device", Patent UK 1425727, Publ. February 1976.
523. Plotnikov, A. F., Seleznev, V. N., Tokarchuk, D. N. and Ferchev, G. P., "Photoelectric reading of information recorded in a MNOS structure", Sov. J. Quantum Electron., $\underline{5}$ (3), 288-90 (1975), CCA11-6385.
524. Raffel, J. I. and Yasaitis, J. A., "Storage experiments with ultra-high density MNOS capacitors", Proc. IEEE, $\underline{64}$ (11), 1629-30 (1976), EEA80-17256.
525. Raffel, J. L. and Yasaitis, J. A., "A selection system for MNOS capacitor memories", 1977 International Solid-State Circuits Conference, Digest of Technical Papers, p. 192-3.
526. Reid, P. R., McLouski, R. M., Cricchi, J. R. and Blaha, F. C., "The effect of Si_3N_4 deposition parameters on MNOS memory transistor response", J. Electrochem. Soc., $\underline{121}$ (3), 85C (1974), EEA77-23055.
527. Ross, E. C., "Variable threshold memory system using minimum amplitude signals", Patent USA 3702990, Publ. November 1972, CCA8-13702.
528. Sewell, F. A., Jr., "The light-sensitive MNOS memory transistor", IEEE Trans. Electron Devices, ED-$\underline{20}$ (6), 563-72 (1973), EEA76-23735.
529. Smart, J., "M.n.o.s. - is this the perfect memory?", Electron, no. 33, 23-4 (1973), CCA9-1234.
530. Tanaka, T. and Nishi, Y., "Nature and mechanism of degradation in MNOS read write memory", Jap. J. Appl. Phys., $\underline{44}$ (Suppl.), 203-9 (1975).
531. Tarui, Y., Hayashi, Y. and Hagai, K., "Electrically reprogrammable nonvolatile semiconductor memory", IEEE J. Solid-State Circuits, SC-7, 369-75 (1972), EEA75-36286.
532. Tickle, A. C., "Electrically alterable non-volatile semiconductor memories", 1972 WESCON Technical Papers, Vol. 16, Paper 4/1, 5 pp., CCA8-7358.
533. Tickle, A. C. and Wanlass, F. M., "Electrically alterable non-volatile semiconductor memory technology", 1972 WESCON Technical Papers, Vol. 16, Paper 4/2, 8 pp., CCA8-7359.
534. Topich, J. A. and Yoh, E. T., "Effects of high temperature annealing of MNOS devices", J. Electrochem. Soc., $\underline{123}$ (4), 535-9 (1976).
535. Uchida, Y., Endo, N., Saito, S., Konaka, M., Nojima, I., Nishi, Y. and Tamaru, K., "A 1024-bit MNOS RAM using avalanche-tunnel injection", IEEE J. Solid-State Circuits, SC-$\underline{10}$ (5), 288-94 (1975), EEA78-37617.

536. Uchida, Y., Endo, N., Saito, S. and Nishi, Y., "Avalanche-tunnel injection in MNOS transistor", IEEE Trans. Electron Devices, ED-24(6), 688-93 (1977), CCA12-15515.
537. Vanstone, G. F., "Metal-nitride-oxide-silicon-capacitor arrays as electrical and optical stores", Electron. Lett., 8, 13-14 (1972), EEA75-9822.
538. Vitanov, P. K., Popova, L. I. and Antov, B. Z., "MNOS memory structure with relatively thick oxide", Electron. Lett., 12 (25), 681 (1976), CCA12-4013.
539. Wegener, H. A. R., Doig, M. B., Marraffino, P. and Robinson, B., "Radiation resistant MNOS memories", IEEE Trans. Nucl. Sci., NS-19 (6), 291-8 (1972), CCA8-9535.
540. White, M. H., Dzimianski, J. W. and Pekkerar, M. C., "Endurance of thin-oxide nonvolatile MNOS memory transistors", IEEE Trans. Electron Devices, ED-24 (5), 577-80 (1977), CCA12-15508.
541. Woods, M. H., "Instabilities in double dielectric structures", 12th Annual Reliability Physics Symposium, Proceedings, 1974, p. 259-66.
542. Yamaguchi, T. and Sato, S., "Investigations of continuously variable threshold voltage devices (CVTD)", 6th Conference on Solid State Devices, Proceedings, 1974, p. 233-42, Publ. 1975.
543. Yamazaki, S., Hatakeyama, K., Kagawa, I. and Yamashita, Y., "Properties of nonvolatile semiconductor memories by using silicon clusters in the gate insulator", 1973 International Electron Devices Meeting, Technical Digest, p. 355-8, CCA9-11974.

c.   MAOS

544. Nakagiri, M., "Discriminating analysis of non-uniform charge distribution from surface states in MAOS diodes", Jap. J. Appl. Phys., 15 (2), 319-26 (1976), CCA11-12331.
545. Pappu, R. V. and Boothroyd, A. R., "Memory behavior of metal-plasma-anodized Al_2O_3 and SiO_2 semiconductor (MAOS) capacitors", Appl. Phys. Lett., 22, 72-4 (1973), EEA76-12634.
546. Sato, S. and Yamaguchi, T., "Study of charge storage behavior in metal-alumina-silicon dioxide-silicon (MAOS) field effect transistor", Solid-State Electron., 367-75 (1974), CCA9-14355.
547. Tarui, Y., Komiya, Y., Sakamoto, T., Suzuki, E., Sigiyama, K. and Fujishiro, T., "Nonvolatile optical pattern memory using MAOS structure", Bull. Electrotech. Lab., 26 (4-5), 32-9 (1976), EEA79-46905.
548. Woods, M. H., "Instabilities in double dielectric structures", Proceedings of the 12th Annual Reliability Physics Symposium, 1974, p. 259-66.

d.   CMOS

549. Alaspa, A. A. and Dingwall, A. G. F., "COS/MOS parallel processor array", 1970 International Solid State Circuits Conference, Digest of Technical Papers, p. 118-19, CCA5-13648.
550. Anon., "C.m.o.s. r.a.m. applications", New Electron., $\underline{7}$ (24),
75, 78, 81, 85 (1974), EEA78-8620.
551. Bangen, G., "CMOS random access memory replaces ROM", Electron.
Engineering, $\underline{48}$ (584), 29 (1976), EEA80-269.
552. Dingwall, A., Lile, W. R. and Scott, J. H., Jr., "Hi-speed SOS
COS/MOS random-access memories", RCA Eng., $\underline{19}$ (4), 70-4 (1973/
1974), CCA9-18413.
553. Gargione, F., Keyzer, W. and Noek, G., "A multilayer hybrid
thin-film 1024 x 3 COS/MOS memory", New Electron., $\underline{6}$ (13),
25-9 (1973), EEA76-38320.
554. Greene, R. and Baxter, B., "A high speed 256 bit memory chip
using complementary N/P MOS transistors", Semiconductor Integrated and Production Conference and Exhibition (abstracts),
1971, 1 pp., CCA6-15250.
555. Huertas, J. L., Acha, J. I. and Carmona, J. M., "Design and
implementation of tristables using CMOS integrated circuits",
IEE J. Electron Circuits & Syst., $\underline{1}$ (3), 88-94 (1977),
EEA80-20387.
556. Lesniewski, R., "A high-speed 256 bit C-MOS memory array",
Semiconductor Integrated and Production Conference and Exhibition (abstracts), 1971, 1 pp., CCA6-15251.
557. Lile, W. H., Scott, J. H. and Dingwall, A. G. F., "SOS CMOS
random-access memories: A mini survey", EDN, $\underline{19}$ (21), 30-5
(1974).
558. Maine, S., "Portable memories and advantages of ion implantation
in m.o.s. memories", Man Made Memoires, Conference Summaries,
1971, 2 pp., CCA6-12754.
559. Miller, H. S., "COS/MOS random access memories", 1971 IEEE
International Convention Digest, p. 34-5, EEA74-22714.
560. Mizokami, H., Ino, M. and Hashimoto, T., "Ion implantation and
HCl treatment simplify SOS/CMOS processing and slash leakage
current", JEE, no. 103, 26-30 (1975), CCA11-2228.
561. Shenton, G., "Non-volatile storage using CMOS RAMS", Electron,
no. 78, 25 (1975), EEA79-1043.
562. Shenton, G., "Wattless storage with CMOS-RAMs", Elektron. J.,
$\underline{10}$ (10), 71-3 (1975), German, CCA11-6379.
563. Shiga, K., Itoh, T. and Anbe, T., "A monostable CMOS RAM with
self-refresh mode", IEEE J. Solid-State Circuits, SC-$\underline{11}$ (5),
609-13 (1976), EEA79-46421.
564. Spampinato, D. P. and Terman, L. M., "High-speed complementary
memory cell", IBM Tech. Disclosure Bull., $\underline{14}$, 1723-5 (1971),
EEA75-10292.
565. Wuttke, G. H., "RAM's in CMOS-technique", Elektron. Anz., $\underline{7}$
(5), 127-32 (1975), EEA78-33200.

e. ESFI MOS

566. Druminiski, M., Kuhl, C., Preuss, E., Schwidefsky, F., Splitt-
     gerber, H. and Takacs, D., "Thin film evaluation techniques for
     the ESFI SOS technology", Jap. J. Appl. Phys., 15 (Suppl. 15-1,
     217-20 (1976), EEA80-9523.
567. Gonauser, E. and Herbst, H., "A master slice design concept
     based on master cells in ESFI-SOS-CMOS technology", Siemens
     Forsch. - & Entwicklungsber., 5 (6), 344-9 (1976), EEA80-3546.
568. Goser, K. and Pomper, M., "Five-transistor memory cells in ESFI
     MOS technology", IEEE J. Solid-State Circuits, SC-8, 324-6
     (1973), EEA76-35356.
569. Goser, K. and Pomper, M., "Static ESFI MOS (SOS) cells for high-
     density memories", Siemens Forsch. Entwicklungsber. Res. Dev.
     Rep., 4 (4), 220-5 (1975).
570. Goser, K., Pomper, M. and Tihanyi, J., "High-density static ESFI
     MOS memory cells", IEEE J. Solid-State Circuits, SC-9 (5),
     234-8 (1974), CCA9-23252.
571. Horninger, K., "Fully decoded MNOS storage arrays in ESFI MOS
     technology", IEEE J. Solid-State Circuits, SC-9 (6), 444-6 (1974).
572. Horninger, K., "High-speed ESFI SOS programmable logic array
     with an MNOS version", IEEE J. Solid-State Circuits, SC-10 (5),
     331-6 (1975).
573. Pomper, M., Horninger, K., Tihanyi, J. and Goser, C., "Fast low-
     power memory devices using ESFI MOS technology", Onde Electr.,
     54 (4), 187-91 (1974), CCA9-16080.
574. Pomper, M. and Tihanyi, J., "Ion-implanted ESFI MOS devices with
     short switching times", IEEE J. Solid-State Circuits, SC-9 (5),
     250-6 (1974), CCA9-23185.
575. Preuss, E., Pomper, M., Raetzel, C. and Splittgerber, H., "Large-
     scale integration in ESFI-SOS technology", Siemens Forsch. - &
     Entwicklungsber., 5 (6), 338-43 (1976), EEA80-3545.
576. Splittgerber, H., Takacs, D. and Schlotterer, H., "ESFI-SOS
     transistors on as-grown sapphire wafers", Siemens Forsch. - &
     Entwicklungsber., 4 (2), 99-102 (1977), EEA80-17273.

f. FAMOS

577. Abbas, S. A. and Barile, C. A., "P-channel rewritable avalanche
     injection device (RAID) operation and degradation mechanisms",
     13th Annual Proceedings of Reliability Physics Symposium, 1975,
     p. 1-5, CCA11-28309.
578. Abbas, S. A., Barile, C. A., Lane, R. D. and Liu, P. T., "Elec-
     trically erasable floating gate FET memory cell", Patent USA
     3836992, Publ. September 1974.
579. Anantha, N. G., Dockerty, R. C. and Lucy, J. M., Sr., "Electri-
     cally erasable floating gate field-effect transistor memory cell",
     IBM Tech. Disclosure Bull., 17 (8), 2311-13 (1975), EEA78-24356.

580. Berger, J., "Improved MOS transducer-drive circuit (piezoelectric transducers)", IEEE J. Solid-State Circuits, SC-8, 92-3 (1973), EEA76-10104.

581. Chang, C. S. and Gokhale, B. V., "Electrically erasable FAMOS memory cell", IBM Tech. Disclosure Bull., 15 (9), 2922-4 (1973), EEA76-19931.

582. Davidson, E. E., Hansen, A. A., Lane, R. D. and Steinberg, G., "Nonvolatile read-mostly memory cell", IBM Tech. Disclosure Bull., 15 (7), 2282-3 (1972), CCA8-11647.

583. DiStefano, T. H. and Laibowitz, R. B., "Electrically reprogrammable floating avalanche injection MOS (FAMOS) structure", IBM Tech. Disclosure Bull., 15, 3264-6 (1973), EEA76-26100.

584. Dockerty, R. C., "N-channel FAMOS cell with a Schottky electrode", IBM Tech. Disclosure Bull., 16 (10), 3244 (1974), EEA77-30338.

585. Dockerty, R. C., "Degradation mechanisms in rewritable N-channel FAMOS devices", 13th Annual Proceedings of Reliability Physics Symposium, 1975, p. 6-9, CCA11-28310.

586. Frohman-Bentchkowsky, D., "Memory behavior in a floating-gate avalanche-injection MOS (FAMOS) structure", Appl. Phys. Lett., 18 (8), 332-4 (1971), EEA74-22724.

587. Frohman-Bentchkowsky, D., "FAMOS - a new semiconductor charge storage device", Solid-State Electron., 17 (6), 517-29 (1974), EEA77-23065.

588. Jeppson, K. O. and Svensson, C. M., "Unintentional writing of a FAMOS memory device during reading", Solid State Electron., 19 (6), 455-7 (1976).

589. Kohler, E., "The floating-gate transistor - a new storage cell to be programmed electrically", Nachrichtentech, 22, 366-7 (1972), German, EEA75-39948.

590. Lochner, D., "Programmable semiconductor memories", Funk-Tech., 30, 320-1 (1975), EEA78-33205.

591. Medwin, L. B. and Ipri, A. C., "Floating transistor-gate field effect transistor memory device", RCA Tech. Not., TN1152, 1-2 (1976), CCA11-28285.

592. Pepper, M., "Memory elements (FAMOS)", Patent USA 3774087, Publ. November 1973.

593. Yu, K. K., Brody, T. P. and Chen, P. C. Y., "Experimental realization of floating-gate-memory thin-film transistor", Proc. IEEE, 63 (5), 826-7 (1975), EEA78-19848.

g.   CCD

594. Baker, R. T., "CCD array forms random-access memory", Electronics, 48 (23), 138-9 (1975), CCA11-2224.

595. Gunsagar, K. C., Guidry, M. R. and Amelio, G. F., "A CCD line addressable random-access memory (LARAM)", IEEE J. Solid-State Circuits, SC-10 (5), 268-72 (1975), CCA10-25846.

596. Hoffmann, K., "The behavior of the continuously charge-coupled random-access memory (C^3RAM)", IEEE J. Solid-State Circuts, SC-<u>11</u> (5), 591-6 (1976), CCA11-31610.

597. Tasch, A. F., Jr., Frye, R. C. and Fu, H. S., "The charge-coupled RAM cell concept", IEEE J. Solid-State Circuits, SC-<u>11</u> (1), 58-63 (1976), CCA11-12327.

h.   SOS

598. Boleky, E. J. and Meyer, J. E., "High-performance low-power CMOS memories using silicon-on-sapphire technology", IEEE J. Solid-State Circuits, SC-<u>7</u>, 135-45 (1972), EEA75-13799.

599. Herbert, P., "SOS and its place in LSI technology", 1974 Electronic Meeting on Microprocessors, p. 139-48, CCA11-9643.

600. King, G., "Silicon-on-sapphire IC's: technology forecast", Circuit Manuf., <u>15</u> (9), 48-52 (1975).

601. Kjar, R. A., Peterson, B. E. and Blandford, J. T., "Radiation hardened 64-bit CMOS/SOS RAM", IEEE Trans. Nucl. Sci., NS-<u>23</u> (6), 1728-31 (1976), CCA12-8898.

602. Kliment, D. C., Ronen, R. S., Nielsen, R. L., Seymour, R. N. and Splinter, M. R., "Architecture and performance of radiation-hardened 64-bit SOS/MNOS memory", IEEE Trans. Nucl. Sci., NS-<u>23</u> (6), 1749-55 (1976), CCA12-8900.

603. Pelletier, H., "Silicon-on-sapphire - A 1024 bits memory", Electron. & Microelectron. Ind., no. 213, 34-5 (1975), French, EEA79-13018.

604. Ronen, R. S. and Micheletti, F. B., "Recent SOS technology, advances and applications", Solid State Technol., <u>18</u> (8), 39-46 (1975).

605. Schwob, P., "SOS technology", Bull. Assoc. Suisse Electr., <u>68</u> (2), 60-5 (1977), French, EEA80-9514.

606. Scrupski, S. E., "Control processor uses C-MOS on sapphire", Electronics, <u>50</u> (14), 128-31 (1977).

i.   Others

607. Angle, R. L. and Talley, H. E., "Characterization of the aluminum-tantalum oxide-silicon dioxide-silicon charge storage (MTOS) device", International Electron Devices Meeting, Technical Digest, 1975, p. 466-8.

608. Anon., "A new type reprogrammable memory", Electron. & Microelectron. Ind., no. 288, 20-1 (1976), EEA80-20371.

609. Barron, M. B. and Butler, W. J., "j.f.e.t. bucket-brigade circuit: some recent experimental results", Electron. Lett., <u>9</u> (25), 603-4 (1973), CCA9-6174.

610. Dion, A., "Study and realisation of a simple capacitor analogue memory", Electron. & Microelectron. Ind., no. 210, 39-41 (1975), French, EEA79-9004.

611. Fagan, J. L., White, M. H. and Lampe, D. R., "A 16-kbit non-
     volatile charge addressed memory", IEEE J. Solid-State Circuits,
     SC-11 (5), 631-6 (1976), EEA79-46424.
612. Ferris-Prabhu, A. V. and Lareau, L. J., "Amnesia in layered
     insulator FET memory devices", 1973 International Electron
     Devices Meeting, Technical Digest, p. 75-7, CCA9-11968.
613. Hart, B. L. and Baker, R. W. J., "'Zero' decay analogue store
     using guarded gate MOSFET's", Radio & Electron. Eng., 42, 114-16
     (1972), EEA75-15966.
614. Heald, R. A. and Hodges, D. A., "Multilevel random-access memory
     using one transistor per cell", IEEE J. Solid-State Circuits,
     SC-11 (4), 519-28 (1976), EEA79-37686.
615. Hollingsworth, R. J., "Memory cell", RCA Tech. Notes, no. 1125,
     2 pp. (1975).
616. Kahng, D., "Developments in dual dielectric charge-storage (DDC)
     ReMM (read mostly memories)", Electrochemical Society Spring
     Meeting, Extended Abstracts, 1975, p. 393, EEA79-4718.
617. Kolankowsky, E. and Nestork, W. J., "Silicon memory cube", IBM
     Tech. Disclosure Bull., 18 (10), 3239-42 (1976), CCA11-17405.
618. Matsushita, R., Aoki, T., Ohtsu, T., Yamoto, H., Hayashi, H.,
     Okayama, M. and Kawana, Y., "Highly reliable high-voltage tran-
     sistors by the use of SIPOS process", IEEE Trans. Electron
     Devices, ED-23 (8), 826-30 (1976).
619. Noda, A., "Leak-compensated analog memory with pair MOSFET's",
     IEEE J. Solid-State Circuits, SC-5, 75-7 (1970), EEA73-25038.
620. Rai, y., Sasami, T. and Hawegawa, Y., "Electrically reprogram-
     mable nonvolatile semiconductor memory with MNMoOs (metal-
     nitride-molybdenum-oxide-silicon) structure", 1975 International
     Electron Devices Meeting, Technical Digest, p. 313-16,
     CCA11-31631.
621. Tarui, Y., Hayashi, Y. and Nagai, K., "Electrically reprogram-
     mable nonvolatile semiconductor memory", Bull. Electrotech.
     Lab., 26 (4-5), 8-15 (1976), Japanese, CCA11-31606.
622. Uzunoglu, V., "An analog memory device", International Tele-
     metering Conference, 1975, p. 574-80, EEA79-28561.
623. White, M. H., Lampe, D. R., Fagan, J. L., Kub, F. J. and Barth,
     D. A., "A nonvolatile charge-addressed memory (NOVCAM) cell",
     IEEE J. Solid-State Circuits, SC-10 (5), 281-7 (1975),
     CCA10-25848.

       4.   Magnetic
         a.   Ferrite Cores

624. Anon., "Interchangeable core memory modules feature 16K and 32K
     sense lines", Comput. Des., 13 (3), 116 (1974), CCA9-16060.
625. Anon., "Bulk core system bridges gap between main memory and
     I/O devices", Comput. Des., 15 (5), 22, 6 (1976), CCA11-22637.

626. Anon., "Curable silicone rubber carries core memories through production - no dropouts along the way", Insul./Circuits, $\underline{22}$ (11), 33-5 (1976), CCA12-10685.
627. Becker, E., "Type DPS 01 print-out buffer storage device", Hernsdorfer Tech. Mitt., $\underline{13}$ (36), 1119-24 (1973), German, CCA9-9204.
628. Brown, J. R., Jr., "Magnetic core memories - a tutorial review", Mod. Data, $\underline{8}$ (8), 51-3 (1975), CCA11-9670.
629. Brune, W. and Dull, E. H., "The power supply to high-speed core memories", Elektrotech. Z. (ETZ) B, $\underline{23}$ (3), 53-6 (1971), German, CCA6-10762.
630. Buhling, D., Petermann, R. and Undeutsch, M., "The effectiveness and stability of ferrite core storage HS 16K/2 and HS 32K/2. I.", Hermsdorfer Tech. Mitt., $\underline{13}$ (38), 1213-17 (1973), German, CCA9-11958.
631. Cardona, G., "Satellite ferrite core memory", Proceedings of the International Conference on Space and Communication, 1971, 1 pp., CCA6-15228.
632. Coats, T., "Hi-G vibration characteristics of core memories", IEEE Proceedings National Aerospace Electronic Conference, 1975, p. 109-15.
633. Cole, J. M. and Sinclair, W., "Method of fabricating a magnetic memory matrix", Patent USA 3676924, Publ. July 1972, CCA8-2653.
634. Dhar, P. and Thyagaraju, D., "Investigation into the thermal hysteresis of ferrite cores used for computer memory", J. Inst. Electron. Telecommun. Eng., $\underline{20}$ (12), 589-92 (1974).
635. Enticknap, N., "Core memory is not on the way out", Comput. Wkly., $\underline{16}$ (384), 10-11 (1974), CCA9-16061.
636. Florjancic, M. and Reiner, H., "Megabit memory for helios space probe", Electr. Commun., $\underline{49}$ (3), 262-7 (1974).
637. Fujitsu Ltd., "Memory", Patent UK 1285974, Publ. August 1972, CCA8-2654.
638. Gaertner, M., Gliem, F., Pittermann, F., Schlenther, M. and Spencer, J., "Modular core memory with low power consumption", Report N76-16854/1SL, Dornier-System G.M.B.H., Friedrichshafen (West Germany), 1975, 135 pp.
639. Gendler, M. B., Rozenblat, M. A. and Rubtchinskii, A. A., "Magnetic states of branched cores of arbitrary configuration", Avtom. & Telemekh., no. 5, 168-75 (1971), Russian, CCA6-15219.
640. Hall, T. P., "Can you improve on core?", Systems, $\underline{3}$ (9), 36-7 (1975), CCA11-17399.
641. Ho, H. H., "Direct storage of decimal numbers by a square loop core", Int. J. Electron., $\underline{39}$ (5), 497-507 (1975), CCA11-4795.
642. Im, H. B. and Wickham, D. G., "Memory-core characteristics of cobalt substituted lithium ferrite", IEEE Trans. Mag., MAG-$\underline{8}$ (4), 765-9 (1972), CCA8-13680.
643. Jensen, J. L., "Read-out and reset control for a memory core", Patent USA 3693188, Publ. September 1972, CCA8-9488.

644. Jones, A., "Ferrite core memories", Compon. Technol., $\underline{4}$ (5),
     7-8 (1970), CCA6-2058.
645. Karpekov, Y. D., Konstantinov, V. F., Makarov, G. P., Prokhorov,
     A. N., Sergeev, V. A., Simonov, Y. N. and Sugonyaev, V. P., "The
     characteristics of wire spark chambers with takeoff of infor-
     mation onto ferrite rings", Instrum. & Exp. Tech., $\underline{16}$ (5), 1350-2
     (1973), CCA9-25091.
646. Kawasaki, J., "Design criteria for switching-core memories",
     Electr. Eng. Jap., $\underline{92}$ (1), 167-77 (1972), CCA8-13662.
647. Kenner, B. A., "Read and write systems for 2 1/2D core memory",
     Patent USA 3693176, Publ. September 1972, CCA8-9487.
648. Kobayashi, S., Torri, M., Yamakawa, F. and Dateki, Z., "Large
     capacity NDRO core memory", IEEE Trans. Magn., MAG-$\underline{8}$ (3), 389
     (1972), CCA8-13678.
649. Koppens, L. J., "Improved ferrite memory cores obtained by a new
     preparation technique", IEEE Trans. Magn., MAG-$\underline{8}$ (3), 303-5
     (1972), CCA8-13664.
650. Ksiazek, J., "Planar ferrite core storages for minicomputers
     made in WZE MERA-ELWRO", Informatyka, $\underline{11}$ (12), 4-8 (1975),
     Polish, CCA12-9668.
651. Kuroda, C. and Kawashima, T., "Temperature-stable low-noise
     ferrite memory cores", IEEE Trans. Magn., MAG-$\underline{5}$ (3), 192-6
     (1969), CCA5-6478.
652. Lighthall, J. T., "Fault detection for a ring core memory",
     Patent USA 3903511, Publ. September 1975, CCA11-6376.
653. Maiwald, W., "The stapelblock: a novel technique for manufactur-
     ing highly realiable ferrite core memory stacks", IEEE Trans.
     Magn., MAG-$\underline{6}$ (3), 684-7 (1970), CCA6-10767.
654. Nicot, J., "A core memory family", Onde Electr., $\underline{52}$ (10), 435-8
     (1972), French, CCA8-5260.
655. Post Office, "Magnetic memory", Patent UK 1294445, Publ.
     October 1972, CCA8-9530.
656. Purcell, C. J., "Control data STAR-100 - performance measure-
     ments", AFIPS National Computer Conference Proceedings, Vol.
     43, 1974, p. 385-7.
657. Rabl, H., "Ferrite-core memory boards for extreme environmental
     conditions", Siemens Rev., $\underline{41}$ (6), 284-6 (1974), CCA9-21991.
658. Riches, E. E., "Ferrites: A review of materials and applica-
     tions", Mills & Boon Ltd., London, England, 1972, 88 pp.,
     CCA8-2649.
659. Samofalov, K. G., Seligei, A. M. and Trostyanetskii, D. S.,
     "Problems in the optimization of the structure of permanent
     core-type memory units", Izv. Vuy. Priborostr., $\underline{15}$ (9), 68-73
     (1972), CCA8-7338.
660. Shchelkanov, S. S., "Magnetic memory device with compound core",
     Prib. & Sist. Upr., no. 5, 49 (1974), Russian, CCA9-25095.
661. Thomson-CSF, "Magnetic memory core manufacture", Patent UK
     1282028, Publ. July 1972, CCA8-2650.

662. Thuras, R. D., "New magnetic materials help core memories stay
     alive", Comput. Des., $\underline{15}$ (7), 100-2 (1976), CCA11-31595.
663. Watson, J. N., "Effects of macroscopic thermal stress in ferrite
     memory cores", Thesis, Lehigh Univ., Bethlehem, Pa., 1974,
     179 pp., Order No. 74-21444, CCA10-15170.
664. Welsh, J. H., "An improved modular core memory for aerospace
     computers", Comput. Des., $\underline{12}$ (11), 106-12 (1973), CCA9-6231.
665. Wiskin, A., "Reduced error rates with core memory", Systems,
     $\underline{3}$ (9), 38-9 (1975), CCA11-17400.

                    b.   Plated Wires

666. Anon., "Fine wire memory nerves", Met. & Mater., $\underline{7}$ (8), 329-30
     (1973), CCA9-1230.
667. Balogh, B. and Nagy, A. B., "A modular plated wire memory plane",
     IEEE Trans. Magn., MAG-$\underline{8}$ (3), 362-3 (1972), CCA8-13670.
668. Brosius, J. P., Jr. and Russell, B. J., "Fault-tolerant plated
     wire memory for long duration space missions", International
     Symposium on Fault-Tolerant Computing 73, Digest of Papers,
     p. 33-8, CCA9-3420.
669. Doyle, W. D. and Josephs, R. M., "The read-write disturb prob-
     lem in plated wire memories", IEEE Trans. Magn., MAG-$\underline{8}$ (3),
     306-9 (1972), CCA8-13665.
670. England, W. A., "Applications of plated wire to the military and
     space environments", IEEE Trans. Magn., MAG-$\underline{6}$ (3), 528-34
     (1970), CCA6-10764.
671. Faktor, Z. and Munz, V., "The properties of memory wires for
     non-destructive readout memories, their manufacture and the
     equipment used", Slaboproudy Obz., $\underline{37}$ (12), 583-8 (1976), Czech,
     CCA12-6666.
672. Fitzwilson, R. L. and Comisar, G. G., "Radiation induced replace-
     ment currents in a plated wire memory", IEEE Trans. Nucl. Sci.,
     NS-$\underline{21}$ (6), 295-304 (1974), CCA10-15167.
673. Fulmer, L. C. and Meilander, W. C., "A modular plated-wire
     associative processor", Memories, Terminals, and Peripherals,
     Proceedings of the 1970 IEEE International Computer Group Con-
     ference, p. 325-35, CCA6-2035.
674. Gibson, A. T., "Plated wire memories", Man Made Memories, Con-
     ference Summaries, 1971, 1 pp., CCA6-12725.
675. Goodyear Aerospace Corp., "Magnetic memory", Patent UK 1299414,
     Publ. December 1972, CCA8-13690.
676. Gruter, O. and Kragh, P., "Homogeneous nickel-iron films for
     plated-wire memories", Siemens Forsch. - & Entwicklungsber.,
     $\underline{1}$ (4), 332-40 (1972), German, CCA8-7337.
677. Harloff, H. J. and Penzel, H. J., "A plated-wire memory of high
     capacity and short cycle time", Elektron. Rechenanlagen, $\underline{15}$ (5),
     208-15 (1973), German, CCA9-6232.

678. Ishii, O., Nitta, M. and Tanaka, T., "A plated wire read-only
     memory using pattern weaving technique", Bull. Electrotech.
     Lab., 36 (6), 469-81 (1972), Japanese, CCA8-7334.
679. Ishii, O. and Shirakura, T., "A plated wire memory using grooved
     ferrite keeper plates", Bull. Electrotech. Lab., 36 (6), 450-68
     (1972), Japanese, CCA8-5257.
680. Kitano, Y. and Yamagata, M., "Plated-memory-wire tester", Rev.
     Electr. Commun. Lab., 21 (5-6), 332-8 (1973), CCA9-6236.
681. Kitano, Y. and Yonezawa, S., "Construction and performance of a
     plated-wire memory plane", IEEE Trans. Magn., MAG-8 (3), 389
     (1972), CCA8-13677.
682. Kitovich, V. V. and Tarayan, I. S., "Concerning different modes
     of plated-wire cutting for memory modules", Avtom. & Telemekh.,
     no. 5, 176-84 (1971), Russian, CCA6-15220.
683. Kobayashi, S., Torii, M. and Jojima, T., "Plated wire memory
     with flexible keeper", IEEE Trans. Magn., MAG-8 (3), 367 (1972),
     CCA8-13673.
684. Lisowski, B., "Pulse method of angular dispersion measurement
     of plated wires", Arch. Elektrotech., 22 (1), 111-41 (1973),
     Polish, CCA8-13660.
685. Littwin, B., "The NDRW-characteristic of a plated wire memory
     element with an interlayer of tin", IEEE Trans. Magn., MAG-11,
     3135 (1975).
686. Lotsch, H. K. V., "Design considerations for a plated-wire
     memory", IEEE Trans. Magn., MAG-5 (3), 312-13 (1969), CCA5-6483.
687. Luborsky, F. E. and Drummond, B. J., "Experimental evaluation
     of cell designs for plated memory wire", Plating, 61 (3), 243-8
     (1974), CCA9-18395.
688. Luborsky, F. E., Drummond, B. J. and Barber, W. D., "Some fac-
     tors controlling the uniformity and properties of NiFe plated
     wires", IEEE Trans. Magn., MAG-9 (3), 495-9 (1973), CCA9-3414.
689. Luborsky, F. E., Skoda, R. E. and Barber, W. D., "Origins of
     nondestructive readout of plated wires", J. Appl. Phys., 42
     (4), 1428-30 (1971), CCA6-15225.
690. Lutes, O. S. and Kooyer, R. L., "Switching mechanisms of plated
     wires in external fields", IEEE Trans. Magn., MAG-13 (1),
     907-13 (1977), CCA12-15500.
691. Makino, K., Kawakami, S., Orihara, S. and Sakai, S., "A highly
     reliable plated wire: Study on corrosion of magnetic films",
     IEEE Trans. Magn., MAG-9 (3), 500-3 (1973), CCA9-3415.
692. Matsuyama, S. and Iwasa, S., "NDRO plated wire for a 100ns cycle
     time EAROM", Fujitsu Sci. & Tech. J., 9 (4), 123-40 (1973),
     CCA9-11954.
693. Murakami, H. and Wada, Y., "A three-dimensional operation in
     plated-wire memory", IEEE Trans. Magn., MAG-5 (3), 422-5 (1969),
     CCA5-6488.
694. McCallister, J. P., "Plated-wire memories: How far can we go?",
     IEEE Trans. Magn., MAG-6 (3), 525-8 (1970), CCA6-10763.

695. North American Rockwell Corp., "Plated wire memory", Patent UK 1284058, Publ. August 1972, CCA8-2651.

696. Odani, Y., Terajima, M. and Moritomo, T., "Plated wire semi-permanent memory for electronic switching system", Jap. Tele-commun. Rev., 13 (2), 57-61 (1971), CCA6-17644.

697. Ooomote, R., Nitta, M. and Ishii, O., "Method of simulation of wave transmission characteristics of woven type plated wire memory matrix, and its application", Bull. Electrotech. Lab., 36 (6), 434-49 (1972), Japanese, CCA8-5256.

698. Orihara, S., Kawakami, S., Hoshi, T. and Segawa, S., "A 50-μm plated-wire mainframe memory", IEEE Trans. Magn., MAG-8 (3), 364-6 (1972), CCA8-13671.

699. Orihara, S., Kawakami, S. and Sakai, S., "A consideration of the bit-density limitations in plated-wire memory", IEEE Trans. Magn., MAG-9 (3), 507-11 (1973), CCA9-3417.

700. Przyluski, J. and Grabowski, R., "Polish-technology of plated wires for memories of IVth-generation computers", Pol. Tech. Rev., no. 8, 17-19 (1976), CCA12-1502.

701. Renard, A. M., "High speed display of the magnetic parameters of plated wires", IEEE Trans. Magn., MAG-8 (3), 309-11 (1972), CCA8-13666.

702. Saitoh, K., Takaba, T., Oguma, S., Eguchi, N. and Nakamura, H., "High-speed plated wire read-only memory", NEC Res. & Dev., no. 27, 50-62 (1972), CCA8-9524.

703. Smale, B. G., "A 15-Mbit random-access plated-wire memory system", IEEE Trans. Magn., MAG-5 (3), 428 (1969), CCA5-6491.

704. Stapleton, D., "Plated wire memories", Compon. Technol., 4 (5), 9-13 (1970), CCA6-2059.

705. Terjima, M., Gomi, Y. and Ikawa, H., "Effect of on-line flash anneal on aging properties of NDRO multilayer plated wire", IEEE Trans. Magn., MAG-6 (3), 716-19 (1970), CCA6-10769.

706. Tvarusko, A., "Current distribution on resistive metal elec-trodes", Plating, 61 (9), 846-9 (1974).

707. Ueoka, Y. and Kitano, Y., "High bit density plated wire memory module", Electr. Commun. Lab. Tech. J., 25 (2), 371-81 (1976), Japanese, CCA11-22656.

708. Ueoka, Y., Kitano, Y. and Hasegawa, K., "256 Kbyte plated wire memory module", Electr. Commun. Lab. Tech. J., 25 (2), 359-79 (1976), Japanese, CCA11-22655.

709. Yamanaka, T. and Toda, T., "Evolution in manufacture of woven plated-wire memory", IEEE Trans. Manuf. Technol., MFG-3 (2), 82-7 (1974).

c.   Thin Films

710. Arnett, P. C., "Magnetization spreading in a thin film memory", IEEE Trans. Magn., MAG-7 (1), 171-4 (1971), CCA6-12718.

711. Ayling, J. K., "Magnetic film memory", IBM Tech. Disclosure Bull., 13 (1), 111-12 (1970), CCA5-22420.

712. Belson, H. S., Desavage, B. F., Tebble, R. S. and Parker, M. R.,
     "NDRO thin-film memory device exhibiting triangular hysteresis
     loop", IEEE Trans. Magn., MAG-6 (3), 722-4 (1970), CCA6-10770.
713. Benrud, V. M., Forslund, G. L., Hanson, M. M. and Others,
     "Oligatomic film memories", J. Appl. Phys., 42 (4), 1364-73
     (1971), CCA6-15224.
714. Blanchet, F., "High density magnetic film memory", Onde Electr.,
     52 (10), 446-51 (1972), French, CCA8-5261.
715. Dangel, J. and Jostan, J. L., "A 90 nanosecond experimental
     storage with thin electroplated NiFeSe storage films", Wiss. Ber.
     AEG-Telefunken, 45 (1-2), 64-71 (1972), German, CCA8-9525.
716. Finzi, L. A. and Fritsch, R. J., "Nondestructive readout in
     nonplanar magnetic films", IEEE Trans. Magn., MAG-5 (3), 300-3
     (1969), CCA5-6480.
717. Franklin, D. W., Hornreich, R. M. and Rubinstein, H., "Magneto-
     statically coupled thin-film magnetic memory devices", Patent
     USA 3701983, Publ. October 1972, CCA8-9529.
718. Karmon, D., "Device and array for a 10^7 bit high density mag-
     netic thin film memory", 8th Convention of the Electrical and
     Electronics Engineers, 1973, 50 pp., CCA9-6238.
719. Kayser, W., "A 32 byte, 200 ns read-write cycle magnetic film
     memory system", IEEE Trans. Magn., MAG-8 (3), 366 (1972),
     CCA8-13672.
720. Lin, C. H. and Fresia, J. F., "Prediction of device performance
     of integrated flat film NDRO devices", IEEE Trans. Magn.,
     MAG-6 (3), 660-2 (1970), CCA6-10765.
721. Manzel, M., Linzen, D. and Heller, I., "Study of the storage
     properties of cylindrical soft-hard magnetic multi-films",
     Hermsdorfer Tech. Mitt., 13 (38), 1202-5 (1973), German,
     CCA9-11957.
722. Oshima, S., "Recent developments of magnetic film memories in
     Japan", IEEE Trans. Magn., MAG-5 (3), 392-8 (1969), CCA5-6484.
723. Overn, W. M. and Root, M. R., "Mated-film memory production –
     a progress report", IEEE Trans. Magn., MAG-8 (3), 361 (1972),
     CCA8-13668.
724. Pelloth, W., Sihling, A. and Stein, K. U., "Read-write-loop with
     integrated circuits for a high-speed planar magnetic-film
     memory with high storage density", Proceedings of the Inter-
     national Conference on Advanced Microelectronics, 1970, 2 pp.,
     CCA6-2067.
725. Pohm, A. V., Wang, J. M., Lee, F. S., Schnasse, W. and Smay,
     T. A., "High-density very efficient magnetic film memory arrays",
     IEEE Trans. Magn., MAG-5 (3), 408-12 (1969), CCA5-6486.
726. Rabinovici, B. M., "Effects of magnetic keepers on permalloy-
     film stores", Proc. Inst. Elec. Eng., 118(6), 750-5 (1971),
     CCA6-15226.

727. Rybinskii, O. A., Orobinskii, S. P., Ivanov, B. D. and Petrenko, O. V., "Integrated memory elements based on single crystal ferrite films", Izv. VUZ Priborostr., 16 (10), 52-6 (1973), Russian, CCA9-9206.

728. Sakai, M., "Large capacity ferromagnetic thin film memory device", Patent USA 3702992, Publ. November 1972, CCA8-13685.

729. Salzmann, G., "A planar magnetic thin-film memory element", IEEE Trans. Magn., MAG-5 (3), 310-11 (1969), CCA5-6482.

730. Salzmann, G., "Influence of eddy currents on the switching and storage characteristics of thin magnetic films", 14th International Scientific Colloquium, Electrical Engineering Theory, 1969, p. 73-8, German, CCA6-15227.

731. Schwee, L. J., Irons, H. R. and Anderson, W. E., "The crosstie memory", IEEE Trans. Magn., MAG-12 (6), 608-13 (1976), CCA12-6648.

732. Schwee, L. J., Irons, H. R. and Anderson, W. E., "Cross tie memory simplified by the use of serrated strips", AIP Conf. Proc., no. 29, 624-5 (1976), CCA12-3989.

733. Sery, R. S., "Dynamic crosstie nucleation thresholds for crosstie memory", IEEE Trans. Magn., MAG-11 (1), 29-30 (1975).

734. Siemens AG, "Correction circuit for magnetic film memory signal", Patent UK 1401523, Publ. July 1975, CCA11-4790.

735. Stein, K. U., "A 1.5 x 10^5 bit low-current high-speed planar-magnetic-film memory with high storage density", IEEE Trans. Magn., MAG-5 (3), 426 (1969), CCA5-6489.

736. Tisone, T. C., Grupen, W. B. and Chin, G. Y., "Stress insensitive permalloys for memory application", IEEE Trans. Magn., MAG-6 (3), 712-16 (1970), CCA6-10768.

737. Valstyn, E. P. and Walker, L. R., "Bit-line interaction in magnetic-film memories", IEEE Trans. Magn., MAG-5 (3), 417-21 (1969), CCA5-6487.

d.   Others

738. Bate, R. T. and West, F. G., Jr., "Magnetic domain memory structure", Patent USA 3702991, Publ. November 1972, CCA8-13684.

739. Bunker Ramo Corp., "Batch fabricated magnetic memory", Patent UK 1291428, Publ. October 1972, CCA8-7341.

740. Bunker Ramo Corp., "Magnetic wire memory", Patent UK 1291429, Publ. October 1972, CCA8-5264.

741. Chamberlayne, J. W. and Gibson, B., "Magnetic materials for integrated cores", IEEE Trans. Magn., MAG-8 (4), 759-64 (1972), CCA8-13679.

742. IBM Corp., "Magnetic storage system", Patent UK 1289371, Publ. September 1972, CCA8-5262.

743. Malinovskii, B. N., Yakovlev, Y. S. and Novikov, B. V., "Design of integrated memory units based on multiaperture ferrite plates", Avtom. & Vychisl. Tekh., no. 3, 73-81 (1972), Russian, CCA8-11622.

744. Mazda, F. F., "The components of computers. VI. Magnetic main-
     frame memories", Electron. Components, 13 (22), 1112-18 (1972),
     CCA8-5259.
745. Philips Ltd., "Magnetic memory with beam addressing", Patent
     UK 1395592, Publ. May 1975, CCA11-4809.
746. Rider, R. E., "Head-per-track memory system", Ind. Ital. Elet-
     trotec. & Elettron., 25 (9), 802-4 (1972), Italian, CCA8-7335.
747. Schouten, G. H., Bohlmeijer, N., van der Steeg, H. and van de
     Maarel, M. C., "Three-dimensional storage device comprising
     magnetic memory elements", Patent USA 3701981, Publ. October
     1972, CCA8-9527.
748. Sperry Rand Corp., "Memory", Patent UK 1292751, Publ. October
     1972, CCA8-7343.
749. Thomson-CSF, "Magnetic memory element", Patent UK 1287287,
     Publ. August 1972, CCA8-2655.

5.   Josephson

750. Anacker, W., "Memory cell using a single Josephson tunneling
     gate", IBM Tech. Disclosure Bull., 15 (2), 449-51 (1972),
     CCA8-2692.
751. Anacker, W., "Superconducting tunneling circuits for computer
     applications", 1974 International Electron Devices Meeting,
     Technical Digest, p. 5-8.
752. Anacker, W., "Superconducting memories employing Josephson
     devices", AFIPS National Computer Conference, Proceedings,
     Vol. 44, 1975, p. 529-34.
753. Broom, R. F., Jutzi, W. and Mohr, T. O., "1.4 mil^2 memory cell
     with Josephson junctions", IEEE Trans. Magn., MAG-11 (2),
     755-8 (1975).
754. Clarke, J., "Application of Josephson junctions to computer
     storage and logic elements and to magnetic measurements", AIP
     Conf. Proc.,no. 29, 17-22 (1976).
755. Fahlenbrach, H., "Superconductors - fundamentals - realization -
     applications", Elektronik, 25 (2), 58-63, 88 (1976), German,
     CCA11-12338.
756. Fang, F. F. and Laibowitz, R. B., "Latent image Josephson
     memory", IBM Tech. Disclosure Bull., 16 (10), 3433-4 (1974),
     CCA9-20773.
757. Gueret, P., "Storage and detection of a single flux quantum in
     Josephson junction devices", IEEE Trans. Magn., MAG-11 (2),
     751-4 (1975).
758. Jutzi, W., "Inductively coupled memory cell for NDRO with two
     Josephson junctions", Cryogenics, 16 (2), 81-8 (1976).
759. Jutzi, W. and Schunemann, C., "A cryogenic memory cell concept
     with two Josephson junctions of very high current density", Sci.
     Electr., 21 (3), 57-67 (1975), CCA11-9695.
760. Matisoo, J., "Superconductive computer technology", IEEE Trans.
     Magn., MAG-12 (6), 623 (1976), CCA12-6676.

761. Zappe, H. H., "Subnanosecond Josephson tunneling memory cell
     with nondestructive readout", IEEE J. Solid-State Circuits,
     SC-10 (1), 12-19 (1975).

6.   Schottky

762. Anon., "4-bit bipolar/LSI processor slice cuts microcycle time
     to 100 ns", Electron. Des., 23 (15), 77-8 (1975), CCA11-2189.
763. Bertolucci, B., "CAMAC magnetostrictive readout system using
     Schottky memories (for use with spark chamber)", IEEE Trans.
     Nucl. Sci., NS-20, 361-6 (1973), PA76-28676.
764. Broom, R. F., Moser, A. and Rideout, A. J., "Vertical Schottky
     diode-memory device", IBM Tech. Disclosure Bull., 15, 2158 (1972).
765. Crowder, B. L., Hovel, H. J. and Urgell, J. J., "Silicon
     Schottky barrier bistable memory element", IBM Tech. Disclosure
     Bull., 15, 891 (1972).
766. Domingo, G., Holm-Kennedy, J. W., Kingsley, W. and Nash, J. G.,
     "Long term memory in junction devices using multivalent trapping
     impurities in silicon", Report AD-A018213/9SL, California Univ.,
     Los Angeles, 1975, 127 pp.
767. Dorler, J. A., Forneris, J. L. and Swietek, D. J., "Schottky
     barrier diode read-only memory", Patent USA 3780320, Publ.
     December 1973.
768. Drangeid, K. E., Broom, R. F., Jutzi, W., Mohr, T. O., Moser,
     A. and Sasso, G., "A fixed-address memory cell with normally-
     off-type Schottky-barrier FETs", 1971 International Solid State
     Circuits Conference, Digest of Technical Papers, p. 68-9.
769. Flassayer, C., Morenza Gil, J. L., Martinez, A. and Esteve, D.,
     "Switching phenomena and memory effects in n-type metal silicon
     Schottky diodes", Report N74-17939/1, Centre National de La
     Rechereche Scientifique, Toulouse, France, 1973, 14 pp.
770. Gaensslen, F. H., "Schottky barrier read-only memory", IBM
     Tech. Disclosure Bull., 14, 252 (1971), EEA74-34116.
771. Guhman, G. F., "Diode-capacitor memory cells supported on a
     sapphire substrate", IBM Tech. Disclosure Bull., 16, 3166-7
     (1974), EEA77-30284.
772. Hashizume, N., Kataoka, S. and Tomizawa, K., "Gun-effect memory
     device using the charge accumulation on a Schottky-trigger
     electrode", Electron. Lett., 13 (2), 60-1 (1977), CCA12-6668.
773. Holm-Kennedy, J. W., Domingo, G. and Kingsley, W., "Long term
     memory in junction devices using multivalent trapping impurities
     in silicon", Report AD-A016939/1SL, California Univ., Los
     Angeles, 1975, 61 pp.
774. Holm-Kennedy, J. W., "Long term memory in junction devices using
     multivalent trapping impurities in silicon", Report AD-A008045/
     7SL, California Univ., Los Angeles, 1975, 67 pp.
775. Householder, V. P., Jr., "Schottky barrier diode transmission
     line termination", IBM Tech. Disclosure Bull., 19 (8), 3133-4
     (1977).

776. Hovel, H. J., "Schottky barrier method sneak path elimination
     in memory devices", IBM Tech. Disclosure Bull., 14, 356-7
     (1971), EEA74-37243.
777. Ingebrigtsen, K. A., "Schottky diode acoustoelectric memory
     and correlator - a novel programmable signal processor", Proc.
     IEEE, 64 (5), 764-9 (1976).
778. Ingebrigtsen, K. A., "Simultaneous storage of spatially ortho-
     gonal acoustic beams in a Schottky-diode memory correlator",
     Electron. Lett., 11 (24), 585-6 (1975), EEA79-6739.
779. Ingebrigtsen, K. A., Cohen, R. A. and Mountain, R. W., "A
     Schottky-diode acoustic memory and correlator", Appl. Phys.
     Lett., 26, 596-8 (1975).
780. Ingebrigtsen, K. A. and Stern, E., "Coherent integration and
     correlation in a modified acoustoelectric memory correlator",
     Appl. Phys. Lett., 27 (4), 170-2 (1975).
781. Koch, S., "Fixed data memory utilizing Schottky diodes", Patent
     USA 3774170, Publ. November 1973.
782. Kurz, B., Barron, M. and Butler, W., "Applications of metal-
     silicon structures in integrated circuits", Eurocon 71 Digest,
     1971, 2 pp., EEA75-3877.
783. Landler, P. F., "Dynamic Schottky cell", IBM Tech. Disclosure
     Bull., 17 (11), 3215-16 (1975), EEA78-28753.
784. Luxton, H. E. G., "Gallium arsenide field effect transistors -
     their performance and application up to X-band frequencies",
     4th European Microwave Conference, 1974, p. 92-6, EEA78-2427.
785. Mette, H. L. and Ahlstrom, E. R., "Magneto-Schottky-diode-
     integrated circuits", Report ECOM-3527, Army Electronics
     Command, Fort Monmouth, New Jersey, 1972, 27 pp., EEA75-35415.
786. Minden, H. T., "Gallium arsenide dual Schottky barrier diodes",
     Solid-State Electron., 16, 1185-8 (1973), EEA76-35299.
787. Philippow, P., "Using the Schottky junction for semiconductor
     memories", Wiss. Z. Tech. Hochsch. Ilmenau, 22 (3), 165-75
     (1976), German, CCA11-17412.
788. Pierre, L., "The evolution of Schottky diode integrated cir-
     cuits", Electronique, no. 134, 415-9 (1970), French, EEA73-30380.
789. Platt, S., "Double emitter-bit line Schottky memory cell", IBM
     Tech. Disclosure Bull., 15, 260 (1972), EEA75-35304.
790. Rattner, J., Cornet, J. C. and Hoff, M. E., Jr., "Bipolar LSI
     computing elements usher in new era of digital design", Elec-
     tronics, 47 (18), 89-96 (1974), EEA78-675.
791. Rein, H. M., Kosak, W. and Kotte, W., "Fast integrated digital
     circuits with Schottky diodes", Wiss. Ber. AEG-Telefunken, 45,
     130-41 (1973), German, EEA76-19962.
792. Trotman, D., "Read while you write r.a.m.s", Electron, no. 21,
     31-3 (1973), CCA8-11642.
793. Waaben, S., "Electronically-variable semiconductor memory using
     two diodes per memory cell", 1970 International Solid State Cir-
     cuits Conference, Digest of Technical Papers, p. 46-7,
     EEA73-24998.

794. Wang, W. C., "Physical processes of s.a.w. Schottky-diode memory correlator", Electron. Lett., 12 (4), 105-6 (1976), EEA79-22749.

7.   Others

795. Anon., "A high-density, low-power memory system", New Electron., 8 (22), 39, 46 (1975), CCA11-6383.
796. Arnett, P. C., "Three-dimensional dual insulator memory", IBM Tech. Disclosure Bull., 16 (11), 3517-18 (1974), CCA9-25100.
797. Barsuhn, H., Gschwendtner, J. and Haug, W. O., "Semiconductor storage circuit utilizing two device memory cells", IBM Tech. Disclosure Bull., 18 (3), 786-7 (1975), CCA11-2226.
798. Baskett, F. and Smith, A. J., "Interference in multiprocessor computer systems with interleaved memory", Commun. ACM, 19 (6), 327-34 (1976), CCA11-22781.
799. Beers, H. R., Dennison, R. T., Keller, D. G., Ludlow, P. J., Marzin, C. and Park, S. J., "Integrated Harper cell", IBM Tech. Disclosure Bull., 18 (9), 2875-7 (1976), CCA11-15068.
800. Berezin, A. S., Vaganov, V. I., Kuzmin, V. A., Mochalkina, O. R. and Pershenkov, V. S., "Thyristor-transistor memory cell", Izv. VUZ Radioelektron., 15 (10), 1270-80 (1972), Russian, CCA8-9537.
801. Chang, W. H., Guhman, G. F., Miller, D. A. and Wang, L. C., "Vertical diode-capacitor memory cells", IBM Tech. Disclosure Bull., 15 (9), 2887-9 (1973), CCA8-13695.
802. Chekmarev, Y. D. and Smolov, V. B., "The use of geometric-code fixed semiconductor memories for the storage of large volumes of instructions and constants in special purpose digital computers", Izu. VUZ Priborostr., no. 1, 54-7 (1973), Russian, CCA8-13696.
803. Chen, A. C. M., "Electron beam heating in amorphous semiconductor beam memory", IEEE Trans. Electron Devices, ED-20 (2), 160-9 (1973), CCA8-7351.
804. Chen, A. C. M. and Wang, J. M., "An amorphous semiconductor electron beam memory", IEEE Trans. Magn., MAG-8 (3), 312-14 (1972), CCA8-13694.
805. Cohn, C. E., "Expanding a memory without plug-compatibility", Comput. Des., 14 (11), 102-4 (1975), CCA11-9653.
806. Filipskii, Y. K. and Mikhailov, Y. P., "Dynamic operating conditions of multistate diode-capacity memories", Izv. VUZ Radioelektron., 15 (10), 1263-9 (1972), Russian, CCA8-9536.
807. Fischle, A. S. and Holmstrom, F. E., "Discrete memory elements are formed by evaporating metal pads onto a silicon dioxide coated silicon wafer. The metal pads surround a central metallized circle", IBM Tech. Disclosure Bull., 15, 741-2 (1972), EEA76-816.
808. Melcher, R. L. and Shiren, N. S., "High-density memory and cinematic 3-D display", IBM Tech. Disclosure Bull., 18 (7), 2356-9 (1975), CCA11-9678.

809. National Cash Register Co., "Semiconductor memory for binary data", Patent UK 1401101, Publ. July 1975, CCA11-9687.
810. Noble, W. P., Jr., "Bulk access-surface storage memory cells", IBM Tech. Disclosure Bull., $\underline{16}$, 3170-2 (1974), EEA77-30335.
811. Novik, A. I. and Sheremet, L. P., "A high-speed memory", Probl. Tekh. Elektrodin., $\underline{54}$, 52-5 (1975), Russian, CCA11-4813.
812. Overington, W. J. G., "A bed-addressed memory with 1000 addresses", Electron. Engineering, $\underline{48}$ (578), 19 (1976), CCA11-17401.
813. Ovshinsky, S. R. and Fritzsche, H., "Amorphous semiconductors for switching, memory and imaging applications", IEEE Trans. Electron Devices, ED-$\underline{20}$ (2), 91-105 (1973), CCA8-7350.
814. Shanko, R. R., "Amorphous semiconductors for electrically alterable memory applications", Comput. Des., $\underline{13}$ (5), 95-100 (1974), CCA9-20767.
815. van de Vaart, H. and Schissler, L. R., "Acoustic surface-wave recirculating memory", IEEE Trans. Microwave Theory & Tech., MTT-$\underline{21}$ (4), 236-43 (1973), CCA8-11653.
816. Winkler, E. and Hartmann, U., "The behaviour of the storing gate-controlled diode (SGGD) and its use in semiconductor memory circuits", Nachrichtentech. Elektron., $\underline{23}$ (12), 459-60 (1973), German, CCA9-11965.
817. Wu, S. Y., "Memory retention and switching behavior of metal-ferroelectric-semiconductor transistors", Ferroelectrics, $\underline{11}$ (1-2), 379-83 (1976), EEA79-28576.

VII. DESIGN

818. Alexander, E. J., "Bipolar-MOS tradeoffs in memory design", 1970 IEEE International Convention Digest, p. 126-7, CCA5-22412.
819. Altman, L., "LSI multiples design options", Electronics, $\underline{48}$ (21), 110-14 (1975), EEA79-4743.
820. Altman, L., "Advances in designs and new processes yield surprising performance (digital MOS LSI)", Electronics, $\underline{49}$ (7), 73-81 (1976), CCA11-14958.
821. Altman, L., "CCD's show influence on design quietly", Electronics, $\underline{50}$ (2), 65-6 (1970).
822. Anon., "A high-density, low-power memory system", New Electron., $\underline{8}$ (22), 39, 46 (1975), EEA79-9014.
823. Artamonov, G. T. and Mozharov, E. A., "Investigation of the throughput capacity of a model for multichannel memory, using a priority analysis block", Prib. & Sist. Upr., no. 9, 13-15 (1976), CCA12-6638.
824. Atterbury, G. F. and Adams, G. F., "Building today's technologies into a large-scale time-sharing system", Comput. Des., $\underline{14}$ (9), 79-85 (1975).
825. Avaev, N. A. and Naumov, Y. E., "Analysis of cell parameters and an operational memory device with injection feed", Izv. VUZ Radioelektron., $\underline{18}$ (11), 15-20 (1975), Russian, CCA11-9677.

826. Basiladze, S. G., "Design of integrated write and readout cir-
     cuits using a type of MOS-transistor", Izv. VUZ Radioelektronika,
     13, 381-8 (1970), Russian, EEA73-27259.
827. Beer, A. F., "A MOS transistor store with discretionary wiring",
     Philips Tech. Rev., 31, 286-95 (1970), EEA74-34117.
828. Bhandarkar, D. P., "Some performance issues in multiprocessor
     system design", IEEE Trans. Comput., C-26 (5), 506-11 (1977).
829. Boehm, R. F. and Van Bogelen, D. W., "Data processing memory
     system with bidirectional data bus", Patent USA 3795901, Publ.
     March 1974, CCA9-23258.
830. Boll, H. J. and Lynch, W. T., "Design of a high-performance
     1024-B switched capacitor p-channel IGFET memory chip", IEEE J.
     Solid-State Circuits, SC-8 (5), 310-18 (1973), EEA76-35355.
831. Boonstra, L., Lambrechtse, C. W. and Salters, R. H. W., "Design
     considerations for a 4K 1 MOS/BIT RAM", Onde Electr., 54 (4),
     175-80 (1974), French, EEA77-26523.
832. Boysel, L. L. and Carter, G. P., "MOS complex array system
     design", Electro-Technology, 83 (2), 35-7 (1969), EEA72-12803.
833. Brown, M. A. C. S., Bounden, J. E. and Vanstone, G. F., "A simple
     equivalent circuit for the MNOST memory element", Solid State
     Electron., 15, 707-19 (1972), EEA75-24367.
834. Bryant, R. W., Tu, G. K., Kwei, T. C. and Robinson, R. H., "A
     high-performance LSI memory system", Comput. Des., 9 (7), 71-7
     (1970), CCA5-22438.
835. Burns, J. R., "Memory subsystem array", Patent USA 3701984,
     Publ. October 1972, CCA8-9545.
836. Capece, R. P., "Memory-oriented designs maximize throughput",
     Electronics, 50 (22), 104-10 (1977).
837. Card, H. C. and Elmasry, M. I., "Functional modelling of non-
     volatile MOS memory devices", Solid-State Electron., 19 (10),
     863-70 (1976), CCA11-28287.
838. Chang, D. Y., Kuck, D. J. and Lawrie, D. H., "On the effective
     bandwidth of parallel memories", IEEE Trans. Comput., C-26 (5),
     480-9 (1977).
839. Cricchi, J. R., Blaha, F. C., Fitzpatrick, M. D. and Sciulli,
     F. M., "Space charge effects in MNOS memory devices and endur-
     ance measurements", 1975 International Electron Devices Meeting,
     Technical Digest, p. 459-62.
840. Critchlow, D. L., Dennard, R. H. and Schuster, S. E., "Design
     and characteristics of n-channel insulated-gate field-effect
     transistors", IBM J. Res. & Dev., 17 (5), 430-42 (1973),
     CCA9-11963.
841. Dunbridge, B., Miller, C. S. and Tsou, H. S., "Building block
     approach to digital signal processing", IEEE Electron. and
     Aerosp. Syst. Conv. (EASCON '74), Rec., 1974, p. 469-76.
842. Ferris-Prabhu, A. V., Lubart, N. D. and Medve, T. J., "Evalua-
     tion of an integral in the theory of nonvolatile semiconductor
     memories", Solid-State Electron., 20 (1), 74-5 (1977), EEA80-6269.

843. Fisher, E. R., "Expand a system's memory capacity without mounting hardware and board space", Electron. Des., 20 (1), 74-5 (1977), CCA11-15063.

844. Frankenberg, R. J. and Cross, D., "Designer's guide to: Semiconductor memories - 8", EDN, 20 (21), 127-37 (1975).

845. Franson, P., "Semiconductors and IC's - new technologies and clever designers promise to solve your every problem", EDN, 21 (13), 20-4 (1976), EEA80-3005.

846. General Instrument Corp., "Memory", Patent UK 1296066, Publ. November 1972, CCA8-13698.

847. Goldberg, J., Levitt, K. N. and Wensley, J. H., "An organization for a highly survivable memory", IEEE Trans. Comput., C-23 (7), 693-705 (1974), CCA9-18399.

848. Hampel, D., "Design and application of electronically programmable LSI arrays", Proceedings, National Computer Conference, 1975, p. 867-76.

849. Hanratty, J. J., "Model RAM's automatically", Electron. Des., 25 (17), 72-4 (1977).

850. Hayashi, Y., Sekigawa, T. and Tarui, Y., "Design and fabrication of ECL-IC", Bull. Electrotech. Lab., 33 (6), 603-10 (1969), Japanese, CCA5-8294.

851. Heightley, J. D. and Waaben, S. G., "Two-terminal dual pnp transistor semiconductor memory", Patent USA 3693170, Publ. September 1972, CCA8-7353.

852. Hilberg, W., "Extending the measure of memory capacity to general digital circuits", Elektron. Rechenanlagen, 18 (3), 111-13 (1976), German, CCA11-22634.

853. Hollock, S. and Clark, A., "Bipolar LSI in high speed systems", New Electron., 9 (7), 14, 19, 22 (1976), CCA11-17353.

854. Howker, J. K. and Procter, B. J., "Systems and the semiconductor memory - the impact and the firm but gentle push", Colloquium Digest on Computer Memories, 1973, p. 1, CCA8-11649.

855. Hughes, K. L., "Application of a new isolation technique to memory circuits", 4th International Microelectronics Conference, 1970, p. 19, CCA6-9323.

856. IBM Corp., "Monolithic data store", Patent UK 1280924, Publ. July 1972, CCA8-2669.

857. IBM Corp., "Transistor memory circuit", Patent UK 1292355, Publ. October 1972, CCA8-5280.

858. IBM Corp., "Bipolar integrated circuit memory", Patent UK 1407847, Publ. September 1975, CCA11-6388.

859. Kaneko, T., "Optimal task switching policy for a multilevel storage system", IBM J. Res. & Dev., 18 (4), 310-15 (1974), CCA9-25086.

860. Kelly, H. J., Kolchak, G. M. and Scott, T. R., "An application of circuit optimization to an FET ROS design", IEEE Trans. Circuits & Syst., CAS-23 (11), 688-90 (1976), CCA11-31616.

861. Kroy, W., Manhart, S. and Mehnert, W. T., "Three-dimensional storage system", Patent USA 3693169, Publ. September 1972, CCA8-2656.

862. Levine, L. and Myers, W., "Timing: a crucial factor in LSI-MOS main-memory design", Electronics, $\underline{48}$ (14), 107-11 (1975), EEA78-33197.

863. Levine, L. and Myers, W., "Design of fault tolerant MOS memories", 11th IEEE Computer Society Conference, COMPCON 75, Digest of Papers, p. 97-100.

864. Liu, P. S. and Mowle, F. J., "Suggestions for efficient utilization of a writable control memory", Proceedings of the IEEE Milwaukee Symposium on Automatic Computation and Control, 1976, p. 155-9, CCA11-28210.

865. Lloyd, R. H. F., "The case for using partially good memory devices", Comput. Des., $\underline{16}$ (4), 93-6 (1977).

866. McCarthy, C. E., Carter, W. C. and White, J. B., "A memory system design which can tolerate multiple storage array faults", Proceedings of the 7th Annual Southeastern Symposium on System Theory, 1975, p. 172-8, CCA11-28269.

867. Maison, F. P., "Logical and technological criteria for LSI-MOS computer partitioning", Proceedings of the International Conference on Advanced Microelectronics, 1970, 2 pp., CCA6-2020.

868. Meusberger, G., "Increased storage time of a single-transistor memory cell", Electron. Lett., $\underline{12}$ (6), 140-1 (1976), German, EEA79-20525.

869. Mouftah, H. T. and Jordan, I. B., "Design of ternary COS/MOS memory and sequential circuits", IEEE Trans. Comput., C-$\underline{26}$ (3), 281-8 (1977).

870. National Cash Register Co., "Transistor memory circuit", Patent UK 1293711, Publ. October 1972, CCA8-9546.

871. Nguyen-Huu, A., "Large bipolar memory arrays", Microelectronics, $\underline{3}$ (5), 30-5 (1970), CCA5-18343.

872. Nutt, G. J., "Memory and bus conflict in an array processor", IEEE Trans Comput., C-$\underline{26}$ (6), 514-21 (1977), CCA12-15443.

873. Onishchenko, Y. M., Pershchenkov, V. S. and Kimarskii, V. I., "Design optimisaticn of an operational integrated circuit memory", Izv. VUZ Radioelektron., $\underline{15}$ (7), 877-85 (1972), Russian, CCA8-2664.

874. Patel, S. and Barry, J. N., "Custom-designed m.o.s. arrays for use in digital systems", Electron. Eng., $\underline{44}$ (536), 64-6 (1972), EEA75-39091.

875. Polkinghorn, R. W., "System of aspects of MOS memory devices", Semiconductor Integrated and Production Conference and Exhibition (abstracts), 1971, p. 11, CCA6-15249.

876. Pooley, D., "Application of radiation damage in information storage", Radiation Damage Processes in Materials, 1973, p. 517-34, Publ. 1975, CCA11-4827.

877. Poppendieck, M. and Desautels, E. J., "Memory extension techniques for minicomputers", Computer, $\underline{10}$ (5), 68-75 (1977).

878. Ruiz-Medrano, C. C., "Memory units and sequential circuits", Metal. & Electr., $\underline{38}$ (440), 149-52 (1974), Spanish, CCA9-25007.

879. Shenton, G., "A systems approach to the design of MOS memory components", Microelectron. & Reliab., 13 (5), 419-24 (1974), EEA78-5369.

880. Torre, E. D. and Roitman, J., "An array of computing memory cells", Proceedings of the 1973 Sagamore Computer Conference on Parallel Processing, p. 102, CCA9-18414.

881. van Beek, H. W., "MOS/LSI circuit design: designing-in reliability", Proceedings of the 10th Annual Conference on Reliability Physics, 1972, p. 36-41, EEA76-4694.

882. Virant, J., "Logical analysis of master-slave memory cells", Elektrotech. Vestn., 43 (3), 109-15 (1976), CCA12-8895.

883. Washburn, J., "Battery backup for minicomputer semiconductor memories", Comput. Des., 16 (4), 108-12 (1977).

884. Ypma, J. E., Gergis, I. S. and Archer, J. L., "64K fast access chip design", AIP Conf. Proc., no. 29, 51-3 (1976), CCA11-22649.

VIII. TESTING

885. Alonso, R. L., "Strategies of memory testing", 1973 Symposium on Semiconductor Memory Testing, Digest of Papers, p. 109-25.

886. Anderson, K., "Device manufacturers test problems", 1972 Symposium on Semiconductor Memory Testing, Digest of Papers, p. 17-26.

887. Anon., "Military tests BORAM module as replacement for discs and drums", Comput. Des., 15 (5), 76, 82 (1976), CCA11-22659.

888. Anon., "V-MOS process gets first test in new memories", Electronics, 50 (13), 29-30 (1977).

889. Balasubramanian, P. S., Grazier, E., Henke, J. D. and Myers, M. J., "Testing LSI memory arrays using on-chip I/O shift register latches", IBM Tech. Disclosure Bull., 17, 2019-20 (1974), EEA78-15409.

890. Barber, M. R., "Techniques for dynamically testing semiconductor memory chips", 1972 Symposium on Semiconductor Memory Testing, Digest of Papers, p. 3-16.

891. Barber, M. R. and Zacharias, A., "Integrated circuit testing", Bell Lab. Record, 55 (5), 124-30 (1977).

892. Barraclough, W., Chiang, A. C. L. and Sohl, W., "Techniques for testing the microcomputer family", Proc. IEEE, 64 (6), 943-50 (1976).

893. Beaston, J., Domenik, S. and Richardson, W., "Development of a testing strategy for a 16 kilobit RAM", 1976 Semiconductor Test Symposium, Digest of Papers, p. 1-2, EEA80-3041.

894. Bladowski, R., "Application and testing of complex integrated circuits", Elektron. Anz., 8 (11), 272-4 (1976), EEA80-9493.

895. Brown, J. R., Jr., "Pattern sensitivity in MOS memories", 1972 Symposium on Semiconductor Memory Testing, Digest of Papers, p. 33-46, EEA76-26129.

896. Campbell, J. F., Jr., "A new real-time function test generation
     system for complex LSI testing", IEEE Trans. Manuf. Technol.,
     MFT-4, 52-5 (1975).
897. Carter, W. C. and McCarthy, C. E., "Implementation of an experi-
     mental fault-tolerant memory system", IEEE Trans. Comput.,
     C-25 (6), 557-68 (1976), CCA11-28270.
898. Cocke, J., "Read-and-test to reduce redundancy requirements in
     memories", IBM Tech. Disclosure Bull., 15 (3), 855-6 (1972),
     CCA8-2667.
899. Cocking, J., "RAM test patterns and test stragedy", 1975 Semi-
     conductor Test Symposium, Digest of Papers, p. 1-8.
900. Cozzi, L., "A universal MOS-LSI testing system", Electron. Prod.
     Methods & Equip., 4 (5), 9, 11 (1975), EEA78-37610.
901. de Jonge, J. H. and Smulders, A. J., "Moving inversions test
     pattern is thorough, yet speedy", Comput. Des., 15 (5), 169-73
     (1976).
902. Dumitrescu, D. and Saucier, G., "Testing an MOS dynamic memory",
     1975 International Symposium on Fault-Tolerant Computing, Digest
     of Papers, p. 66-71, French, CCA11-28271.
903. Evans, W. H., "Testing IC memory devices - the user's viewpoint",
     1975 International Microelectronics Conference, Proceedings of
     the Technical Program.
904. Feldmann, D. and Healy, J. E., "Problems with sample- and 'goods-
     in' testing of semi-conductor stores (memories)", Elektro-Anz.,
     29 (18), 429-32 (1976), German, CCA12-1509.
905. Fischer, J. E., "4K RAM's: Increased densities bring difficult
     testing problems", EDN, 19 (22), 47-51 (1974).
906. Fitzpatrick, J. F., "Today's three test environments", 1972
     Symposium on Semiconductor Memory Testing, Digest of Papers,
     p. 47-59.
907. Foss, R. C. and Harland, R., "MOS dynamic RAM - design for test-
     ability", 1976 Semiconductor Test Symposium, Digest of Papers,
     p. 9-12, EEA80-3043.
908. Foster, R. C., "Why consider screening, burn-in, and 100-percent
     testing for commercial devices?", IEEE Trans. Manuf. Technol.,
     MFT-5 (3), 52-8 (1976).
909. Fulcomer, E. J. and Verma, S. P., "Modelling and test generation
     for a custom LSI chip", 1976 Semiconductor Test Symposium, Digest
     of Papers, p. 41-4, EEA80-3049.
910. Ganesan, S., "Testing unit for memory integrated circuits",
     Electro-Technol., 19 (1), 19-24 (1975), EEA79-13012.
911. Gillow, G., "Simplifying processor maintenance with a carefully
     designed maintenance panel", Comput. Des., 14 (7), 95-6, 99
     (1975).
912. Goldblatt, R. C., "How computers can test their own memories",
     Comput. Des., 15 (7), 69-75 (1976), CCA11-31586.

913. Gonzales, A. J. and Powell, M. W., "Internal waveform measurements of the MOS three-transistor, dynamic RAM using S.E.M. stroboscopic techniques", 1975 International Electron Devices Meeting, Technical Digest, p. 119-22, EEA79-37707.

914. Grimm, A. V., "LSI integrated circuit test development", Report BDX-613-1017, Bendix Corp., Kansas City, Mo., 1974, 62 pp., EEA78-24509.

915. Healy, J., "Testing 1k, 4k and 16k RAMs", Electron. Ind., $\underline{3}$ (2), 24-5, 27 (1977), CCA12-15504.

916. Hnatek, E. R., "4-kilobit memories present a challenge to testing", Comput. Des., $\underline{14}$ (5), 117-25 (1975).

917. Hnatek, E. R., "Test results from screening 4000 series CMOS IC's", Eval. Eng., 14 (3), 20-1 (1975), EEA78-33167.

918. Hnatek, E. R. and Schmitt, R. G., "A 2048 bit RAM characterization program", 1975 Semiconductor Test Symposium, Digest of Papers, p. 23-9.

919. Hodge, W. A. and Flaningam, D. H., "What you need to know about testing 4k MOS memories", Tekscope, $\underline{6}$ (5), 3-7 (1974), EEA78-41699.

920. Huston, R. E., "Testing semiconductor memories", 1973 Symposium on Semiconductor Memory Testing, Digest of Papers, p. 27-62.

921. Jackson, K. and Sear, B., "Need for time testing in memories and LSI circuits", 1974 Semiconductor Test Symposium: Memory and LSI, Digest of Papers, p. 7-32.

922. Jackson, K. and Sear, B. E., "Measurement techniques for testing high speed memories", Comput. Des., $\underline{14}$ (5), 127-32 (1975).

923. Johnson, B., "Semiconductor memory testing", New Electron., $\underline{9}$ (21), 17-18 (1976), EEA80-3022.

924. King, J. L., "Method for testing MOS memory store device", Patent USA 3813032, Publ. May 1974.

925. Knaizuk, J., Jr. and Hartmann, C. R. P., "An algorithm for testing random access memories", IEEE Trans. Comput., C-$\underline{26}$ (4), 414-16 (1977), CCA12-10682.

926. Knowlton, D., "Testing problems associated with a 4096 bit MOS memory", 1975 International Microelectronics Conference, Proceedings of the Technical Program, p. 8-15.

927. Kodama, K. and Kasai, Y., "Test methods for some failure modes of 1K MOS RAM", 1974 International Electron Devices Meeting, Technical Digest, p. 104-7.

928. Koplowitz, J., "Necessary and sufficient memory size for m-hypothesis testing", IEEE Trans. Inf. Theory, IT-$\underline{21}$ (1), 44-6 (1975).

929. Lattin, B., "Testing dynamic RAM's", 1971 WESCON Technical Papers, Paper 3/4, 5 pp., EEA75-1183.

930. Marshall, M., "Through the memory cells - further explorations of IC's in testingland", EDN, $\underline{21}$ (4), 77-85 (1976).

931. Merryman, V. A., "Memories that forget", 1973 Symposium on Semiconductor Memory Testing, Digest of Papers, p. 77-91.

932. Mitchell, T., "Problems encountered in testing bipolar semicon-
     ductor memories at the chip and card level", 1972 Symposium on
     Semiconductor Memory Testing, Digest of Papers, p. 27-31.
933. Morgan, M. K. M., "New patterns for testing 4K RAMs", Electron,
     no. 89, 12, 14 (1976), CCA11-15066.
934. Morgan, M. K. M., "A new approach to memory testing", Micro-
     electron. & Reliab., 15 (4), 351-3 (1976), EEA79-46438.
935. Mow, W. C. W., "Systems approach to device testing", 1972 Sym-
     posium on Semiconductor Memory Testing, Digest of Papers,
     p. 61-81.
936. Mow, W. C. W. and Chiang, A. C. L., "20 MHz semiconductor mem-
     ory test system", 1973 Symposium on Semiconductor Memory Test-
     ing, Digest of Papers, p. 93-107.
937. Muller, K., "The strategy of the electrical functional test of
     integrated semiconductor stores", Z. Elektr. Inf. - & Energie-
     tech., 5, 137-43 (1975), German, EEA78-37565.
938. O'Brien, T., "Tester requirements for CMOS LSI", 1975 Semicon-
     ductor Test Symposium, Digest of Papers, p. 92-5.
939. Owen, R. W., "When testing 16-k dynamic RAM's", Electron. Des.,
     25 (19), 86-9 (1977).
940. Palmquist, S. and Chapman, D., "Expanding the boundaries of LSI
     testing with an advanced pattern controller", 1976 Semiconductor
     Test Symposium, Digest of Papers, p. 70-5, EEA80-3052.
941. Palphi, T., "Dynamic memories", 1973 Symposium on Semiconductor
     Memory Testing, Digest of Papers, p. 1-6.
942. Panigrahi, G., Woo, B. and Chu, B., "Charge coupled memory test
     philosophy", 1975 Semiconductor Test Symposium, Digest of
     Papers, p. 9-18.
943. Perlmutter, D., "Handle complex testing with word generators",
     Electron. Prod., 17 (12), 51-3, 55 (1975).
944. Pound, A. E., "Designing MOS for maximum testability", 1971
     WESCON Technical Papers, Paper 3/2, 3 pp., EEA75-436.
945. Reitan, A., "Testing of LSI components", Elektro, 89 (21), 13,
     15 (1976), Norwegian, EEA80-9495.
946. Richardson, W. S., "Diagnostic testing of MOS random access
     memories", 1973 Symposium on Semiconductor Memory Testing,
     Digest of Papers, p. 7-20.
947. Richardson, W. S., "Diagnostic testing of MOS random access
     memories", Solid State Technol., 18 (3), 31-4 (1975).
948. Routh, W. S., "Relationship of MOS circuit properties to LSI
     testing", Solid State Technol., 15 (3), 41-5 (1972), EEA75-27664.
949. Schuermeyer, F. L., "Test results on an MNOS memory array",
     IEEE Trans. Electron Devices, ED-24 (5), 564-8 (1977),
     EEA80-21017.
950. Seltzer, R., "Test strategies for LSI and memories", Circuit
     Manuf., 17 (3), 40-6 (1977).
951. Shaw, A. A., "Design and application of semiconductor memory
     test hardware", GTE Autom. Electr. Tech. J., 15 (2), 65-79 (1976).

952. Sklyarevich, A. N., "Conditional checking capabilities of memory
     element input arrays", Autom. Control Comput., $\underline{9}$ (5), 31-7 (1975).
953. Srini, V. P., "API tests for RAM chips", Computer, $\underline{10}$ (7),
     32-5 (1977).
954. Williams, B., "L.S.I. automatic test equipment applied to
     dynamic microprocessor testing", New Electron., $\underline{7}$ (21), 50-2
     (1974), EEA78-2912.
955. Zimmer, B. A., "Test techniques for circuit boards containing
     large memories and microprocessors", 1976 Semiconductor Test
     Symposium, Digest of Papers, p. 16-21, EEA80-3045.

IX. RELIABILITY

956. Anon., "Cost, performance, reliability tradeoffs grow with
     increased (semiconductor) memory-system use", Digital Des.,
     $\underline{6}$ (12), 10 (1976), CCA12-15502.
957. Anon., "Switch to conductive polyolefin PCB racks eliminates
     CMOS/FET failures due to static electricity", Insul./Circuits,
     $\underline{23}$ (1), 25-6 (1977), EEA80-9521.
958. Baranyai, A., Kerek, L. and Barsony, I., "Experimental method
     for non-destructive analysis of MOS IC's", Hiradstech. Ipari
     Kut. Intez. Kozl., $\underline{14}$, 16-37 (1974), Hungarian, EEA77-38880.
959. Barnes, D. E. and Thomas, J. E., "Reliability assessment of a
     semiconductor memory by design analysis", 12th Annual Reliability
     Physics Symposium, Proceedings, 1974, p. 74-81.
960. Barrett, C. R. and Smith, R. C., "Failure modes and reliability
     of dynamic RAMs", 1976 International Electron Devices Meeting,
     Technical Digest, p. 319-22, EEA80-17279.
961. Basham, G. R., "New error-correcting technique for solid-state
     memories saves hardware", Comput. Des., $\underline{15}$ (10), 1110-13 (1976),
     CCA12-4010.
962. Beattie, C. G., Adams, J. D., Carrell, S. L., George, T. D.
     and Valek, M. H., "Elements of semiconductor device reliability",
     Proc. IEEE, $\underline{62}$ (2), 149-68 (1974), EEA77-9969.
963. Behera, S. K. and Speer, D. P., "A procedure for the evaluation
     and failure analysis of m.o.s. memory circuits using the scan-
     ning electron microscope in potential contrast mode", Proceed-
     ings of the 10th Annual Reliability Physics Conference, 1972,
     p. 5-11, EEA76-4250.
964. Brooks, J., "Error correcting stores for a small computer", New
     Electron., $\underline{9}$ (19), 30, 34 (1976), CCA11-31620.
965. Brucker, G. J. and Parson, B. R., "Radiation test and simulation
     of CMOS/SOS/Si-gate ALU and ROM devices", IEEE Trans. Nucl.
     Sci., NS-$\underline{23}$ (6), 1720-7 (1976), CCA12-8832.
966. Cheney, G. T., Freyman, R. L. and Mammele, A. A., "Reliability
     of $Al_2O_3SiO_2$ IGFET integrated circuits", Proceedings of the 9th
     Annual Reliability Physics Conference, 1971, p. 62-6, EEA75-3982.

967. Child, J. W. M. and de Voil, G. F., "Destruction of m.o.s.
     device by static", New Electron., 7 (5), 16-17 (1974),
     EEA77-26483.
968. Clelford, A. P. and Payne, J. R., "Package development for a
     large multi-chip array", Electron. Packag. Prod., 14 (9), 47,
     50, 52, 54 (1974).
969. Colbourne, E. D., Coverley, G. P. and Behera, S. K., "Reli-
     ability of MOS LSI circuits", Proc. IEEE, 62 (2), 244-59
     (1974), EEA77-9705.
970. Elmer, B., Tchon, W. E., Denboer, A. J., Frommer, R., Hira-
     bayashi, K., Kohyama, S. and Nojima, I., "Fault tolerant 92160-
     bit multiphase CCD memory", 1977 International Solid-State Cir-
     cuits Conference, Digest of Technical Papers, p. 116-17.
971. Feeleston, W. and Pepper, M., "Modes of failure of MOS devices
     (in integrated circuits)", Microelectron. & Rel., 10, 32-8
     (1971), EEA75-4708.
972. Fitzgerald, D. J., Parker, G. H. and Spiegel, P., "Reliability
     studies of MOS Si-gate arrays", Proceedings of the 9th Annual
     Reliability Physics Conference, 1971, p. 57-61, EEA75-4724.
973. Forsythe, D. D., "Surface-charge induced failures observed on
     MOS integrated circuits", Microelectron. & Rel., 8 (4), 339-47
     (1969), EEA73-14319.
974. Franklin, P., "A reliability assessment of bipolar PROMs", 14th
     Annual Proceedings Reliability Physics Conference, 1976,
     p. 201-18, CCA12-15523.
975. Franklin, P. and Burgess, D., "Reliability aspects of nichrome
     fusible link PROM's", Proceedings of the 12th Annual Reliability
     Physics Sympoisum, 1974, p. 82-6.
976. Fujiwara, E., "Reliability of main memories", Inf. Process.
     Soc. Jap. (joho Shori), 16 (4), 295-304 (1975), Japanese,
     CCA10-25827.
977. Gear, G., "FAMOS PROM reliability studies", 14th Annual Pro-
     ceedings Reliability Physics Conference, 1976, p. 198-201,
     CCA12-15521.
978. Horiuchi, M. and Itoh, Y., "Fatigue phenomena of FTMIS memory
     transistors", IEEE Trans. Electron Devices, ED-24 (5), 587-90
     (1977), CCA12-15509.
979. Jeppson, K. O., "Loss of memory: only one memory", Te. Tidskr.,
     107 (5), 45, 47, 51 (1977), Swedish, EEA80-21019.
980. Kenney, G. B., Kinzy, W. and Ogilvie, R. E., "Fusing mechanism
     of nichrome-linked programmable read-only memory devices", 14th
     Annual Proceedings Reliability Physics Conference, 1976,
     p. 164-72, CCA12-15539.
981. Kodama, K. and Kasai, Y., "Test methods for some failure modes
     of 1K MOS RAM", 1974 International Electron Devices Meeting,
     Technical Digest, p. 104-7.
982. Kostychev, G. I., "Physical aspects of the reliability of inte-
     grated microcircuits employing MIS transistors", Telecommun. &
     Radio Eng. Pt. 2, 29, 104-8 (1974), EEA78-37656.

983. Krause, D., "The reliability analysis of semiconductor memories",
     Wiss. Ber. AEG-Telefunken, 48 (5), 197-212 (1975), EEA79-24922.
984. Kuo, C., Kitagawa, N., Ogden, D. and Hewkin, J., "16-k RAM
     built with proven process may offer high start-up reliability",
     Electronics, 49 (10), 81-6 (1976), CCA11-17403.
985. Lattin, W. and DeMassa, T., "Geometric and temperature compen-
     sating effects of the MOS enhanced capacitor", 1974 Interna-
     tional Electron Devices Meeting, Technical Digest, p. 50-2.
986. Levine, L. and Meyers, W., "Semiconductor memory reliability
     with error detecting and correcting codes", Computer, 9 (10),
     43-50 (1976), CCA12-6667.
987. Levine, S., "Performance and reliability in large systems - the
     best PR", 8th IEEE Computer Society International Conference,
     COMPCON 74, Digest of Papers, p. 23-5.
988. Lindner, R. and Otto, J., "On-wafer failure analysis of LSI-MOS
     memory circuits by scanning electron microscopy", Siemens
     Forsch. - & Entwicklungsber., 6 (1), 39-46 (1977), CCA12-6672.
989. Lindwedel, J. H., "LSI reliability screen developments", 1972
     IEEE Region Six (US Western region) Conference Record, p. 129-31,
     CCA8-2690.
990. Majumder, D. D., Das, J. and Chatterjee, N., "On the properties
     of error correcting codes for improving the computer's memory
     reliability", J. Inst. Electron Telecommun. Eng., 21 (5),
     253-8 (1975).
991. Maloney, M. J. and Murahashi, S., "Design considerations for
     ruggedized memories", Comput. Des., 13 (5), 114-16 (1974).
992. Marques, A. M. and Partridge, J., "Progress report on nichrome
     link PROM reliability studies", 14th Annual Proceedings Reli-
     ability Physics Conference, 1976, p. 182-91, CCA12-15541.
993. Moeenuddin, S. K., "Semiconductor memory maintenance techniques
     applicable to on-line system failure detection", GET Autom.
     Electr. Tech. J., 15 (2), 80-8 (1976).
994. Nichols, E. D., "MOS/LSI reliability prediction", Proceedings
     of the 1972 Reliability and Maintainability Symposium,
     p. 474-8, EEA75-15993.
995. Palfi, T. L., "MOS memory system reliability", 1975 Semicon-
     ductor Test Symposium, Digest of Papers, p. 37-46.
996. Pinon, M., "Qualification and supervision of custom design MOS
     LSI production line", International Conference on Large Scale
     Integrated Circuits, 1974, p. 159-66, French, EEA78-11751.
997. Piwczyk, B. and Siu, W., "Specialized scanning electron micro-
     scopy voltage contrast techniques for LSI failure analysis",
     12th Annual Proceedings, Reliability Physics Conference, 1974,
     p. 49-53.
998. Raisanen, W., Hillis, D., Marley, J. and Trolsen, G., "Large
     scale integrated circuit bipolar memories (MOS)", Onde Elec.,
     50 (9), 763-7 (1970), French, EEA74-4917.

999. Rao, T. R. N. and Chawla, A. S., "Asymmetric error codes for some LSI semiconductors memories", Proceedings of the 7th Annual Southeastern Symposium on System Theory, 1975, p. 170-1, CCA11-28307.

1000. Reynolds, F. H., Parrott, R. W. and Braithwaite, D., "Use of tests at elevated temperatures to accelerate the life of an MOS integrated circuit", Proc. Inst. Elec. Eng., 118 (3-4), 475-85 (1971), EEA74-18808.

1001. Ross, E. C., Goodman, A. M. and Duffy, M. T., "The effects of SiO_2 thickness variation on the operation of direct-tunneling mode m.n.o.s. memory transistors", 8th Annual Reliability Physics Symposium, 1970, p. 6, EEA73-28225.

1002. Schlegel, E. S., Keen, R. S. and Schnable, G. L., "The effects of insulator surface-ion migration on MOS and bipolar integrated circuits", 8th Annual Reliability Physics Symposium, 1970, p. 4-5, EEA73-25055.

1003. Schnable, G. L., Ewald, H. J. and Schlegel, E. S., "MOS integrated circuit reliability", IEEE Trans. Rel., R-21, 12-19 (1972), EEA75-9569.

1004. Smith, R. C., Rosenberg, S. J. and Barrett, C. R., "Reliability studies of polysilicon fusible link PROM's", 14th Annual Proceedings Reliability Physics Conference, 1976, p. 193-7, CCA12-15520.

1005. Spivak, H. A., Hawkins, L. E. and Nasuti, A. J., "An MOS-LSI memory reliability program for a commercial computer application", Proceedings of the 1971 Annual Symposium on Reliability, p. 331-7, EEA74-18818.

1006. Steventon, A. G., "Some aspects of reliability in amorphous chalcogenide memory switches", J. Phys. D., 8 (15), 1869-81 (1975).

1007. Szep, I. C. and Tihanyi, J., "Low-cost fabrication of MOS-IC's", Proceedings of the International Conference on Advanced Microelectronics, 1970, 1 pp., EEA74-1616.

1008. Tanaka, T. and Nishi, Y., "The nature and mechanism of degradation in MNOS read write memory", Oyo Buturi, 44 (Suppl.), 203-9 (1975), EEA78-41615.

1009. Topich, J. A. and Yon, E. T., "The effects of high temperature annealing of MNOS devices", J. Electrochem. Soc., 123 (4), 535-9 (1976), CCA11-17407.

1010. Toschi, E. A. and Watanabe, T., "An all-semiconductor memory with fault detection, correction, and logging", Hewlett-Packard J., 27 (12), 8-13 (1976), CCA11-28282.

1011. Vincoff, M. N. and Schnable, G. L., "Relability of complementary MOS integrated circuits", IEEE Trans. Reliab., R-24 (4), 255-60 (1975), EEA78-37570.

1012. Wolfgang, E., Otto, J. and Kantz, D., "Use of scanning elec-
      tron microscope for stroboscopic voltage contrast studies for
      electrical failure analysis of dynamic 4096-bit MOS memory
      devices", Siemens Forsch. - & Entwicklungsber., 5 (1), 33-8
      (1976), CCA11-9686.
1013. Wolfgang, E., Otto, J., Kantz, D. and Lindner, R., "Strobo-
      scopic voltage contrast of dynamic 4096 bit MOS RAMs: failure
      analysis and function testing (in SEM)", Scanning Electron
      Microscopy 1976. I., p. 625-32, EEA79-46481.
1014. Woods, M. H. and Tuska, J. W., "Degradation of M.N.O.S. memory
      transistor characteristics and failure mechanism model", Pro-
      ceedings of the 10th Annual Conference on Reliability Physics,
      1972, p. 120-5, EEA76-4675.

X.   FUTURE TRENDS

1015. Alewijnse, C. P. J., "Memory requirements for future computer
      systems", Onde Electr., 54 (6), 277-80 (1974), CCA10-20747.
1016. Allan, R., "Semiconductor memories", IEEE Spectrum, 12 (8),
      40-5 (1975), CCA11-6381.
1017. Allan, R., "Amorphous semiconductors revisited", IEEE Spectrum,
      14 (5), 41-5 (1977).
1018. Allison, J. and Thompson, M. J., "Amorphous semiconductor
      devices and components", Radio & Electron. Eng., 40 (1), 11-22
      (1976).
1019. Altman, L. and Cohen, C. L., "The gathering wave of Japanese
      technology", Electronics, 50 (12), 99-122 (1977).
1020. Anon., "Computer mainframe demand 1975-1980 - analysis of the
      market for general purpose computer systems and memory utili-
      zation", Report IDC-1612, International Data Corp., Waltham,
      Mass., 1975, 101 pp.
1021. Anon., "The trends in semiconductors", Toute Electron., no.
      414, 37-41 (1976), French, EEA80-3370.
1022. Anon., "Semiconductor forecast: study tech growth", Circuits
      Manufac., 17 (2), 8, 10 (1977).
1023. Anon., "VLSI - new technologies for future digital integrated
      circuits", Nachrichtentech. Z. (NTZ), 30 (3), 212-16 (1977),
      German, EEA80-32461.
1024. Anon., "Magnetic bubble memories: past, present and future",
      Digital Des., 7 (5), 50-66 (1977).
1025. Anon., "Computer memory technology: A state-of-the-art and
      patent activity analysis", Digital Des., 1 (8), 22-34 (1977).
1026. Avery, L. R., "Mixed-discipline monolithic integrated cir-
      cuits - bipolar, MOS and COS/MOS technologies and future
      trends", Microelectron. & Reliab., 13 (5), 249-61 (1974).
1027. Avery, L. R., "Present technologies and future trends in inte-
      grated circuits", Mundo Electron., no. 56, 98-108 (1976),
      Spanish, EEA80-3526.

1028. Berlekamp, E. R., Garwin, R. L., Knuth, D. E., Lederberg, J. and Leibler, R. A., "Report of the ARPA study group on advanced memory concepts", Report AD-A021274/6SL, Science Applications Inc., Arlington, Va., 1976, 54 pp.

1029. Bernstein, H., "Programmable logic array (PLA) - a new type of memory", Elektron. Ind., 7 (1-2), 2-5 (1976), German, CCA11-17342.

1030. Bloch, E. and Galage, D., "Component progress: Its effect on high-speed computer architecture and machine organization", Presented at Symposium on High-speed Computer and Algorithm Organization, University of Illinois, April 1977.

1031. Bradford, P. V., "Semiconductor memories as seen from Wall Street", Mod. Data, 8 (9), 46-7, 49-53 (1975), CCA11-9683.

1032. Branscomb, L. M., "Promising areas of research in computer science and technology", Proceedings of the 2nd USA-Japan Computer Conference, AFIPS & IPST, 1975, p. 1-7.

1033. Bremer, J. W., "Hardware technology in the year 2001", Computer, 9 (12), 31-6 (1976).

1034. Bremermann, H. J., "The challenges of biocybernetics and artificial intelligence", IEEE Trans. Syst. Man Cybernetics, SMC-5 (3), 362-5 (1975).

1035. Brown, A., "The next 20 years in microelectronics", New Sci., & Sci. J., 53 (787), 590-3 (1972), CCA7-14652.

1036. Brown, R. J., Jr., "The status and outlook of semiconductor memories", 1976 International Solid-State Circuits Conference, Digest of Technical Papers, p. 170, CCA11-28297.

1037. Dance, M., "Digital devices", Electron. Ind., 2 (11), 31-3, 35 (1970), EEA80-9015.

1038. Davis, S., "Selection and applications of semiconductor memories", Comput. Des., 13 (1), 65-77 (1974), CCA9-9212.

1039. Evans, J., "Digital devices", Electron. Ind., 1 (2), 31-2, 34-5, 37 (1975), EEA79-1045.

1040. Ferris-Prabhu, A. V., "Future memories", Presented at IEEE-India Sect, Annual Convention, Bombay, India, 1973, 9 pp.

1041. Feth, G. C., "Memories: smaller, faster, and cheaper", IEEE Spectrum, 13 (6), 36-43 (1976), CCA11-28267.

1042. Frohman-Bentchkowsky, D., "Trends in memory technologies", Proceedings of the 6th International Congress on Microelectronics, 1974, 3 pp., EEA78-24513.

1043. Greenblott, B. J. and Hsiao, M. Y., "Where is technology taking us in data processing systems?", Proceedings of the National Computer Conference, Vol. 44, 1975, p. 623-8.

1044. Haddad, J. A., "Implications of LSI", 1974 International Solid-State Circuits Conference, Digest of Technical Papers, p. 50-1.

1045. Hobbs, L. C., "A look at the future", Computer, 9 (12), 9-10 (1976).

1046. Hodges, D. A., "Computer memories", IEEE Stud. J., 8 (4), 15-20 (1970), CCA6-3754.

1047. Hodges, D. A., "Alternative component technologies for advanced memory systems", Computer, $\underline{6}$ (9), 35-7 (1973), CCA9-3384.
1048. Hodges, D. A., "The future of integrated storage technology", 8th IEEE Computer Society International Conference, Digest of Papers, 1974, p. 291-2.
1049. Hodges, D. A., "A review and projection of semiconductor components for digital storage", Proc. IEEE, $\underline{63}$ (8), 1136-47 (1975), EEA78-28869.
1050. Hodges, D. A., "Trends in computer hardware technology", Comput. Des., $\underline{15}$ (2), 77-85 (1976).
1051. Ishii, O., "Progress of memory technology", J. Acoust. Soc. Jap., $\underline{28}$ (11), 612-6 (1972), Japanese, CCA8-11620.
1052. Jacobsen, H., "New technologies and circuits gaining ground", Elektro, $\underline{88}$ (18), 18, 20-1, 23-5 (1975), EEA79-4751.
1053. Kakihana, S., "Current status and trends in high frequency transistors", Microelectronics, $\underline{5}$ (4), 7-12 (1974), EEA78-37520.
1054. Kodama, K., "Current state and future directions of semiconductor memories", JEE, no. 86, 48-50 (1972), Japanese, CCA9-11964.
1055. Koehler, H. F., "Advances in memory technology", Comput. Des., $\underline{13}$ (6), 71-7 (1974), CCA9-21994.
1056. Kohler, R., "Computer memory of the future", Elektro-Anz., $\underline{28}$ (7), 163-6 (1975), German, CCA10-17624.
1057. Koppel, R. J. and Maltz, I., "Predicting the real costs of semiconductor-memory systems", Electronics, $\underline{49}$ (24), 117-22 (1976), CCA12-4011.
1058. Koppel, R. L. and Maltz, I. R., "Cost, performance, and reliability tradeoffs in semiconductor memory systems", 13th IEEE Computer Society International Conference, COMPCON 76, Digest of Papers, p. 218-22.
1059. Landauer, R., "Optical logic and optically accessed digital storage", In: Optical Information Processing, Nesterikhin, Y. E., Stroke, G. W., Kock, W. E., (Eds.), Plenum Press, New York, p. 219-51 (1976).
1060. Landauer, R. and Hall, J. T., "Solid state physics as a source of modern electronics", Science, $\underline{160}$, 736-41 (1968).
1061. Lindgren, N., "Semiconductors face the 80's", IEEE Spectrum, $\underline{14}$ (10), 42-8 (1977).
1062. Longo, T. A., "Recent advances in semiconductor logic and memory", Proceedings of International Conference on Automotive Electronics and Electric Vehicles, 1976, 20/1-6, EEA80-9035.
1063. Luecke, G., "Overview of semiconductor technology trends", 13th IEEE Computer Society International Conference, COMPCON 76, Digest of Papers, p. 52-5, EEA80-681.
1064. Mackintosh, I. M., "Semiconductor devices - portrait of a technological explosion", Radio Electron Eng., $\underline{45}$ (10), 515-24 (1975).
1065. Madland, G. R., "The future of silicon technology", Solid State Technol., $\underline{20}$ (8), 91-5 (1977).

1066. Markkula, A. C., Jr., "Semiconductor memory costs present and
      future", 1974 IEEE Intercom Technical-Program Papers,
      No. 10-3.
1067. Marston, A. D., "Impact of recent advancements of memory
      technologies on products", 1973 WESCON Technical Papers,
      Paper 16/4, 8 pp.
1068. Martin, R. R. and Frankel, H. D., "Forecast of computer memory
      technology", 8th IEEE Computer Society International Conference,
      COMPCON 74, Digest of Papers, p. 287-90.
1069. Marvel, O. E., "Advanced memory technologies", Proceedings of
      8th Asilomar Conference on Circuits, Systems and Computers,
      1974, p. 726-32, Publ. 1975, CCA10-17629.
1070. Mayo, J. S., "The role of microelectronics in communication",
      Sci. Am., $\underline{237}$(3), 192-209 (1977).
1071. Moore, G. E., "Progress in digital integrated electronics",
      1975 International Electron Devices Meeting, Technical Digest,
      p. 11-13, EEA79-37705.
1072. Myers, W., "Key developments in computer technology: A survey",
      Computer, $\underline{9}$ (11), 48-77 (1976).
1073. Neukom, H., "The development of storage technologies in elec-
      tronic data processing", Bull. Assoc. Suisse Electr., $\underline{66}$ (10),
      520-5 (1975), German, CCA10-19928.
1074. Noyce, R. N., "Probable evolution of techniques and prices, of
      semiconductor memories", Onde Elec., $\underline{50}$ (9), 757-62 (1970),
      French, EEA74-4916.
1075. Noyce, R. N., "From relays to MPU's", Computer, $\underline{9}$ (12), 26-9
      (1976).
1076. Noyce, R. N., "Large-scale-integration: What is yet to come?",
      Science, $\underline{195}$, 1102-6 (1977).
1077. Noyce, R. N., "Microelectronics", Sci. Am., $\underline{237}$(3), 62-9 (1977).
1078. Oguchi, B., "Promising areas of research in computer science
      and technology", Proceedings of the 2nd USA-Japan Computer
      Conference, AFIPS & IPSJ, 1975, p. 8-13.
1079. Oliphant, J. M., "Technological impact on semiconductor mem-
      ories", 13th IEEE Computer Society International Conference,
      COMPCON 76, Digest of Papers, p. 174-7, EEA80-9506.
1080. Pullen, E. W. and Simko, R. G., "Our changing industry", Data-
      mation, $\underline{23}$ (11), 49-55 (1977).
1081. Rege, S. I., "Cost, performance, and size tradeoffs for dif-
      ferent levels in a memory hierarchy", Computer, $\underline{9}$ (4), 43-51
      (1976), CA11-22633.
1082. Reiner, H., "Semiconductor memories - status and future out-
      look", VDE Fachber, $\underline{29}$, 155-64 (1976), German, CCA12-8902.
1083. Ruth, R. N., "Integrated injection logic and linear bipolar
      present and future", 1975 WESCON Technical Papers, Paper 19/1,
      4 pp., CCA11-12267.
1084. Sautter, D., "Focal points of modern semiconductor technology",
      Elektron. Anz., $\underline{8}$ (12), 293-5 (1976), German, EEA80-9494.

1085. Schiedewitz, W., "On further developments in memories for digital computer equipment", Radio Fernsehen Elektron., $\underline{24}$ (7), 239-42 (1975), CCA10-22910.
1086. Shimp, A. C., "Can core survive?", Digital Des., $\underline{7}$ (11), 31-2, 34, 36, 38 (1977).
1087. Slana, M. F., "Semiconductor memories: New dimensions in storage technology", Computer, $\underline{10}$ (7), 12-13 (1977).
1088. Slana, M. F., "Workshop report: A computer element technology update", Computer, $\underline{10}$ (7), 37-9 (1977).
1089. Soni, G., "LSI: technological options and implications", Electron. Inf. & Plann., $\underline{3}$ (7), 521-32 (1976), EEA79-33112.
1090. Spruth, W. G., "Trends in computer system structure and architecture", Proceedings of the 1st Conference of the European Cooperation in Informatics, 1976, p. 12-32, CCA12-15593.
1091. Stehlin, R. and Sloan, B. J., "Advances in I^2L development", 13th IEEE Computer Society International Conference, COMPCON 76, Digest of Papers, p. 56-8.
1092. Stein, K. U., "The present state of and future developments in microelectronics", Elektrotech. & Maschinenbau, $\underline{93}$ (6), 240-8 (1976), EEA79-33073.
1093. Striz, V., "MOS integrated circuits - the technology of future?", Automatizace, $\underline{19}$ (3), 82-3 (1976), Czech.
1094. Sugano, T., "Very large scale integrated circuits", Solid State Phys., $\underline{11}$ (8), 459-64 (1976), Japanese, EEA80-13281.
1095. Tarui, Y., "LSI and VLSI research in Japan", 1977 International Electron Devices Meeting, Technical Digest, p. 2-6.
1096. Thim, H., "New techniques in ultra-fast microelectronics", Elektrotech & Maschinenbau, $\underline{93}$ (6), 275-81 (1976), German, EEA79-33075.
1097. Tomash, E., "SOS: the next 11111 years (EDP industry outlook)", Computer, $\underline{9}$ (12), 11-21 (1976), CCA12-5100.
1098. Triebwasser, S., "Future trends in solid state electronics", Proceedings of the 5th International Conference on Solid State Devices, 1973, p. 3-12, Publ. 1974.
1099. Verhofstadt, P. W. J., "Evaluation of technology options for LSI processing elements", Proc. IEEE, $\underline{64}$ (6), 842-51 (1976), CCA11-22814.
1100. Wickham, R. F., "Projections of data processing memory usage", Proc. IEEE, $\underline{63}$ (3), 1096-1103 (1975).
1101. Wilcock, J. D., "Semiconductor memories 1975", New Electron, $\underline{9}$ (4), 18-19, 21-2 (1976), CCA11-15069.
1102. Withington, F. G., "Beyond 1984: a technology forecast", Datamation, $\underline{31}$ (1), 54-61, 63, 65, 67, 71, 73 (1975), CCA10-9661.
1103. Withington, F. G., "Innovations in the operation of future computers", Proceedings of the National Computer Conference, Vol. 44, AFIPS, 1975, p. 633-5.
1104. Withington, F. G., "Future computer technology", Data Base, 7-4 (Spring 1976).

1105. Withington, F. G., "The world computer industry 1976-1981",
      Report R770301, Arthur D. Little, Inc., Cambridge, Mass.,
      1977, 37 pp.
1106. Wolf, P., "Josephson junctions in digital technology",
      Bussl. Assoc. Suisse Electr., 68 (2), 66-70 (1977), CCA12-6571.

XI. LIMITATIONS

1107. Abramov, V. A., "Optimization of the reliability of complex
      systems with technical and economic limitations", Sov. Micro-
      electron., 5 (4), 253-7 (1976), EEA80-36411.
1108. Anon., "Limitations on the maximum operating voltage of CMOS
      integrated circuits", 1975 International Electron Devices
      Meeting, Technical Digest, p. 551-4, EEA80-3551.
1109. Anon., "Towards the limits of microelectronics", Technol.
      Elettr., no. 7-8, 40-3 (1976), Italian, EEA80-679.
1110. Bajenesco, T. I., "The reliability and limitations of micro-
      circuits", Rev. Polytech., no. 11, 1051-9, 1095 (1976),
      French, EEA80-9503.
1111. Bennett, C. H., "Logical reversibility of computation", IBM
      J. Res. & Develop., 17 (6), 525-32 (1973).
1112. Broers, A. N. and Dennard, R. H., "Impact of electron beam
      technology on silicon device fabrication", Proceedings of the
      2nd Symposium of Silicon Material Science and Technology,
      1973, p. 830-41.
1113. Dill, F. H., Tuttle, J. A. and Neureuther, A. R., "Limits of
      optical lithography", 1974 International Electron Devices
      Meeting, Technical Digest, p. 13-16.
1114. Folberth, O. G. and Bleher, J. H., "Fundamental limitations
      of digital semiconductor technology", Nachrichtentech. Z.
      (NTZ), 30 (4), 307-14 (1977), German, EEA80-32375.
1115. Freiser, M. J. and Marcus, P. M., "A survey of some physical
      limitations on computer elements", IEEE Trans. Magn., MAG-5
      (2), 82-90 (1969).
1116. Garbrecht, K. and Stein, K. U., "Perspectives and limitations
      of large-scale integration", Siemens Forsch- & Entwicklungsber,
      5 (6), 312-18 (1976).
1117. Georgiev, V. K. and Tomeva, R. S., "On the physical limitations
      in MOS technology", C. R. Acad. Bulg. Sci., 26 (8), 1021-3
      (1973), EEA77-14523.
1118. Hara, H., "Note on limitations in MOS LSIs from a design view-
      point", Jap. J. Appl. Phys., 15, Suppl. 15-1, 333-4 (1976),
      EEA80-9524.
1119. Hoeneisen, B. and Mead, C. A., "Fundamental limitations in
      microelectronics. I. MOS technology", Solid-State Electron,
      15, 819-29 (1972), EEA75-23335.
1120. Hoeneisen, B. and Mead, C. A., "Limitations in microelectronics-
      II Bipolar Technology", Solid-State Electron., 15, 891-7 (1972).

1121. Hook, H. O., "The capabilities and limitations of photo-lithography", J. Vac. Sci. & Technol., 11, 94 (1974), EEA77-38865.
1122. Keyes, R. W., "Physical problems and limits in computer logic", IEEE Spectrum, 6 (5), 36-45 (1969).
1123. Keyes, R. W., "Power and energy limitations in magnetic bubble devices", Proc. IEEE, 59 (10), 1528-30 (1971), EEA75-33582.
1124. Keyes, R. W., "Physical limits in digital electronics", Proc. IEEE, 63 (5), 740-67 (1975).
1125. Keyes, R. W., "Physical limits in semiconductor electronics", Science, 195, 1230 (1977).
1126. Keyes, R. W., "Microstructure fabrication", Science, 196 (4293), 945-9 (1977), EEA80-36409.
1127. Keyes, R. W. and Armstrong, J. A., "Thermal limitations in optical logic", Appl. Opt., 8 (12), 2549-52 (1969).
1128. Landauer, R., "Fundamental limitations in computer processes", Ber. Bunsenges. Phys. Chen., 80 (11), 1047-59 (1976).
1129. Laquer, H. L., "Superconductivity, energy storage and switching", Report LA-UR-75-758, Los Alamos Scientific Lab., New Mexico, 1974, 27 pp.
1130. Mallinson, J. C., "A theoretical limit to digital pulse resolution in saturation recording", IEEE Trans. Magn., MAG-5 (2), 91-7 (1969).
1131. Muller, R., Pfeiderer, H. J. and Stein, K. U., "A figure of merit for ICs-definition and determination of minimum energy per binary operation", Siemens Forsch. - & Entwicklungsber, 5 (6), 368-70, EEA80-3528.
1132. Ono, K., "Limit on the microelectronic fabrication", Jap. J. Appl. Phys., 15, Suppl. 15-1, 331-2 (1976), EEA80-9501.
1133. Phillips, A., Jr., O'Brien, R. R. and Joy, R. C., "IGFET hot electron emission model", 1975 International Electron Devices Meeting, Technical Digest, p. 39-42.
1134. Stein, K. U., "Noise-induced error rate as a limiting factor for energy per operation in digital IC's", IEEE J. Solid-State Circuits, SC-12 (5), 527-9 (1977).
1135. Sutherland, I. E., Mead, C. A. and Everhart, T. E., "Basic limitations in microcircuit fabrication technology", Report AD-A035149/4SL, Rand Corp., Santa Monica, CA, 1976, 56 pp.
1136. Wallmark, J. T., "Fundamental physical limitations in integrated electronic circuits", Proceedings of the European Conference on Solid State Devices, 1974, p. 133-67, EEA78-24530.
1137. Ware, W. H., "The ultimate computer", IEEE Spectrum, 9 (3), 84-91 (1972).

C. MASS STORAGE

I. BOOKS

1. Butters, J. N., "Holography and its technology", Peter Peregrinus
   Ltd., London, England, 1971.
2. Casasent, D. and Sawchuk, A. A., (Eds.), "Optical information
   processing: Real-time devices and novel techniques", SPIE
   Seminar Proceedings, Vol. 83; SPIE, Palos Verdes Estate, California, 1977.
3. Cathey, W. T., "Optical information processing and holography",
   Wiley, New York, 1973.
4. Chang, H., (Ed.), "Magnetic bubble technology: Integrated-circuit magnetics for digital storage and processing", IEEE Press,
   New York, 1975.
5. Compiled by Committee for Bubble Materials, "Handbook of bubble
   technology", Ohm Book Co., Tokyo, Japan, 1976.
6. Fuller, S. H., "Analysis of drum and disk storage units", Springer-
   Verlag, New York, 1975.
7. Matick, R., "Computer storage systems and technology", Wiley-
   Interscience, New York, 1977.
8. Melen, R. and Buss, D., (Eds.), "Charge-coupled devices: technology and applications", IEEE Press, New York, 1977.
9. Renwick, W. and Cole, A. J., "Digital storage systems", Chapman
   & Hale, Ltd., London, England, 1971.
10. Sebestyen, L. G., "Digital magnetic tape recording for computer
    applications", Chapman & Hale, London, England, 1973.
11. Seitzer, D., "Computer storage devices", Springer-Verlag, New
    York, 1975.
12. Sequin, C. H. and Tompsett, M. F., "Charge transfer devices",
    Academic Press, Inc., New York, 1975.
13. Seraphin, B. O., (Ed.), "Optical properties of solids: New
    developments", North-Holland Publishing Co., Amsterdam,
    Netherlands, 1976.
14. Smith, A. B., (Ed.), "Bubble-domain memory devices", Artech
    House, Inc., Dedham, Massachusetts, 1974.

II. REVIEW ARTICLES

15. Alsberg, H. and Nathan, R., "Man machine interactive imaging
    and data processing using high speed digital mass storage",
    18th Annual Meeting of the Human Factors Society, Proceedings,
    1974, p. 498-505.

16. Anacker, W., "Possible uses of charge transfer devices and magnetic domain devices in memory hierarchies", Presented at the International Magnetics (Intermag.) Conference, Denver, Colorado, 1971, CCA6-17637.
17. Anon., "CCDs are potential replacements for moving magnetic memories", Comput. Des., $\underline{15}$ (5), 72-4 (1976), CCA11-22658.
18. Anon., "Computer memory technology: Part II", Digital Des., $\underline{7}$ (9), 86-98 (1977).
19. Axford, J. G., "Matching computer storage to the demands of the CPU", Presented at Colloquium Digest on Computer Memories, London, England, February 1973, CCA8-11621.
20. Barrekette, E. S., "Trends in storage of digital data", Appl. Opt., $\underline{13}$ (4), 749-54 (1974), CCA9-16049.
21. Bate, G., "The future of recording", Proceedings of the Conference on Video and Data Recording, Birmingham, England, July 1976, p. 97, CCA11-31590.
22. Bousky, S., "Optical and magnetic digital recording - a comparative review", Opt. Eng., $\underline{15}$ (2), 109-14 (1976).
23. Bowers, D. M., "The charge-coupled device - is this the end for core, drum and disk?", Mod. Data, $\underline{8}$ (8), 54-6 (1975), CCA11-6382.
24. Branscomb, L. M., "Technology trends in peripheral devices", 8th IEEE Computer Society International Conference, COMPCON 74, Digest of Papers, p. 17-20, 1974.
25. Brown, A. J., "Considerations in selecting a mass memory system", Systems, $\underline{2}$ (2), 22-4 (1974), CCA9-14344.
26. Burt, D. J., "Charge coupled devices, today and for the future", Microelectron. & Reliab., $\underline{14}$ (4), 379 (1975), CCA11-9682.
27. Chang, H., "Capabilities of the bubble technology", AFIPS National Computer Conference, Proceedings, Vol. 43, 1974, p. 847-55.
28. Chang, H., "Bubbles - status and prospects", 13th IEEE Computer Society International Conference, COMPCON 76, Digest of Papers, p. 324-5.
29. Chen, D. and Zook, J. D., "Overview of optical data storage technology", Proc. IEEE, $\underline{63}$(8), 1207-30 (1975).
30. Cohen, M. S. and Chang, H., "Frontier of magnetic bubble technology", IEEE Proc., $\underline{63}$ (8), 1196-206 (1975).
31. Curtis, D. A., "The status of memory technologies under development in Europe and their use in scientific and earth resources observation satellites, volumes 1 and 2", Report N75-20465/1SL, Cambridge Consultants Ltd., England, 1974, 150 pp.
32. Eward, R. S., "Optical memories are alive and well and growing", Opt. Spectra, $\underline{9}$ (8), 25-8 (1975), CCA11-9693.
33. Eschenfelder, A. H., "Promise of magneto-optic storage systems compared to conventional magnetic technology", J. Appl. Phys., $\underline{41}$, 1372-6 (1970).
34. Gilder, J. H., "Mass memories raising speed, dropping cost and storing billions of bytes", Electron. Des., $\underline{23}$ (22), 46-50 (1975), CCA11-9660.

35. Guidry, M., "Directions in CCD digital memory", 13th IEEE Computer Society International Conference, COMPCON 76, Digest of Papers, p. 326-8.
36. Hoagland, A. S., "Mass storage: past, present and future", Computer, 6 (9), 29-33 (1973), CCA9-3389.
37. Hoagland, A. S., "Magnetic recording storage", IEEE Trans. Comp., C-25 (12), 1283-8 (1976).
38. Houston, G. B., "Trillion bit memories", Datamation, 19 (10), 52-8 (1973), CCA9-1224.
39. Ishii, O., "Magnetic bubble and its competing technologies", Jap. J. Appl. Phys., 15, 141-2 (1976), CCA12-4007.
40. Kelly, J., "The development of an experimental electron-beam-addressed memory module", Computer, 8 (2), 32-42 (1975), CCA10-15203.
41. Mallinson, J. C., "Tutorial review of magnetic recording", Proc. IEEE, 64 (2), 196-208 (1976), CCA11-15050.
42. Martin, R. R. and Frankel, H. D., "Electronic disks in the 1980's", Computer, 8 (2), 24-30 (1975), CCA10-15202.
43. Mee, C. D., "A comparison of bubble and disk storage technologies", IEEE Trans. Magn., MAG-12 (1), 1-6 (1976), CCA11-6367.
44. Panigrahi, G., "Charge-coupled memories for computer systems", Computer, 9 (4), 33-42 (1976), CCA11-22661.
45. Panigrahi, G., "The implications of electronic serial memories", Computer, 10 (7), 18-25 (1977).
46. Pohm, A. V., "Cost/performance perspectives of paging with electronic and electromechanical backing stores", Proc. IEEE, 63 (8), 1123-8 (1975).
47. Pohm, A. V., "Electronic replacements for head-per-track drums or disks", Computer, 9 (3), 16-20 (1976), CCA11-22638.
48. Robinson, C. R., "Charge-coupled devices. How will they affect future systems?", Can. Datasyst., 7 (10), 49-54 (1975), CCA11-15058.
49. Rogge, H., "Present state and evolution of mass memories", Onde Elect., 54 (6), 273-6 (1974), CCA9-20746.
50. Sallet, H. W., "Magnetic tapes: a high performer", IEEE Spectrum, 14 (7), 26-31 (1977).
51. Schneidewind, N. and Syms, G., "Mass memory system peripherals", 8th IEEE Computer Society International Conference, COMPCON 74, Digest of Papers, p. 87-91.
52. Schneidewind, N. F., Syms, G. H., Grainger, T. L. and Carden, R. J., "A survey and analysis of high density magnetic storage devices", Report AD-743049, Naval Postgraduate School, Monterey, California, 1972, 76 pp., CCA8-2639.
53. Schneidewind, N., Syms, G., Grainger, T. and Carden, R., "High density mass memories", EE/Syst. Eng. Today, 32 (8), 46-51 (1973), CCA9-16051.
54. Schneidewind, N., Syms, G., Grainger, T. and Carden, R., "High density mass memories. TII", EE/Syst. Eng. Today, 32 (10), 39-42 (1973), CCA9-14341.

55. Schramm, H. F. W., "In the laboratory: optical data storage in the near future", Online-ADL-Nachr., no. 1-2, 62-8 (1977), German, CCA12-15536.
56. Shahbender, R., "Magnetic memories – present status and future trends", 1971 IEEE International Convention Digest, p. 274-5, CCA6-15209.
57. Speliotis, D. E., "EBAM – electron beam addressable memories", 13th IEEE Computer Society International Conference, COMPCON 76, Digest of Papers, p. 329-31.
58. Terman, L. M. and Heller, L. G., "Overview of CCD memory", IEEE J. Solid-State Circuits, SC-$\underline{11}$ (1), 4-10 (1976), EEA79-16877.
59. Thompson, D. A., Romankiw, L. T. and Mayadas, A. F., "Thin film magnetoresistors in memory, storage, and related applications", IEEE Trans. Magn., MAG-$\underline{11}$ (4), 1039-50 (1975).
60. Tschudin, H., "A new storage technique – magnetic bubble memories", Micomp, $\underline{1}$ (2), 27-8, 30-1 (1976), German, CCA12-4008.
61. Tufte, O. N. and Chen, D., "Optical techniques for data storage", IEEE Spectrum, $\underline{10}$ (2), 26-32 (1973), CCA8-13714.
62. Villard, J. C. and Randet, D., "Mass memories", Onde Elec., $\underline{51}$ (11), 886-92 (1971), French, CCA7-6286.
63. Weil, J. W., "An introduction to massive stores", Honeywell Comput. J., $\underline{5}$ (2), 88-92 (1971).
64. Weiss, H., "Physics and technology of data storage", Solid-State Electron., $\underline{19}$ (5), 347-56 (1976), CCA11-15071.
65. Wensley, J. H., "The impact of electronic disks on system architecture", Computer, $\underline{8}$ (2), 44-8 (1975), CCA10-15204.
66. Ypma, J. E., "Bubble domain memory systems", Proceedings of the AFIPS National Computer Conference, Vol. 44, 1975, p. 523-8.

III. BUBBLE DOMAIN STORAGE
    1. General

67. Anon., "Magnetic bubble attract attention", Electron. & Power, $\underline{18}$, 205 (1972), EEA75-23632.
68. Anon., "Magnetic bubbles – latest discoveries herald exciting performance improvements", JEE, no. 108, 28-31 (1975), CCA11-15054.
69. Bailey, R. F. and Ypma, J. E., "Bubble domain memories – present achievements and future applications", Solid State Technol., $\underline{19}$ (9), 74-82 (1976), CCA12-1503.
70. Bhandarkar, D. P. and Juliussen, J. E., "Tutorial: computer system advantages of magnetic bubble memories", Computer, $\underline{8}$ (11), 35-40 (1975), CCA11-9665.
71. Bobeck, A. H., Bonyhard, P. I. and Geusic, J. E., "Magnetic bubbles – an emerging new memory technology", Proc. IEEE, $\underline{63}$ (8), 1176-95 (1975).
72. Bobeck, A. H., Fischer, R. E. and Smith, J. L., "An overview of magnetic bubble domains material device interface", AIP Conf. Proc., $\underline{5}$ (Pt. I), 45-55 (1971), EEA75-19524.

73. Bobeck, A. H. and Scovil, H. E. D., "Magnetic bubbles", Sci. Amer., $\underline{224}$(6), 78-9 (1971), EEA74-30218.

74. Chang, H., "Magnetic bubble technology - present and future", Jap. J. Appl. Phys., $\underline{15}$ (Suppl. 15-1), 3-10 (1976), CCA12-4004.

75. Coeure, P., Jouve, H., Manduit, D. and Randet, D., "Magnetic bubble domain technology and devices", Onde Electr., $\underline{54}$ (4), 165-74 (1974), French, CCA9-16065.

76. Druyvesteyn, W. F. , van d Enden, A. W. M., Kuijpers, F. A., de Niet, E. and Verhulst, A. G. H., "Magnetic bubbles", 4th European Solid State Physics Device Research Conference, ESSDERC, 1975, p. 37-74, CCA10-17639.

77. Fairholme, R. J., "An introduction to magnetic bubbles", New Electron., $\underline{6}$ (7), 36-8, 40 (1973), CCA8-13681.

78. Furuoya, T., "Magnetic bubbles", JEE, no. 68, 38-42 (1972), EEA75-39423.

79. Kanyo, H., "A technology of magnetic bubbles", J. Inst. Electron. Commun. Eng. Japan, $\underline{53}$ (5), 673-5 (1970), Japanese, CCA6-2063.

80. Lapidus, G., "The domain of magnetic bubbles", IEEE Spectrum, $\underline{9}$ (9), 58-62 (1972), EEA75-35670.

81. Lock, R. D. and Lucas, J. M., "Magnetic bubbles and their applications", Radio & Electron. Eng., $\underline{42}$ (10), 435-46 (1972), PA75-8162.

82. Torrero, E. A., "Bubbles rise from the lab", IEEE Spectrum, $\underline{13}$ (9), 28-31 (1976), CCA12-1500.

83. VDC, "Bubble domain memory markets, 1978-1983", Report, Venture Development Corporation, 1 Washington Street, Wellesley, MA, 1977.

84. Warnar, R. B. J. and Calomeris, P. J., "Foreign and domestic accomplishments in magnetic bubble device technology", Natl. Bur. Stand. Spec. Publ., no. 500-1, 1977, 50 pp.

2.   Materials

85. Ahn, K. Y. and Lin, Y. S., "Composite cylindrical magnetic domain materials", IBM Tech. Disclosure Bull., $\underline{13}$ (11), 3220 (1971), EEA74-30125.

86. Albert, P. A. and Griest, A. J., Jr., "Cylindrical domain materials for crystal growth", IBM Tech. Disclosure Bull., $\underline{13}$ (11), 3229 (1971), EEA74-30089.

87. Baszynski, J., Sulkowska, S. and Szymanski, B., "Bubble domains in remanence state in monocrystalline, epitaxial ferrite layers /MgMnFe/$_3$0$_4$", IEEE Trans. Magn., MAG-$\underline{13}$ (5), 1098-101 (1977).

88. Besser, P. J., Mee, J. E., Glass, H. L., Hemz, D. M., Austerman, S. B., Elkins, P. E., Hamilton, T. N. and Whitcomb, E. C., "Film/substrate matching requirements for bubble domain formation in CVD garnet films", AIP Conf. Proc., $\underline{5}$ (Pt. 1), 125-9 (1971), PA75-36750.

89. Blank, S. L., Biolsi, W. A. and Nielsen, J. W., "The effect of melt composition on the Curie temperature and flux spin-off from lutetium containing LPE garnet films", IEEE Trans. Magn., MAG-13 (5), 1095-7 (1977).

90. Blank, S. L., Hewitt, B. S., Shick, L. K. and Nielsen, J. W., "Kinetics of LPE growth and its influence on magnetic properties", AIP Conf. Proc., 10, 256 (1972).

91. Blank, S. L. and Nielsen, J. W., "The growth of magnetic garnets by liquid phase epitaxy", J. Cryst. Growth, 17, 302-11 (1972), PA75-2667.

92. Blank, S. L., Shick, L. K. and Nielsen, J. W., "Single crystal growth of yttrium orthoferrite by a seeded Bridgman technique", J. Appl. Phys., 42 (4), 1556-8 (1971), PA74-48319.

93. Bobeck, A. H., Schmidt, P. H., Spencer, E. G., Van Uitert, L. G. G. and Walters, E. M., "Magnetic devices utilizing garnet compositions", Patent USA 3613056, Publ. October 1971.

94. Bobeck, A. H., Sherwood, R. C. and Van Uitert, L. G. G., "Magnetic devices utilizing garnet compositions", Patent USA 3665427, Publ. May 1972.

95. Bobeck, A. H., Smith, D. H., Spencer, E. G. and Van Uitert, L. G. G., "Magnetic devices utilizing garnet compositions", Patent USA 3646529, Publ. February 1972.

96. Bobeck, A. H., Spencer, E. G., Van Uitert, L. G., Abrahams, S. C., Barns, R. L., Grodkiewicz, W. H., Sherwood, R. C., Schmidt, P. H., Smith, D. H. and Walters, E. M., "Uniaxial magnetic garnets for domain wall 'bubble' devices", Appl. Phys. Lett., 17 (3), 131-4 (1970), EEA73-34525.

97. Bobeck, A. H. and Van Uitert, L. G. G., "Magnetic devices", Patent USA 3643238, Publ. February 1972.

98. Bonner, W. A., Geusic, J. E., Smith, D. H., Rossol, E. C., Van Uitert, L. G. and Vella-Coleiro, G. P., "Characteristics of temperature-stable Eu-based garnet films for magnetic bubble applications", J. Appl. Phys., 43 (7), 3226-8 (1972), EEA75-28058.

99. Bourne, H. C., Jr., Goldfarb, R. B., Wilson, W. L., Jr. and Zwingman, R., "Effects of dc bias on the fabrication of amorphous GDCO rf sputtered films", IEEE Trans. Magn., MAG-11 (5), 1332-4 (1975).

100. Brandle, C. D. and Valentino, A. J., "Czochralski growth of rare earth gallium garnets", J. Crystal Growth, 12, 3 (1972).

101. Breed, D. J., Stacy, W. T., Voermans, A. B., Logmans, H. and van der Heijden, A. M. J., "New bubble materials with high peak velocity", IEEE Trans. Magn., MAG-13 (5), 1087-91 (1977).

102. Brouchier, A., Coeure, P., Ferrand, B., Gay, J. C., Joupert, J. C., Mateschal, J., Viguie, J. C., Martin-Binachon, J. C. and Spitz, J., "Heteroepitaxy of yttrium iron garnet thin films by the flux method and hydrothermal synthesis", J. Cryst. Growth, 13-14, 571-5 (1972), French, PA75-50513.

103. Challeton, D., Coeure, P. and Mareschal, J., "Preparation and study of rare-earth orthoferrites for memory application", Proceedings of the International Conference on Advanced Microelectronics, 1970, 2 pp., CCA6-2068.

104. Chaudhari, P., Cuomo, J. J. and Gambino, R. J., "Amorphous metallic films for bubble domain applications", IBM J. Res. Develop., 17, 66 (1973).

105. Cronemeyer, D. C., Giess, E. A., Klokholm, E., Argyle, B. E. and Plaskett, T. S., "Annealing of LPE garnet films", AIP Conf. Proc., 5 (Pt. I), 115-19 (1971), PA75-36752.

106. Damen, J. P. M., Pistorius, J. A. and Robertson, J. M., "Calcium gallium germanium garnet as a substrate for magnetic bubble application", Water Res. Bull., 12 (1), 73-8 (1977).

107. de Jone, F. A. and Druyvesteyn, W. F., "Bubble lattices", AIP Conf. Proc., 5 (Pt. I), 130-4 (1971), PA75-36735.

108. Enz, U. and Rooijmans, G., "Research into magnetic bubbles and bubble materials", Polytech. Tijdschr. Elektrotech. Elektron., 28 (21), 713-16 (1973), Dutch, CCA9-6234.

109. Ferrand, B., Daval, J. and Joubert, J. C., "Heteroepitaxial growth of single crystal films of YIG on GdGaG substrates by hydrothermal synthesis", J. Cryst. Growth, 17, 312-14 (1972), PA75-1922.

110. Freyhardt, H. C., "'Magnetic bubbles' in uniaxial magnetic oxides - a review of new carriers of information for computers", Z. Metallk., 64 (1), 13-25 (1973), German, CCA8-9526.

111. Geusic, J. E., Levinstein, H. J., Licht, S. J., Shick, L. K. and Brandle, C. D., "Cylindrical magnetic domain epitaxial films with low defect density", Appl. Phys. Lett., 19 (4), 93-5 (1971), PA74-69805.

112. Gianola, U. F., Smith, D. h., Thiele, A. A. and van Uitert, L. G., "Material requirements for circular magnetic domain devices", IEEE Trans. Magn., MAG-5 (3), 558-61 (1969), EEA73-13266.

113. Giess, E. A., Argyle, B. E., Calhoun, B. A., Cronemeyer, D. C., Klokholm, E., McGuire, T. R. and Plaskett, T. S., "Rare earth yttrium iron-gallium garnet epitaxial films for magnetic bubble domain applications", Mater. Res. Bull., 6 (11), 1141-50 (1971), PA75-22690.

114. Giess, E. A., Argyle, B. E., Cronemeyer, D. C., Klokholm, E., McGuire, T. R., O'Kane, D. F., Plaskett, T. S. and Sadagopan, V., "Europium-yttrium iron-gallium garnet films grown by liquid phase epitaxy on gadolinium gallium garnet", AIP Conf. Proc., 5 (Pt. I), 110-14 (1974), PA75-36743.

115. Giess, E. A., Calhoun, B. A., Klokholm, E., McGuire, T. R. and Rosier, L. L., "Garnet compositions for bubble domain systems utilizing stress-induced uniaxial anisotropy", Mater. Res. Bull., 6 (5), 317-28 (1971), PA74-45461.

116. Giess, E. A., Kuptsis, J. D. and White, E. A. D., "Liquid phase epitaxial growth of magnetic garnet films by isothermal dipping in a horizontal plane with axial rotation", J. Cryst. Growth, 16 (1), 36-42 (1972), PA75-1912.

117. Glass, H. L. and Hamilton, T. N., "Replication of substrate growth band and core structures by epitaxial CVD garnet films", Mater. Res. Bull., 7 (8), 761-8 (1972), PA75-71338.

118. Griest, A. J., Jr., "Cobalt-platinum group alloys whose anisotropy is greater than their demagnetizable field for use as cylindrical memory elements", Patent USA 3755796, Publ. August 1973.

119. Gyorgy, E. M., Sherwood, R. C. and Van Uitert, L. G. G., "Magnetic devices", Patent USA 3648260, Publ. March 1972.

120. Hagedorn, F. B., "Characterization of magnetic materials and circuits for bubble mass memories", IEEE Trans. Magn., MAG-8 (3), 293 (1972), CCA8-13663.

121. Heinz, D. M., Besser, P. J., Eliins, P. E., Glass, H. L. and Mee, J. E., "Epitaxial film growth of bubble domain materials", Report AD-727070, Autonetics, Anaheim, California, 1971, 39 pp., PA75-29823.

122. Heinz, D. M., Besser, P. J. and Mee. J. E., "Bubble domains in CVD films of gallium-substituted erbium iron garnet", AIP Conf. Proc., 5 (Pt. I), 96-100 (1971), PA75-36742.

123. Kaveggia, F. S., "Chemical lapping of yttrium aluminum", Patent USA 3632513, Publ. January 1972.

124. Kempter, K. and Beogner, W., "Preparation of epitaxial magnetic garnet films by chemical vapor deposition", Thin Solid Films, 12 (1), 35-40 (1972), EEA75-39435.

125. Kolb, E. D. and Laudise, R. A., "Hydrothermal growth of bubble-domain memory materials", J. Appl. Phys., 42 (4), 1552-4 (1971), EEA74-25907.

126. Laudise, R. A., "Single crystals for bubble domain memories", J. Cryst. Growth, 13-14, 27-33 (1972), EEA75-23607.

127. Laudise, R. A. and Van Uitert, L. G., "Materials for magnetic bubbles", Bull. Lab. Rec., 49 (8), 238-43 (1971), PA74-73856.

128. LeCraw, R. C., Pierce, R. D., Blank, S. L. and Wolfe, R., "Effect of lanthanum on growth-induced anisotropy in LPE bubble garnet films", IEEE Trans. Magn., MAG-13 (5), 1092-4 (1977).

129. Lessoff, H. and Webb, D. C., "Magnetic epitaxial films: applications and potential", Thin Solid Films, 39 (1-2-3), 185-94 (1976).

130. Levinstein, H. J., "Controlled epitaxial growth from supercooled nutrient-flux solution", Patent USA 3790405, Publ. February 1974.

131. Mendel, E., "Method for polishing magnetic oxide materials", Patent USA 3662500, Publ. May 1972.

132. Mendel, E., "Method for polishing magnetic oxide materials", Patent USA 3662501, Publ. May 1972.

133. Moody, J. W., Sandfort, R. M. and Shaw, R. W., "Magnetic bubble materials. Initial characterization report", Report AD-741390, Monsanto Res. Corp., St. Louis, Mo., 1972, 58 pp., CCA8-2659.

134. Nielsen, J. W., "Growth of garnet substrates and epitaxial films for bubble devices", AIP Conf. Proc., $\underline{5}$ (Pt. I), 56 (1971), EEA75-19525.

135. Nielsen, J. W., Blank, S. L., Smith, D. H., Vella-Coleiro, G. P., Hagedorn, F. B., Barns, R. L. and Biolsi, W. A., "Three garnet compositions for bubble domain memories", J. Electron Mater., $\underline{3}$ (3), 693-707 (1974).

136. Pistorius, J. A., Robertson, J. M. and Stacy, W. T., "Perfection of garnet bubble materials", Philips Tech. Rev., $\underline{35}$ (1), 1-10 (1975).

137. Quon, H. H. and Potvin, A. J., "Effect of volatilization loss in flux crystallization of $YFeO_3$ or Y_3FeO_{12}", J. Cryst. Growth, $\underline{10}$ (1), 124-6 (1971), PA74-48320.

138. Quon, H. H., Potvin, A. and Entwistle, S. D., "Growth of ortho-ferrite single crystals from flux melts", Mater. Res. Bull., $\underline{6}$ (11), 1175-84 (1971), PA75-22114.

139. Rooymans, C. J. M., "Current problems in bubble memory materials", Presented at 2nd European Electro-Optics Conference, Montreux, Switzerland, April 1974, CCA9-16066.

140. Shick, L. K. and Nielsen, J. W., "Liquid phase homoepitaxial growth of rare-earth orthoferrites", J. Appl. Phys., $\underline{42}$ (4), 1554-6 (1971), PA74-48271.

141. Simsa, Z., Simsova, J., Suk, K., Kratochvilova, E. and Marysko, M., "Magnetic bubbles in spinel ferrite films", Phys. Status Solidi A, $\underline{34}$ (2), 639-44 (1976).

142. Smith, D. H. and Van Uitert, L. G. G., "Magnetic devices", Patent USA 3638207, Publ. January 1972.

143. Swartz, J. C., Siegel, B., Morrison, A. D. and Lingertat, H., "Growth of ribbon-shaped crystals of gadolinium gallium garnet for bubble memory substrates", J. Electron. Mater., $\underline{3}$ (2), 309-26 (1974).

144. Tanasoiu, C., Florescu, V. and Rosenberg, M., "Crystal growth and bubble domain properties of some aluminium substituted hexa-ferrites", Mater. Res. Bull., $\underline{6}$ (11), 1257-64 (1971), PA75-22103.

145. Telesnin, R. V. and Dudorov, V. N., "Periodicity of hexagonal bubble domain lattices in iron garnet films", Sov. Phys. Solid State, $\underline{17}$ (6), 1063-5 (1975).

146. Tolksdorf, W., "Growth and properties of garnet films for stor-age application", IEEE Trans. Magn., MAG-$\underline{11}$ (5), 1074-8 (1975), CCA11-4797.

147. Van Uitert, L. G., Bonner, W. A., Grodkiewicz, W. H., Pictroski, L. and Zydzik, G. J., "Garnets for bubble domain devices", Mater. Res. Bull., $\underline{5}$ (9), 825-35 (1970), PA74-1823.

148. Van Uitert, L. G., Sherwood, R. C., Bonner, W. A., Grodkiewicz, W. H., Pictorski, L. and Zydzik, G., "Rare earth orthoferrites for bubble domain devices", Mater. Res. Bull., $\underline{5}$(2), 153-61 (1970), PA73-59340.

149. Van Uitert, L. G., Smith, D. H., Bonner, W. A., Grodkiewicz,
     W. H. and Zydzik, G. J., "Hexagonal ferrites for bubble domain
     devices", Mater. Res. Bull., 5 (6), 455-63 (1970), PA73-72256.
150. White, E. A. D. and Wood, J. D. C., "Bubble materials - com-
     position, growth and evaluation", Radio Electron. Eng., 45
     (12), 711-24 (1975).

3.   Overlay Fabrication

151. Ahn, K. Y., "All-amorphous film single-level masking bubble
     devices and fabrication method", IBM Tech. Disclosure Bull.,
     18 (3), 940-1 (1975), CCA11-9667.
152. Ahn, K. Y., Chang, T. H. P., Hatzakis, M., Kryder, M. H. and
     Luhn, H., "Electron-beam fabrication of high-density amorphous
     bubble film devices", IEEE Trans. Magn., MAG-11 (5), 1142-4
     (1975), CCA11-4801.
153. Ahn, K. Y. and Hatzakis, M., "Recessed overlay structure", IBM
     Tech. Disclosure Bull., 17 (10), 3172 (1975), CCA10-15169.
154. Ahn, K. Y. and McGouey, R. P., "Conductor metallization for
     bubble devices", IBM Tech. Disclosure Bull., 18 (3), 938-9
     (1975), CCA11-9666.
155. Almasi, G. S., Canavello, B. J., Giess, E. A., Hendel, R. J.,
     Horstmann, R. E., Jamba, T. F., Keefe, G. E., Powers, J. V.
     and Rosier, L. L., "Fabrication and operation of a self contained
     bubble domain memory chip", AIP Conf. Proc., 5 (Pt. I), 220-4
     (1971), EEA75-19575.
156. Bailey, P. T., "Magnetic bubble circuit with hard-soft overlay",
     Patent USA 3875568, Publ. April 1975.
157. Bobeck, A. H., Ciak, F. J. and Strauss, W., "Single wall domain
     arrangement", Patent USA 3702995, Publ. November 1972.
158. Chang, T. H. P., Hatzakis, M., Wilson, A. D., Speth, A. J.,
     Kern, A. and Luhn, H., "Scanning electron beam lithography for
     fabrication of magnetic bubble circuits", IBM J. Res. & Dev.,
     20 (4), 376-88 (1976), CCA11-31600.
159. DeFabritis, R. P., Fefferman, G. B., Gesner, B. D. and Radner,
     R. J., "Magnetic bubble device and method of manufacture",
     Patent USA 3699552, Publ. October 1972.
160. Genovese, E. R., "Method and apparatus for creation of cylin-
     drical, single wall domains", Patent USA 3662359, Publ. May 1972.
161. Harris, R. A., Clegg, W. W., Pickard, R. M., Gourley, S. F. and
     Hardy, C. J., "Electron beam techniques for magnetic bubble
     device fabrication", Proceedings of the Conference on Computers
     - Systems and Technology, 1972, p. 57-64, Publ. 1974, CCA8-2661.
162. Kita, Y., Inose, F., Homma, N. and Yasuda, M., "Magnetic domain
     circuit arrangement", Patent USA 3821725, Publ. June 1974.
163. Littwin, B., "Bubble circuit fabrication by electrodeposition",
     IEEE Trans. Magn., MAG-11 (5), 1139-42 (1975), CCA11-4800.

164. Littwin, B., "Fabrication of magnetic circuits for bubble memory chips", J. Magn. & Magn. Mater., $\underline{4}$ (1-4), 166-73 (1977), German, CCA12-6662.

165. Mee, J. E., Heinz, D. M., Hamilton, T. N., Besser, P. J. and Pulliam, G. R., "Method of forming multiple-layer structures including magnetic domains", Patent USA 3645788, Publ. February 1972.

166. Reekstin, J. P., "Fabrication of 'bubble' propagating circuits by electroless deposition of nickel cobalt phosphorus", J. Appl. Phys., $\underline{42}$ (4), 1362-3 (1971), EEA74-21945.

167. Reekstin, J. P. and Kowalchuk, R., "Fabrication of large bubble circuits", IEEE Trans. Magn., MAG-$\underline{9}$ (3), 485-8 (1973), CCA9-3411.

168. Reekstin, J. P., Lehner, A. G., Vratny, F. and Kammlott, G. W., "Fabrication of 10^4 bit permalloy-first magnetic bubble circuits on epitaxial garnet chips", J. Vac. Sci. & Technol., $\underline{10}$ (5), 847-51 (1973), CCA9-9207.

169. Rose, D. K., "Planar processing for magnetic bubble devices", IEEE Trans. Magn., MAG-$\underline{12}$ (6), 618-21 (1976), CCA12-3994.

170. Spiller, E., Feder, R., Topalian, J., Castellani, E., Romankiw, L. and Heritage, M., "X-ray lithography for bubble devices", Solid State Technol., $\underline{19}$ (4), 62-7, 78 (1976).

171. Sweet, J. N., Schroeder, D. H., Maurin, J. K. and Adams, J. R., "Fabrication of 2-4 μm magnetic bubble memories with electron beam negative resist lithography", AIP Conf. Proc., no. 29, 56-7 (1976), CCA11-22650.

172. Takahashi, M., Nishida, H., Kasai, T. and Sugita, Y., "Fabrication of bubble memory chips", IEEE Trans. Magn., MAG-$\underline{10}$ (4), 1067-71 (1974).

173. Webb, D., "Bubble device overlay fabrication using scanned electron beams", Microelectronics, $\underline{7}$ (1), 22-6 (1975).

4.   Characteristics

174. Argyle, B. E. and DeLuca, J. C., "Magnetic domain systems using different types of domains", Patent USA 3911411, Publ. October 1975.

175. Bar'yakhtar, V. G. and Gorobets, Y. L., "Magnetic bubbles in ferrites near the compensation point", Sov. Phys. Solid State, $\underline{18}$ (8), 1386-8 (1976).

176. Bobeck, A. H., Geusic, J. E. and Hagedorn, F. B., "Magnetic bubble structure for suppression of dynamic bubble conversion", Patent USA 3836898, Publ. September 1974.

177. Bokov, V. A., Volkov, V. V., Trofimova, T. K. and Sher, E. S., "Dynamic transformation of magnetic bubbles in translational motion", Sov. Phys. Solid State, $\underline{17}$ (12), 2338-40 (1975).

178. Calhoun, B. A., Eggenberger, J. S., Rosier, L. L. and Shew, L. F., "Column access of a bubble lattice: column translation and lattice translation", IBM J. Res. & Dev., $\underline{20}$ (4), 368-75 (1976), CCA11-31599.

179. Carr, W. J., Jr., "Magnetic domain computational arrangement", Patent USA 3845478, Publ. October 1974.
180. Chen, T. T., Tocci, L. R. and Archer, J. L., "Device component margin evaluation using generalised field interruption technique", IEEE Trans. Magn., MAG-12 (6), 677-9 (1976), CCA12-3999.
181. DeBonte, W. J. and Zappulla, R., "Relationship of bias field setting procedure to field stability against external field perturbations for magnetic bubble memory bias-field structures", IEEE Trans. Magn., MAG-12 (6), 645-7 (1976), CCA12-6654.
182. Dorleijn, J. W. F. and Druyvesteyn, W. F., "Mutual interaction between bubbles in different sheets", AIP Conf. Proc., 5 (Pt. I), 135-9 (1971), PA75-36736.
183. Doyle, W. D. and Flannery, W. E., "The effect of operating conditions on margin decay for half disk bubble propagation elements", IEEE Trans. Magn., MAG-13 (5), 1592-4 (1977).
184. Gal, F. A., Kulikov, Y. S. and Malyutin, V. I., "Threshold characteristics of plane magnetic domains", Autom. & Remote Control, 36 (4), 657-60 (1975), CCA11-6370.
185. Gal, L., "Wall field interpretation of magnetic bubble behaviour", Phys. Status Solidi A, 28 (1), 181-6 (1975).
186. Gal, L., Zimmer, G. J. and Humphrey, F. B., "Transient magnetic bubble domain configurations during radial wall motion", Phys. Status Solidi A, 30 (2), 56-9 (1975).
187. Gergis, I. S., Chen, T. T. and Tocci, L. R., "The effect of DC in plane field on the operation of field access bubble memory devices", IEEE Trans. Magn., MAG-12 (1), 7-14 (1976), CCA11-6374.
188. Geusic, J. E. and Van Uitert, L. G. G., "Magnetic single wall domain arrangement", Patent USA 3711841, Publ. January 1973.
189. Heinz, D. M., "Method of forming bubble domain system", Patent USA 3728153, Publ. April 1973.
190. Heinz, D. M., "Bubble domain system", Patent USA 3728697, Publ. April 1973.
191. Heinz, D. M., Besser, P. J., Owens, J. M., Mee, J. E. and Pulliam, G. R., "Mobile cylindrical magnetic domains in epitaxial garnet films", J. Appl. Phys., 42, 1243 (1971).
192. Hsu, T. L., "Stabilizing buffer domain stripes in bubble lattice file", IBM Tech. Disclosure Bull., 17 (10), 3025 (1975), CCA10-17633.
193. Josephs, R. M., "Characterization of the magnetic behavior of bubble domains", AIP Conf. Proc., 10, 286 (1972).
194. Keyes, R. W., "Power and energy limitations in magnetic bubble devices", Proc. IEEE, 59 (10), 1528-9 (1971), EEA75-33582.
195. Kohara, H., "Magnetic domain splitter", Patent USA 3820091, Publ. June 1974.
196. Kryder, M. H., Tao, L. J. and Wilts, C. H., "Dynamics of amorphous film bubble devices", IEEE Trans. Magn., MAG-13 (5), 1626-30 (1977).

197. LeCraw, R. C., Levinstein, H. J. and Wolfe, R., "Method for controlling magnetization in garnet material and devices so produced", Patent USA 3845477, Publ. October 1974.
198. Licht, S. J., "Technique for controlled adjustment of bubble collapse field in epitaxial garnet films by etching", J. Electron. Mater., $\underline{4}$ (4), 757-68 (1975).
199. Luff, P. B., "Bubble in low coercivity channel", Patent USA 3790935, Publ. February 1974.
200. Majumder, D. D. and Das, J., "On the formation of magnetic bubbles and their application as computer storage element", J. Sci. & Ind. Res., $\underline{35}$ (3), 149-52 (1976), CCA12-8888.
201. Malozemoff, A. P., "Magnetic bubble domain system using different types of domains", Patent USA 3899779, Publ. August 1975.
202. Malozemoff, A. P. and Slonczewski, J. C., "Angular momentum compensation in bubble domain devices", IBM Tech. Disclosure Bull., $\underline{18}$ (5), 1622-4 (1975), CCA11-4807.
203. Matsuyama, S., Orihara, S., Iwasa, S. and Yamagishi, K., "Margin degradation in the long-term testing of 3 µm bubble devices", IEEE Trans. Magn., MAG-$\underline{13}$ (5), 1586-91 (1977).
204. Mee. J. E., Besser, P. J., Pulliam, G. R., Heinz, D. M. and Elkins, P. E., "Method for producing bubble domains in magnetic film-substrate structures", Patent USA 3745046, Publ. July 1973.
205. Michaelis, P. C., "Asynchronous magnetic circuit", Patent USA 3480925, Publ. November 1969.
206. Mukhortov, Y. P., "Effect of propagation of a magnetic bubble on its shape", Sov. Phys. Solid State, $\underline{18}$ (5), 777-81 (1976).
207. Pulliam, G. R., Besser, P. J., Mee, J. E. and Heinz, D. M., "Confinement of bubble domains in film-substrate structures", Patent USA 3753814, Publ. August 1973.
208. Slonczewski, J. C., "Magnetic domain systems using domains having different properties", Patent USA 3890605, Publ. June 1975.
209. Sugiyama, Y. and Kataoka, S., "Direct measurement of stray magnetic field distributions from a magnetic bubble domain in yttrium orthoferrite with a micro hall element", Proc. IEEE, $\underline{64}$ (11), 1643-4 (1976).
210. Tomas, I., "Frequency dependence of bubble domain oscillations", Phys. Status Solidi A, $\underline{30}$ (2), 587-91 (1975).
211. Vella-Coleiro, G. P., "Domain wall mobility in epitaxial garnet films", AIP Conf. Proc., $\underline{10}$, 424 (1972).
212. Vella-Coleiro, G. P. and Tabor, W. J., "Measurement of magnetic bubble mobility in epitaxial garnet films", Appl. Phys. Lett., $\underline{21}$, 7 (1972).
213. Wolfe, R. and North, J. C., "Suppression of hard bubbles in magnetic garnet films by ion implantation", Bell Syst. Tech. J., $\underline{51}$, 1436 (1972).

214. Yamaguchi, K., Asama, K. and Namikata, T., "Hard bubble proper-
     ties in garnet films suppressed by ion implantation, permalloy
     film and thin garnet film", Fujitsu Sci. Tech. J, 11 (2),
     49-70 (1975).

5.   Drive Circuits
a.   Generation

215. Almasi, G. S., Argyle, B. E. and Deluca, J. C., "Method for
     generating hollow magnetic bubble domains", Patent USA 3811119,
     Publ. May 1974.
216. Anon., "Stripe-line coil for magnetic-field generation in bubble
     memory devices", Comput. Des., 15 (7), 116 (1976), CCA11-31596.
217. Archer, J. L., Tocci, L. R. and Chen, T. T., "Multiple bar
     bubble domain", Patent USA 3824565, Publ. July 1974.
218. Ashkin, A. and Dziedzic, J. M., "Interaction of laser light with
     magnetic domains", Appl. Phys. Lett., 21 (6), 253-5 (1972),
     PA75-75230.
219. Bobeck, A. H. and Danylchuk, I., "Single wall domain generator",
     Patent USA 3706082, Publ. December 1972.
220. Bobeck, A. H., Kowalchuk, R. and Reekstin, J. P., Jr., "Fail-
     safe domain generator for single wall domain arrangements",
     Patent USA 3706081, Publ. December 1972.
221. Bonyhard, P. I., "Single wall domain source", Patent USA
     3611331, Publ. October 1971.
222. Brown, B. R. and Henry, G. R., "Magnetic bubble nucleation with
     controlled chirality", IBM Tech. Disclosure Bull., 18 (2), 553-4
     (1975), CCA10-25840.
223. Bullock, D. C. and Epstein, D. J., "A new technique for gener-
     ating magnetic bubbles", Proc. IEEE, 59 (12), 1713-14 (1971),
     EEA75-634.
224. Chang, H., "Magnetic means for collapsing and splitting of
     cylindrical domains", Patent USA 3727197, Publ. April 1973.
225. Chen, Y. S., Geusic, J. E., Nelson, T. H. and Shapiro, H. M.,
     "Single wall domain nucleator", Patent USA 3789375, Publ.
     January 1974.
226. Chen, Y. S., Geusic, J. E. and Smith, J. L., "Characterization
     of magnetic bubble generators", IEEE Trans. Magn., MAG-10 (1),
     23-7 (1974), CCA9-18389.
227. Clover, R. B., Jr. and Waites, R. F., "Magnetic bubble genera-
     tion", Patent USA 3824571, Publ. July 1974.
228. Danylchuk, I., "Single-wall domain generator", Patent USA
     3633185, Publ. January 1972.
229. Danylchuk, I., "Single wall domain fanout circuit", Patent USA
     3713118, Publ. January 1973.
230. George, P. K., "Disk generator", Patent USA 3925769, Publ.
     December 1975.
231. Geusic, J. E., "Single wall magnetic domain generator", Patent
     USA 3781833, Publ. December 1973.

232. Geusic, J. E., "Single wall domain generator", Patent USA 3786452, Publ. January 1974.
233. Mee, J. E. and Besser, P. J., "Method for producing bubble domains in magnetic film-substrate structures", Patent USA 3788896, Publ. January 1974.
234. Mee, J. E., Besser, P. J., Pulliam, G. R., Heinz, D. M. and Elkins, P. E., "Method for producing bubble domains in magnetic film-substrate structures", Patent USA 3728152, Publ. April 1973.
235. Michaelis, P. C., "Input for single-wall domain arrangement", Patent USA 3735370, Publ. May 1973.
236. Navratil, F., "Generation and fast switching of high frequency rotating fields for bubble memories", IEEE Trans. Magn., MAG-11 (5), 1154-6 (1975), CCA11-4803.
237. Olivei, A., "Generating and detecting circuits in the magnetic bubble memory devices", Ind. Ital. Elettrotec. & Elettron., 25 (2), 128-35 (1972), Italian, EEA75-23634.

b.   Propagation

238. Almasi, G. S., "Bubble domain propagation and sensing", Proc. IEEE, 61 (4), 438-44 (1973), CCA8-13682.
239. Almasi, G. S. and Keefe, G. E., "Two-phase propagation of cylindrical magnetic domains", Patent USA 3893089, Publ. July 1975.
240. Aziz, Z., Clegg, W. and Pickard, R. M., "Theoretically-computed bubble motion and experimental data", Electron. Engineering, 48 (580), 48-50 (1976), CCA11-22653.
241. Bailey, P. T. and Doerr, L. J., III, "Mutually exclusive magnetic bubble propagation circuits with discrete elements", Patent USA 3899716, Publ. April 1975.
242. Battarel, C., "Propagation register for magnetic domains", Patent USA 3889246, Publ. June 1975.
243. Bobeck, A. H., "Domain propagation arrangement", Patent USA 3529303, Publ. September 1970.
244. Bobeck, A. H., "Magnetic domain propagation arrangement", Patent USA 3534346, Publ. October 1970.
245. Bobeck, A. H., "Single wall domain propagation arrangement", Patent USA 3534347, Publ. October 1970.
246. Bobeck, A. H., "Domain-propagation arrangement", Patent USA 3644908, Publ. February 1972.
247. Bobeck, A. H., "Single wall domain transfer circuit", Patent USA 3676870, Publ. July 1972.
248. Bobeck, A. H., "Single wall domain propagation arrangement", Patent USA 3714640, Publ. January 1973.
249. Bobeck, A. H., "Magnetic domain propagation arrangement having channels defined by straight line boundaries", Patent USA 3811120, Publ. May 1974.

250. Bobeck, A. H., Copeland, J. A., III and Wolfe, R., "Single wall domain propagation arrangement", Patent USA 3778788, Publ. December 1973.
251. Bobeck, A. H. and Della Torre, E., "Magnetic single wall domain propagation device", Patent USA 3523286, Publ. June 1970.
252. Bobeck, A. H., Della Torre, E. and Scovil, H. E. D., "Magnetic propagation device wherein pole patterns move along the periphery of magnetic disks", Patent USA 3516077, Publ. June 1970.
253. Bobeck, A. H., Della Torre, E. and Scovil, H. E. D., "Domain propagation device", Patent USA 3530444, Publ. September 1970.
254. Bobeck, A. H., Della Torre, E. and Thiele, A. A., "Single domain wall propagation in magnetic sheets", Patent USA 3513452, Publ. May 1970.
255. Bobeck, A. H. and Fischer, R. F., "Conductor arrangement for propagation of single wall domains in magnetic sheets", Patent USA 3506975, Publ. April 1970.
256. Bobeck, A. H. and Fischer, R. F., "Single wall domain device", Patent USA 3540019, Publ. November 1970.
257. Bobeck, A. H. and Fischer, R. F., "Strip domain propagation arrangement", Patent USA 3710356, Publ. January 1973.
258. Bobeck, A. H., Fischer, R. B. and Scovil, H. E. D., "Inverted mode domain propagation device", Patent USA 3540021, Publ. November 1970.
259. Bobeck, A. H. and Gianola, U. F., "Magnetic domain propagation arrangement", Patent USA 3541534, Publ. November 1970.
260. Bobeck, A. H., Gianola, U. F., Sherwood, R. C. and Shockley, W., "Magnetic domain propagation circuit", Patent USA 3460116, Publ. August 1969.
261. Bobeck, A. H. and Levinstein, H. J., "Magnetic domain propagation arrangement including medium with graded magnetic properties", Patent USA 3701127, Publ. October 1972.
262. Bobeck, A. H. and Nelson, T. J., "Transfer circuit for single wall", Patent USA 3832701, Publ. August 1974.
263. Bobeck, A. H. and Scovil, H. E. D., "Magnetic domain shifting arrangement employing moveable strip domain", Patent USA 3887905, Publ. June 1975.
264. Bobeck, A. H., Scovil, H. E. D. and Thiele, A. A., "Magnetic domain propagation arrangement", Patent USA 3603939, Publ. September 1971.
265. Bobeck, A. H. and Smith, J. L., "Domain propagation media organized for repertory dialer operation", Patent USA 3479654, Publ. November 1969.
266. Bogholtz, W. E., Hendel, R. J. and Lin, Y. S., "Gapless propagation element for magnetic bubble domains", IBM Tech. Disclosure Bull., 16 (10), 3395 (1974), CCA9-20763.
267. Bonyhard, P. I., "Magnetic, single wall domain, or logic using chevron domain propagating elements", Patent USA 3789373, Publ. January 1974.

268. Bonyhard, P. I., Chen, Y. S. and Smith, J. L., "Magnetic bubble
     passive replicator", Patent USA 3868661, Publ. February 1975.
269. Bonyhard, P. I. and Danylchuk, I., "Single wall domain fast
     transfer circuit", Patent USA 3623034, Publ. November 1971.
270. Bonyhard, P. I., Kish, D. E. and Smith, J. L., "Domain propa-
     gation arrangement", Patent USA 3597748, Publ. August 1971.
271. Bonyhard, P. I., Kisk, D. E. and Smith, J. L., "Magnetic domain
     propagation arrangement", Patent USA 3613058, Publ. October 1971.
272. Bonyhard, P. I. and Michaelis, P. C., "Single-wall domain
     arrangement", Patent USA 3713116, Publ. January 1973.
273. Bonyhard, P. I. and Nelson, T. J., "Dynamic data reallocation
     in bubble memories", Bell Syst. Tech. J., $\underline{52}$ (3), 307-17 (1973),
     CCA8-13661.
274. Carr, W. J., Jr. and Miller, R. C., "Propagation of magnetic
     domains by self-induced drive fields", Patent USA 3825910,
     Publ. July 1974.
275. Chang, H., Chen, T. C., Lin, C. C. and Tung, C., "Stoppage and
     diversion of magnetic bubble streams", IBM Tech. Disclosure
     Bull., $\underline{17}$ (8), 2493-4 (1975), CCA10-15168.
276. Chen, Y. S., Geusic, J. E. and Nelson, T. J., "Single wall
     domain propagation arrangement", Patent USA 3797001, Publ.
     February 1974.
277. Chow, W. F., "Domain propagation circuit", Patent USA 3680067,
     Publ. July 1972.
278. Clover, R. B., Jr., "Bubble domain sonic propagation device",
     Patent USA 3673582, Publ. June 1972.
279. Cohen, M. S., "Contiguous bubble domain propagation structures",
     IBM Tech. Disclosure Bull., $\underline{18}$ (9), 3082-4 (1976), CCA11-15053.
280. Copeland, J. A., III, "Magnetic domain propagation arrangement",
     Patent USA 3631413, Publ. December 1971.
281. Copeland, J. A., III, "Magnetic domain propagation arrangement",
     Patent USA 3638205, Publ. January 1972.
282. Copeland, J. A., III, "Domain propagation arrangement", Patent
     USA 3636531, Publ. January 1972.
283. Copeland, J. A., III, "Multilevel domain propagation arrange-
     ment", Patent USA 3668677, Publ. June 1972.
284. Copeland, J. A., III, "Single wall domain fanout circuit",
     Patent USA 3680066, Publ. July 1972.
285. Copeland, J. A., III, "Conductor propagation circuits for high-
     density bubble-domain memories", IEEE Trans. Magn., MAG-$\underline{8}$ (3),
     378 (1972), CCA8-13676.
286. Copeland, J. A., III, "Single wall domain information transfer
     arrangement", Patent USA 3696347, Publ. October 1972.
287. Copeland, J. A., III, "Self-biasing single wall domain arrange-
     ment", Patent USA 3701129, Publ. October 1972.
288. Copeland, J. A., III, "Write circuit using enhanced propagation
     pulses for lateral displacement coding of patterns of single-wall
     magnetic domains", Patent USA 3711840, Publ. January 1973.

289. Copeland, J. A., III, "Method and apparatus for selectively accessing a serial string of enhanced propagation pulse write circuits", Patent USA 3745542, Publ. July 1973.
290. Copeland, J. A., III, "Conductor-pattern apparatus for controllably inverting the sequence of a serial pattern of single-wall magnetic domains", Patent USA 3774182, Publ. November 1972.
291. Copeland, J. A., III, Elward, J. P., Johnson, W. A. and Ruch, J. G., "Single-conductor magnetic-bubble propagation circuits", J. Appl. Phys., 42 (4), 1266-7 (1971), CCA6-15223.
292. de Jonge, F. A., Druijvesteijn, W. F., Verhulst, A. G. H. and Enz, U. E., "Magnetic domain propagation device", Patent USA 3866190, Publ. February 1975.
293. Della Torre, E., "Oscillating transverse field bubble domain propagation", Patent USA 3705394, Publ. December 1972.
294. Della Torre, E. D. and Dimyan, M., "Magnetic domain propagation plate with minimized temperature sensitivity", Patent USA 3827035, Publ. July 1974.
295. Della Torre, E. and Kinsner, W., "Bubble domain propagation circuit", Patent USA 3895364, Publ. July 1975.
296. Fischer, R. F., "Domain propagation arrangement", Patent USA 3699551, Publ. October 1972.
297. Fischer, R. F., North, J. C. and Wolfe, R., "Single wall domain propagation arrangement", Patent USA 3838329, Publ. August 1974.
298. Fischer, R. F., Schmidt, P. H. and Spencer, E. G., "Single wall domain propagation arrangement", Patent USA 3824568, Publ. July 1974.
299. Flannery, W. E., "Bi-directional magnetic domain transfer circuit", Patent USA 3896421, Publ. July 1975.
300. Fresia, J. F. and Lin, C. C., "Control field conductor for bubble memory", IBM Tech. Disclosure Bull., 14 (4), 1262-3 (1971), EEA75-9779.
301. George, P. K., "Complementary corner structures for magnetic domain propagation", Patent USA 3924249, Publ. December 1975.
302. Ho, C. P. and Chang, H., "Field-access bubble-lattice propagation devices", IEEE Trans. Magn., MAG-13 (2), 945-52 (1977), CCA12-6656.
303. Hsin, C. H., Mateovich, T. L. and Coren, R. L., "Measurements of bubble drive fields in a T bar configuration", AIP Conf. Proc., 5 (Pt. I), 244-8 (1971), EEA75-19502.
304. Hu, H. L., Keefe, G. E., Lin, Y. S. and Rosier, L. L., "Half-circle permalloy circuits for bubble domain devices", IBM Tech. Disclosure Bull., 16 (10), 3393 (1974), CCA9-20761.
305. Keefe, G. E., Lin, Y. S. and Rosier, L. L., "Gapless multi-thickness propagation structure for magnetic domain devices", Patent USA 3914751, Publ. October 1975.
306. Kefalas, J. H., "Magnetic device for domain wall propagation", Patent USA 3711838, Publ. January 1973.

307. Khodenkov, H. E. and Raev, V. K., "Dynamics of normal magnetic bubble domains in propagation circuit", Phys. Status Solidi A, $\underline{28}$ (1), K29-32 (1975).

308. Kish, D. E. and Smith, J. L., "Domain propagation arrangement", Patent USA 3702994, Publ. November 1972.

309. Kleparsky, V. G., Rozenblat, M. A. and Romanov, A. M., "Bubble domain propagation dynamics in field-access permalloy driving circuits", IEEE Trans. Magn., MAG-$\underline{11}$ (5), 1130-2 (1975).

310. Kohara, H., "Magnetic domain propagating circuit", Patent USA 3891978, Publ. June 1975.

311. Krupp, R. S. and Tomko, L. A., "Electrically controllable steering arrangement for magnetic single-wall domain propagation paths", Patent USA 3723985, Publ. March 1973.

312. Kurtzig, A. J., "Faraday rotation devices", Patent USA 3560944, Publ. February 1971.

313. Kurtzig, A. J., "Single wall magnetic domain propagation arrangement", Patent USA 3602911, Publ. August 1971.

314. Lee, F. S., "Magnetic bubble crossover circuit", Patent USA 3676873, Publ. July 1972.

315. Lin, Y. S., "High-density bubble domain propagation structure", IBM Tech. Disclosure Bull., $\underline{16}$ (10), 3405 (1974), CCA8-20765.

316. Lin, Y. S., "Gapless double-sided propagation structure for bubble domain devices", Patent USA 3925768, Publ. December 1975.

317. Lock, R. D., "Propagation of magnetic bubble domains", Patent USA 3676872, Publ. July 1972.

318. Luff, P. P. and Lucas, J. M., "Cylindrical magnetic domain propagation in $Sm_{0.55}Tb_{0.45}FeO_3$ platelets possessing high surface coercivity", J. Appl. Phys., $\underline{42}$ (12), 5173-5 (1971), PA75-6967.

319. Marsh, A., "Circular magnetic domain devices", Patent USA 3736579, Publ. May 1973.

320. Matsuyama, S., Kinoshita, R. and Segawa, M., "Two layer permalloy circuits for 1.5 μm diameter bubble propagation", AIP Conf. Proc., no. 24, 645-6 (1974), CCA11-6373.

321. Michaelis, P. C., "Magnetic domain propagation device", Patent USA 3454939, Publ. July 1969.

322. Minnick, R. C., Bailey, P. T. and Sandfort, R. M., "Magnetic bubble transmission circuit", Patent USA 3921155, Publ. November 1975.

323. Morrow, R. H., "Domain propagation arrangement", Patent USA 3577131, Publ. May 1971.

324. Morrow, R. H. and Perneski, A. J., "Single wall domain apparatus having intersecting propagation channels", Patent USA 3543255, Publ. November 1970.

325. Oshima, S., Watanabe, T. and Ishihara, H., "Control system for magnetic bubbles", Patent USA 3772661, Publ. November 1973.

326. Owens, J. M., "Conductor arrangement for propagation in magnetic bubble domain systems", Patent USA 3678479, Publ. July 1972.

327. Parzefall, F., "Cylindrical domain propagation pattern", Patent
     USA 3828330, Publ. August 1974.
328. Parzefall, F., Lill, A., Littwin, B., Metzdorf, W. and Mavratil,
     F., "A 4k bit bubble memory chip using X-bar propagation patterns
     fabricated by electrode position", Microelectronics, $\underline{7}$ (1),
     27-30 (1975), CCA11-6375.
329. Parzefall, F., Lill, A., Metzdorf, W. and Navratil, F., "A 4k
     bit bubble memory chip using X-bar propagation patterns", IEEE
     Trans. Magn., MAG-$\underline{11}$ (5), 1160-3 (1975), CCA11-4805.
330. Perneski, A. J., "Propagation of cylindrical magnetic domains
     in orthoferrites", IEEE Trans. Magn., MAG-$\underline{5}$, 554 (1969).
331. Perneski, A. J., "Magnetic domain propagation arrangement",
     Patent USA 3518643, Publ. June 1970.
332. Perneski, A. J., "Magnetic domain fanout circuit", Patent USA
     3530446, Publ. September 1970.
333. Perneski, A. J., "Domain propagation arrangement having repeti-
     tive patterns of overlay material of different coercive forces",
     Patent USA 3541535, Publ. November 1970.
334. Perneski, A. J., "Domain propagation arrangement", Patent USA
     3543252, Publ. November 1970.
335. Perneski, A. J., "Domain propagation arrangement", Patent USA
     3555527, Publ. January 1971.
336. Perneski, A. J., "Single wall domain switching arrangement",
     Patent USA 3573765, Publ. April 1971.
337. Perneski, A. J., "Single wall domain switching arrangement",
     Patent USA 3596261, Publ. July 1971.
338. Rossol, F. C., "Stroboscopic observation of cylindrical domain
     propagation in a T-bar structure", IEEE Trans. Magn., MAG-$\underline{7}$
     (1), 142-5 (1971), PA74-41409.
339. Sandfort, R. M. and Bailey, P. T., "Field-accessed magnetic
     bubble replicator", Patent USA 3922652, Publ. November 1975.
340. Sherwood, R. C. and Thiele, A. A., "Domain propagation device
     with high domain mobility", Patent USA 3534341, Publ. October
     1970.
341. Shick, L. K., Nielsen, J. W., Bobeck, A. H., Kurtzig, A. J.,
     Michaelis, P. C. and Reekstin, J. P., "Liquid phase epitaxial
     growth of uniaxial garnet films: circuit deposition and bubble
     propagation", Appl. Phys. Lett., $\underline{18}$ (3), 89-91 (1971),
     EEA74-15009.
342. Smith, J. L., "Memory device employing a propagation medium",
     Patent USA 3508225, Publ. April 1970.
343. Smith, R. M., "Magnetic domain switching matrix and control
     arrangement", Patent USA 3753253, Publ. August 1973.
344. Tabor, W. J., "Faraday effect readout of magnetic domains in
     magnetic materials exhibiting birefringence", Patent USA
     3585614, Publ. June 1971.
345. Thiele, A. A., "Magnetic device propagation sheet", Patent
     USA 3508221, Publ. April 1970.

346. Thiele, A. A., "Domain propagation arrangement", Patent USA
     3638206, Publ. January 1972.
347. Watanabe, T. and Ishihara, H., "High-speed transmission system
     for magnetic bubbles", Patent USA 3778789, Publ. December 1973.
348. Watanabe, T. and Ishihara, H., "Magnetic bubble transmission
     system using a rotating magnetic field", Patent USA 3899781,
     Publ. August 1975.
349. Yamagishi, K., Maegawa, H., Komatsu, M. and Takai, S., "A new
     proposal on field-access bubble drives", IEEE Trans. Magn.,
     MAG-11, 16-20 (1975).
350. Yamauchi, F., "Cylindrical magnetic domain propagating circuit
     and logic circuit", Patent USA 3753250, Publ. August 1973.
351. Yamauchi, F., "Information propagation path switching device",
     Patent USA 3786447, Publ. January 1974.
352. Yoshizawa, S., Mikami, I., Kamoshita, G. and Saito, N., "Driving
     system in magnetic single wall domain device", Patent USA
     3763478, Publ. October 1973.

c.   Detection

353. Ahearn, W. E., Almasi, G. S. and Genovese, E. R., "High-density
     magneto-optic readout apparatus", Patent USA 3729724, Publ.
     April 1973.
354. Almasi, G. S., Chang, H., Keefe, G. E. and Thompson, D. A.,
     "Integrated magneto-resistive sensing of bubble domains", Patent
     USA 3691540, Publ. September 1972.
355. Almasi, G. S., Hendel, R. J. and Keefe, G. E., "Magneto-
     resistive sensing of magnetic bubble domains using expansion",
     Patent USA 3781832, Publ. December 1973.
356. Almasi, G. S. and Keefe, G. E., "Magnetoresistive sensing of
     bubble domains with noise suppression", Patent USA 3736419,
     Publ. May 1973.
357. Almasi, G. S., Keefe, G. E., Lin, Y. S. and Thompson, D. A.,
     "Magnetoresistive detector for bubble domains", J. Appl. Phys.,
     42 (4), 1268-9 (1971), EEA74-21884.
358. Almasi, G. S., Keefe, G. E., Lin, Y. S. and Thompson, D. A.,
     "Magnetoresistive sensing device for detection of magnetic
     fields having a shape anisotropy field and uniaxial anisotropy
     field which are perpendicular", Patent USA 3716781, Publ.
     February 1973.
359. Almasi, G. S., Keefe, G. E. and Terlep, K. D., "High-speed
     sensing of small bubble domains", AIP Conf. Proc., 10, 207
     (1972).
360. Argyle, B. E. and DiStefano, T. H., "Inductive sensor for
     magnetic bubble domain detection", Patent USA 3842407, Publ.
     October 1974.
361. Bailot, N. P., Enright, C. J., Kiseda, J. T. and Taylor, D. M.,
     "Magnetic bubble sensing", IBM Tech. Disclosure Bull., 13
     (10), 3100 (1971), EEA74-17865.

362. Bajorek, C. H., Levannoni, M. and Thompson, D. A., "Scanning magnetoresistive sensors used in recording and bubble sensing applications", IBM Tech. Disclosure Bull., 18 (7), 274-5 (1975), CCA10-25839.

363. Bate, R. T., "Magnetic field sensor", Patent USA 3714523, Publ. January 1973.

364. Beausoleil, W. F., Keefe, G. E. and Walker, E. L., "Magneto resistive signal multiplier for sensing magnetic bubble domains", Patent USA 3858189, Publ. December 1974.

365. Bierlein, J. D., "Optical scanner", Patent USA 3760385, Publ. September 1973.

366. Bobeck, A. H., "Magnetic domain detector arrangement", Patent USA 3689901, Publ. September 1972.

367. Bobeck, A. H., "Magnetoresistance detector for single wall domains", Patent USA 3713117, Publ. January 1973.

368. Bobeck, A. H., "Magnetic single wall domain expansion circuit", Patent USA 3786453, Publ. January 1974.

369. Bobeck, A. H., "Integrated bubble expansion detector and dynamic guard rail arrangement", Patent USA 3810132, Publ. May 1974.

370. Bobeck, A. H., Ciak, F. J. and Strauss, W., "Magnetoresistance detector for single wall magnetic domains", Patent USA 3713120, Publ. January 1973.

371. Bobeck, A. H., Ciak, F. J. and Strauss, W., "Magnetic domain detector arrangement", Patent USA 3820092, Publ. June 1974.

372. Brown, B. R. and Henry, G. R., "Magnetic bubble chirality detector", IBM Tech. Disclosure Bull., 18 (2), 535 (1975), CCA10-25841.

373. Brown, B. R. and Henry, G. R., "Magnetic bubble chirality detector", IBM Tech. Disclosure Bull., 18 (2), 536 (1975), CCA10-25842.

374. Buhrer, C. F., "An rf magneto-resistive magnetic domain detector", Patent USA 3813660, Publ. May 1974.

375. Chang, H., "Sensing of cylindrical magnetic domains", Patent USA 3720928, Publ. March 1973.

376. Cohen, L. D., Shortt, B. A. and Urban, M. J., "Magneto-resistive magnetic domain detector", Patent USA 3838406, Publ. September 1974.

377. Copeland, J. A., III, "Detector for magnetic domain arrangements", Patent USA 3701128, Publ. October 1972.

378. Holtzberg, F., Mayadas, A. F., Thompson, W. A. and Von Molnar, S., "Detection of magnetic domains by tunnel junctions", Patent USA 3840865, Publ. October 1973.

379. IBM Corp., "Magnetic bubble domain apparatus", Patent UK 1400961, Publ. July 1975, CCA11-4810.

380. Katoaka, S., Yamada, H., Iida, S. and Sugiyama, Y., "Semiconductor functional arrayed detector for magnetic bubble domains", Proc. IEEE, 60 (4), 460 (1972), EEA75-19582.

381. Kinsner, W. and Della Torre, E., "Magnetic bubble domain sensing device", Patent USA 3909809, Publ. September 1975.

382. Kryder, M. H., Ahn, K. Y. and Powers, J. V., "Amorphous film
     magnetic bubble domain devices", IEEE Trans. Magn., MAG-11 (5),
     1145-7 (1975), CCA11-4802.
383. Lama, U. G., "Magnetic bubble domain detection", Patent USA
     3820089, Publ. June 1974.
384. Lill, A., "Magnetoresistive detectors for bubble memories",
     J. Magn. & Magn. Mater., 4 (1-4), 159-65 (1977), German,
     CCA12-6661.
385. Marsh, A., "Circular magnetic domain devices", Patent USA
     3836897, Publ. September 1972.
386. Myer, J. H., "Magnetoresistive readout for domain addressing
     interrogator", Patent USA 3806899, Publ. April 1974.
387. Ooomote, R., Sakamot, K., Nitta, M. and Ishii, O., "A considera-
     tion of optical detection system for magnetic bubble-domain
     memories", Bull. Electrotech. Lab., 37 (7), 718-26 (1973),
     Japanese, CCA9-9202.
388. Owens, J. M., "Conductor arrangement for propagation in
     magnetic bubble domain systems", Patent USA 3693177, Publ.
     September 1972.
389. Scarzello, J. F., "Magnetic bubble resonance sensor", Patent
     USA 3824573, Publ. July 1974.
390. Strauss, W., "A Permalloy detector for orthoferrite domains",
     Proc. IEEE, 58 (9), 1386-7 (1970), EEA73-37239.
391. Strauss, W., "Detection of cylindrical magnetic domains", J.
     Appl. Phys., 42 (4), 1251-7 (1971), EEA74-21905.
392. Strauss, W., "Magnetic domain detector", Patent USA 3609720,
     Publ. September 1971.
393. Strauss, W., Bobeck, A. H. and Ciak, F. J., "Characteristics
     of a detection-propagation structure for bubble-domain devices",
     AIP Conf. Proc., 10, 202 (1972).
394. Strauss, W., Shumate, P. W., Jr. and Ciak, F. J., "Magneto-
     resistance sensors for garnet bubble domains", AIP Conf. Proc.,
     5 (Pt. I), 235-9 (1971), EEA75-19527.
395. Strauss, W. and Smith, G. E., "Hall-effect domain detector",
     J. Appl. Phys., 41 (3), 1169-70 (1970), EEA73-24150.
396. Tabor, W. J., "Optical readout implementation", Patent USA
     3515456, Publ. June 1970.
397. Thompson, D. A., Romankiw, L. T. and Mayadas, A. F., "Thin film
     magnetoresistors in memory, storage, and related applications",
     IEEE Trans, Magn., MAG-11 (4), 1039-50 (1975).
398. Yoshizawa, S., Shigeta, T. O. J., Mikami, I. and Kamoshita, G.
     K., "Detection of bubble domains by Hall effect of evaporated
     indium antimonide", AIP Conf. Proc., 5 (Pt. I), 230-4 (1971),
     EEA75-19526.
399. Yoshizawa, S., Yamamoto, N., Ota, H., Shigeta, T. O. J. and
     Kamoshita, G., "Small bubble domain detection by Hall effect in
     InSb films", IEEE Trans. Magn., MAG-8, 454 (1972).

6. Chip Design

400. Almasi, G. S., "Bubble domain chip arrangement", Patent USA
     3750154, Publ. July 1973.
401. Almasi, G. S. and Lin, Y. S., "Analytical design theory for
     field-access bubble domain devices", IEEE Trans. Magn., MAG-12
     (3), 160-202 (1976).
402. Anon., "Conceptual design of a 10 to the 8th power bit magnetic
     bubble domain mass storage unit and fabrication, test and
     delivery of a feasibility model", Report NASA-CR-123577, IBM,
     Huntsville, Ala., 1973, 283 pp., CCA8-7344.
403. Anon., "Design, fabrication and test of 16k bubble memory chips",
     IEEE Trans. Magn., MAG-11 (5), 1157-9 (1975).
404. Anon., "A 1-1/2-level on-chip-decoding bubble memory chip
     design", Comput. Des., 16 (2), 116 (1977), CCA12-10694.
405. Beck, L., Brennian, R., Casey, M. and 16 others, "The design
     fabrication and test of 16K bubble memory chips", IEEE Trans.
     Magn., MAG-11 (5), 1157-9 (1975), CCA11-4804.
406. Bhandarkar, D. P., "On the performance of magnetic bubble mem-
     ories in computer systems", IEEE Trans. Comput., C-24 (11),
     1125-9 (1975).
407. Bobeck, A. H., "Switching crosspoint arrangement", Patent USA
     3470547, Publ. September 1969.
408. Bobeck, A. H., "Magnetic memory arrangement comprising domain
     wall propagation channels", Patent USA 3470546, Publ. September
     1969.
409. Bonyhard, P. I., Chen, Y. S. and Smith, J. L., "Device design
     and system organization for a decoder accessed magnetic bubble
     memory chip", AIP Conf. Proc., no. 18, 100-3 (1973), CCA9-20755.
410. Bonyhard, P. I., Chen, Y. S. and Smith, J. L., "High perform-
     ance magnetic bubble replicate gate design", IEEE Trans. Magn.,
     MAG-13 (5), 1258-60 (1977).
411. Bonyhard, P. I., Gianola, U. F. and Perneski, A. J., "Magnetic
     domain storage organization", Patent USA 3618054, Publ.
     November 1971.
412. Bonyhard, P. I. and Smith, J. L., "68k bit capacity 16 μm-period
     magnetic bubble memory chip design with 2 μm minimum features",
     IEEE Trans. Magn., MAG-12 (6), 614-17 (1976), CCA12-6649.
413. Bouricius, W. G., Carter, W. C., Hsieh, E. P., Jessep, D. C.,
     Jr. and Wadia, A. B., "Modeling of a bubble-memory organization
     with self-checking translators to achieve high relability",
     IEEE Trans. Comput., C-22 (3), 269-75 (1973), CCA8-7336.
414. Boyarchenkov, M. A., Abramov, V. V. and Rosental, J. D., "Cal-
     culation and design of magnetic bubble logic elements", Micro-
     electronics, 7 (1), 31-4 (1975), CCA11-6335.
415. Brown, E. T., "Design criteria with magnetic-bubble memories",
     Elektronik, 25 (9), 45-50 (1976), German, CCA11-31598.
416. Chang, H., "Domain transfer between adjacent magnetic chips",
     Patent USA 3736577, Publ. May 1973.

417. Chang, H. and Fan, G. J., "Magnetic bubble domain system using
     multiple magnetic sheets", Patent USA 3760387, Publ. September
     1973.
418. Chen, T. T., Gergis, I. S. and Tocci, L. R., "Multiple level
     storage organization for bubble memories", IEEE Trans. Magn.,
     MAG-12 (6), 629 (1976), CCA12-6651.
419. Cohen, M. S., Beall, G. W., Hsieh, W. J. and Chang, H., "The
     Y-bar switch - a single-level- masking switch", IEEE Trans.
     Magn., MAG-13 (5), 1264-6 (1977).
420. Copeland, J. A., Josenhans, J. G. and Spiwak, R. R., "Circuit
     and module design for conductor-groove bubble memories", IEEE
     Trans. Magn., MAG-9 (3), 489-92 (1973), CCA9-3412.
421. Dekker, E. H. L. J., van Mierloo, K. L. L. and de Werdt, R.,
     "Combination of field and current access magnetic bubble cir-
     cuits", IEEE Trans. Magn., MAG-13 (5), 1261-3 (1977).
422. Geusic, J. E., "Bubble memory design and performance", IEEE
     Trans. Magn., MAG-12 (6), 622 (1976), CCA12-6650.
423. Halatsis, C., Philokyprou, G. and Maritsas, D., "Designing
     block-structured bubble-up buffers", Electron. Eng., 48 (578),
     45-8 (1976).
424. Hiroshima, M., Asano, A., Yoshizawa, S., Saito, N. and Kasai,
     M., "Design and evaluation of 64k bit magnetic bubble memory
     chip", Jap. J. Appl. Phys., 15 (Suppl. 15-1), 113-17 (1976),
     CCA12-4006.
425. Hoffman, E. J., Moore, R. C. and McGovern, T. L., "Designing
     a magnetic bubble data recorder - the component level", Comput.
     Des., 15 (3), 77-85 (1976).
426. Igarashi, S., Igarashi, K., Hirano, A., Orihara, S. and
     Yamagishi, K., "Design of a 3 µm bubble 80-k bit memory chip",
     AIP Conf. Proc., no. 29, 48-50 (1976), CCA11-22647.
427. Judd, F. F., Young, I. H. and Michaelis, P. C., "Analysis of
     rotating fields in magnetic bubble drive coils", IEEE Trans.
     Magn., MAG-12 (6), 642-4 (1976), CCA12-3998.
428. Kamoshita, G. and Sugita, Y., "Some approaches to high density
     bubble memory chips", Jap. J. Appl. Phys., 15 (Suppl. 15-1),
     145 (1976), CCA12-6658.
429. Kinberg, C., "Managing dummy bubbles in bubble lattice file
     storage systems", IBM Tech. Disclosure Bull., 19 (7), 2702-5
     (1976).
430. Kujipers, F. A., "Single-mask bubble memories: bit/chip organi-
     zation with decoding", IEEE Trans. Magn., MAG-11 (5), 1136-8
     (1975), CCA11-4799.
431. Matsuda, J. and Kinoshita, K., "The two-dimensional bubble
     memory with switching function", Syst. Comput. Control, 6 (2),
     29-35 (1975), CCA11-28276.
432. Nunotani, M., Masumori, T. and Furukawa, S., "Module and cir-
     cuit design for a magnetic bubble memory", Rev. Electr.
     Commun. Lab., 24 (3-4), 266-73 (1976), EEA79-4951.

433. O'Donnell, C. F. and Pulliam, G. R., "Magnetic bubble domain
     system", Patent USA 3827036, Publ. July 1974.
434. Sakai, S., Orihara, S. and Matsuyama, S., "10 kbit bubble memory
     chip: design and fabrication", Jap. J. Appl. Phys., $\underline{15}$ (Suppl.
     15-1), 107-12 (1976), CCA12-4005.
435. Sandfort, R. M., "Nonuniform spacing layer for magnetic bubble
     circuits", Patent USA 3921157, Publ. November 1975.
436. Takahashi, K., Kohara, H., Wada, Y. and Yamada, H., "Magnetic
     bubble memory - a pilot model", NEC Res. & Dev., no. 42, 61-70
     (1976), CCA11-31601.
437. Takasu, M., Maegawa, H., Furuichi, S., Okada, M. and Yamagishi,
     K., " A fast access memory design using 3 μm bubble 80K chip",
     IEEE Trans. Magn., MAG-$\underline{12}$ (6), 633-5 (1976), CCA12-3997.
438. Wong, C. K. and Coppersmith, D., "The generation of permutations
     in magnetic bubble memories", IEEE Trans. Comput., C-$\underline{25}$ (3),
     265-62 (1976), CCA11-9657.
439. Yamaguchi, N., Tsuzuki, N. Yamamoto, M. and Kato, K., "Magnetic
     bubble memory chip design and fabrication", Rev. Electr. Commun.
     Lab., $\underline{24}$ (3-4), 258-65 (1976).

7.   Testing

440. Chang, H., Chen, T. C. and Tung, C., "Error detection and ac-
     commodation in magnetic bubble storage loops", IBM Tech. Dis-
     closure Bull., $\underline{16}$ (10), 3396-7 (1974), CCA9-20764.
441. Ferrio, T., Kenman, R. and Naden, R., "Magnetic bubble memory
     testing", Digital Des., $\underline{7}$ (6), 69-78 (1977).
442. Hagedorn, F. B., "Long-term testing of 68 kbit bubble device
     chips", IEEE Trans. Magn., MAG-$\underline{12}$ (6), 680-2 (1976), CCA12-4000.
443. Hagedorn, F. B., Rago, L. F., Kish, D. E., Chen, Y. S., Hess,
     W. E., Beurrier, H. R. and Wagner, W. D. P., "Magnetic bubble
     device testing", IEEE Trans. Magn., MAG-$\underline{13}$ (5), 1364-9 (1977).
444. Naden, R. A., Keenan, W. R. and Lee, D. M., "Electrical char-
     acterization of a packaged 100 kbit major/minor loop bubble
     device", IEEE Trans. Magn., MAG-$\underline{12}$ (6), 685-7 (1976), CCA12-4001.
445. Orihara, S., Iwasa, S., Majima, T., Nogiwa, K. and Yamagishi,
     K., "Diagnostic testing of a 10-kbit bubble memory chip", IEEE
     Trans. Magn., MAG-$\underline{11}$ (6), 1685-8 (1975), CCA11-2220.
446. Tsuboya, I., Saito, M., Hattanada, T., Yamaguchi, N. and Arai,
     Y., "2 Mbit magnetic bubble memory", IEEE Trans. Magn., MAG-$\underline{13}$
     (5), 1360-3 (1977).
447. Wada, Y., Takahashi, K. and Suga, S., "Development of bubble
     memory chip test system", NEC Res. Dev., no. 41, 61-9 (1976).

8.   Reliability

448. Almasi, G. S., Bouricius, W. G. and Carter, W. C., "Reliability
     and organization of a 10^8-bit bubble domain memory", AIP Conf.
     Proc., 5 (Pt. I), 225-9 (1971), EEA75-19576.

449. Baird, D. H., Buhrer, C. F., Vytal, J. J., Archer, J. L., Toccio, L. R. and Bohning, O. D., "Reliability of magnetic bubble domain memories", AIP Conf. Proc., no. 18, 139 (1973), CCA9-20758.

450. Bouricius, W. G., "Reliability modeling: application to bubble domain memories", AIP Conf. Proc., no. 18, 132-8 (1973), CCA9-20757.

451. Chen, T. T., Tocci, L. R. and Archer, J. L., "Data unreliability in bubble memory devices caused by spontaneous annihilation", AIP Conf. Proc., no. 29, 50 (1976), CCA11-22648.

452. Chen, T. T., Tocci, L. R. and Archer, J. L., "Experimental techniques for studying the reliability of bubble memory devices", J. Appl. Phys., 48 (1), 373-8 (1977), CCA12-10686.

453. DeBonte, W. J. and Butherus, A. D., "Temperature dependence of Ba-ferrite bubble memory bias magnets as a function of magnet geometry", IEEE Trans. Magn., MAG-12 (6), 648-50 (1976), CCA12-6655.

454. Nelson, W. F. and Pulliam, G. R., "Reliability of components in magnetic bubble technology", Communications Systems and Technology Conference, 1974, p. 29-40, CCA9-21992.

455. Shumate, P. W. and Peirce, R. J., "Lifetime characterization of propagated bubble-data streams", Appl. Phys. Lett., 23 (4), 204-5 (1973), CCA8-22809.

456. Strauss, W., "Some temperature characteristics of a bubble memory", IEEE Trans. Magn., MAG-12 (6), 688-90 (1976), CCA12-4002.

457. Yamagishi, K., "Consideration on bubble memory as practical device", Jap. J. Appl. Phys., 15 (Suppl. 15-1), 147-8 (1976), CCA12-6659.

458. Yoshimi, K. and Fujiwara, S., "Information stability in bubble memory elements", AIP Conf. Proc., no. 18, 157-61 (1973), CCA9-20759.

9.   Packaging

459. Anon., "Open coil structure for bubble-memory-device packaging", Comput. Des., 15 (9), 112 (1976), CCA12-3992.

460. Anon., "Bubble-memory packaging alternatives: producibility vs. interchangeability", Digital Des., 7 (1), 54-5 (1977), CCA12-15497.

461. Bobeck, A. H. and Danylchuk, I., "Characterization and test results for a 272K bubble memory package", IEEE Trans. Magn., MAG-13 (5), 1370-2 (1977).

462. Brumley, R. L., "Bubble memory chip mounting", IBM Tech. Disclosure Bull., 16 (10), 3128-9 (1974), CCA9-18394.

463. Carlo, J. T., Stephenson, A. D., Jr. and Hayes, D. J., "Chip packing efficiency in bubble memories", IEEE Trans. Magn., MAG-12 (6), 624-8 (1976), CCA12-3995.

464. DeBonte, W. J. and Butherus, A. D., "Magnetically permeable adhesives and adhesive-jointed shield structures", IEEE Trans. Magn., MAG-13 (5), 1376-8 (1977).

465. Kiseda, J. R., "High speed stripline coils for magnetic bubble devices", IEEE Trans. Magn., MAG-9 (3), 425-9 (1973).
466. Kowalchuk, R., Bobeck, A. H., Butherus, A. D., Ciak, F. J. and Smith, D. H., "Magnetic bubble dual-in-line package (DIP) functional and reliability testing", IEEE Trans. Magn., MAG-12 (6), 691-3 (1976).
467. Lyons, W. A., "Permanent magnet bias schemes for bubble memory applications", 1973 Digest of the Intermag Conference, Paper 26-6, 1 pp.
468. Maegawa, H., Matsuda, J. and Takasu, M., "Flat packaging of magnetic bubble devices", IEEE Trans. Magn., MAG-10 (5), 753-6 (1974).
469. Rifkin, A. A., "A practical approach to packaging magnetic bubble devices", IEEE Trans. Magn., MAG-9 (3), 429-33 (1973), EEA77-6244.
470. Rifkin, A. A., "Non volatile magnetic domain device having binary value bias field excitation", Patent USA 3836896, Publ. September 1974.
471. Rifkin, A. A., Bagholtz, W. B., Bosch, L. S., Downing, R. A., Kiseda, J. R., Lennon, A. L., Jr. and Scott, E. W., "Packaged magnetic domain device having integral bias and switching magnetic field means", Patent USA 3958155, Publ. May 1976.
472. Takasu, M., Maegawa, H., Sukeda, T. and Yamagishi, K., "Reflection coil packaging for bubble devices", IEEE Trans. Magn., MAG-11 (5), 1151-3 (1975), EEA79-4981.
473. Uberbacher, E. C., "Bubble memory package via eggerate", IBM Tech. Disclosure Bull., 14 (11), 3378 (1972), EEA75-28072.
474. Yu, E. Y., "Thermal characteristics of a four-chip magnetic bubble package", IEEE Trans. Magn., MAG-13 (5), 1373-5 (1977).

10. Applications
    a. General

475. Ashkin, A. and Dziedzic, J. M., "Devices employing the interaction of laser light with magnetic domains", Patent USA 3810131, Publ. May 1974.
476. Bobeck, A. H., "Magnetic bubble devices", J. Vac. Sci. & Technol., 9 (4), 1145-50 (1972), EEA75-35674.
477. Bogar, J. E., Bragg, J. K. and Hare, G. H., "System for overcoming faults in magnetic anisotropic material", Patent USA 3792450, Publ. February 1974.
478. Bonner, W. A., Geusic, J. E. and Van Uitert, L. G., "Magnetic devices utilizing garnet epitaxial material", Patent USA 3886533, Publ. May 1975.
479. Bonyhard, P. I., Danylchuk, I., Kish, D. E. and Smith, J. L., "Applications of bubble devices", IEEE Trans. Magn., MAG-6 (3), 447-51 (1970), EEA74-14973.
480. Chang, H., "Bubble domain composer, display, and printer", IBM Tech. Disclosure Bull., 15 (3), 902-3 (1972), CCA8-2644.

481. DeStefanis, G., "Magnetic bubble: logic and memory devices",
     Electron. & Telecommun., <u>21</u> (4), 129-36 (1972), Italian,
     EEA75-39457.
482. Ho, I. T. and Riseman, J., "Integrated magnetic bubble and
     semiconductor device", Patent USA 3786445, Publ. January 1974.
483. Juliussen, J. E., "Magnetic bubble systems approach practical
     use", Comput. Des., <u>15</u> (10), 81-191 (1976), CCA12-3993.
484. Mavity, W. C. and Davis, J. P., "Applications of magnetic
     bubbles", 1972 WESCON Technical Papers, Vol. 16, Paper 8/5,
     6 pp., CCA8-7346.
485. Mikami, I., "Bubble domain apparatus", Patent USA 3742471,
     Publ. June 1973.
486. Murakami, H., "Multimatch processing system with cylindrical
     magnetic domain elements", Patent USA 3803564, Publ. April 1974.
487. Owens, J. M. and Heinz, D. M., "Magnetic bubble domain system",
     Patent USA 3699547, Publ. October 1972.

                          b.   Memory

488. Anon., "A single screen 'bubble' type memory", Rev. Polytech.,
     no. 2, 103-4 (1976), French, CCA11-15057.
489. Anon., "92K bit, 14-pin D.I.L. bubble memory", New Electron.,
     <u>10</u> (5), 20, 22-3 (1977).
490. Bate, R. T. and West, F. G., Jr., "Magnetic domain memory
     structure", Patent USA 3702991, Publ. November 1972.
491. Baxter, P., "Introduction to magnetic bubble stores", Presented
     at Colloquium Digest on Computer Memories, London, England,
     February 1973, CCA8-11636.
492. Beausoleil, W. F., Brown, D. T. and Walker, E. I., "Dynamically
     ordered magnetic bubble shift register memory", Patent USA
     3670313, Publ. June 1972.
493. Beausoleil, W. F. and Phelps, B. E., "Multidimensional dynami-
     cally ordered bubble memory store", IBM Tech. Disclosure Bull.,
     <u>16</u> (9), 2973-5 (1974), CCA9-16062.
494. Beausoleil, W. F. and Phelps, B. E., "Multidimensional dynami-
     cally ordered bubble memory store (with no distinct I/O regi-
     ster)", IBM Tech. Disclosure Bull., <u>16</u> (9), 2976-9 (1974),
     CCA9-16063.
495. Bobeck, A. H., "NDRO of single domain walls in a magnetic sheet
     memory", Patent USA 3471840, Publ. October 1969.
496. Bobeck, A. H. and Danylchuk, I., "Magnetic domain replicator
     arrangement", Patent USA 3810133, Publ. May 1974.
497. Bobeck, A. H., Fischer, R. F., Geusic, J. E. and Nelson, T. J.,
     "Single wall domain memory arrangement", Patent USA 3879585,
     Publ. April 1975.
498. Bobeck, A. H., Michaelis, P. C. and Shockley, W., "Readout
     implementation for magnetic memory", Patent USA 3508222, Publ.
     April 1970.
499. Bobeck, A. H. and Scovil, H. E. D., "Mass memory organization",
     Patent USA 3703712, Publ. November 1972.

500. Bohnlein, A., "Magnetic bubble domain memories", Bull. Assoc.
     Suisse Electr., <u>68</u> (2), 71-6 (1976), CCA12-6646.
501. Bonyhard, P. I. and Nelson, T. J., "Dynamic reallocation of
     information on serial storage arrangements", Patent USA 3701132,
     Publ. October 1972.
502. Bosch, L. J., Downing, R. A., Keefe, G. E., Rosier, L. L. and
     Terlep, K. D., "1024 bit bubble memory chip", IEEE Trans. Magn.,
     MAG-<u>9</u> (3), 481-4 (1973), CCA9-3410.
503. Buhrer, C. F., "Rapid access cylindrical magnetic domain memory",
     Patent USA 3806901, Publ. April 1974.
504. Caron, L., "Magnetic domain serial-to-parallel arrangement",
     Patent USA 3765004, Publ. October 1973.
505. Chang, H., "Bubble domain memory chips", IEEE Trans. Magn.,
     MAG-<u>8</u>, 564 (1972).
506. Chang, H., Fox, J., Lu, D. and Rosier, L. L., "A self-contained
     magnet bubble-domain memory chip", IEEE Trans. Magn., MAG-<u>8</u>
     (2), 214-22 (1972), EEA75-28075.
507. Chang, H. and Genovese, E. R., "Self-contained magnetic bubble
     domain memory chip", Patent USA 3701125, Publ. October 1972.
508. Chen, T. T., Oeffinger, T. R. and Gergis, I. S., "A hybrid
     decoder bubble memory organization", IEEE Trans. Magn., MAG-12
     (6), 630-2 (1976), CCA12-3996.
509. Copeland, J. A., III "Stacking arrangement which provides self-
     biasing for single wall domain organizations", Patent USA
     3678478, Publ. July 1972.
510. Cutler, L. S. and Lacey, R. F., "Memory protect for magnetic
     bubble memory", Patent USA 3744042, Publ. July 1973.
511. DeJonge, F. A., "Magnetic domain store", Patent USA 3787825,
     Publ. January 1974.
512. DeJonge, F. A., "Opto-magnetic memory", Patent USA 3836895,
     Publ. August 1974.
513. Doo, V. and Chang, H., "Random-access bubble domain memory",
     IBM Tech. Disclosure Bull., <u>15</u> (1), 223-4 (1972), EEA75-35672.
514. Dorleijn, J. W. F., Jr. and Janssen, G. A. M., Jr., "Applica-
     tion of cylindrical magnetic domains in memories", Umsch. Wiss.
     & Tech., <u>73</u> (20), 633-4 (1973), German, CCA9-1231.
515. Druyvesteyn, W. F., Kuijpers, F. A., Verhulst, A. G. H. and
     Witmer, C. H. M., "Single-mask bubble memory with rotating-
     field control", Philips Tech. Rev., <u>36</u> (6), 149-59 (1976),
     CCA12-6664.
516. Enz, U. E. and DeJonge, F. A., "Device for the magnetic storage
     of data", Patent USA 3793639, Publ. February 1974.
517. Fox, J. H., "Single-line decoder for bubble memories", IBM
     Tech. Disclosure Bull., <u>15</u> (6), 2027-9 (1972), CCA8-9523.
518. Fugere, D. G. and Lee, J. M., "Bubble domain storage", IBM
     Tech. Disclosure Bull., <u>13</u> (11), 3453-4 (1971), CCA6-20140.
519. Furuoya, T., "Cylindrical magnetic domain memory apparatus",
     Patent USA 3737882, Publ. June 1973.

520. Furuoya, T., "Problems on high speed bubble devices (memory applications)", Jap. J. Appl. Phys., 15 (Suppl. 15-1), 143 (1976), CCA12-6657.

521. Genovese, E. R. and Lin, Y. S., "Cylindrical magnetic domain random access memory", IBM Tech. Disclosure Bull., 13 (11), 3297 (1971), CCA6-20139.

522. Geusic, J. E., "Magnetic bubble devices: moving from lab to factory", Bell Lab. Rec., 54 (10), 262-8 (1976), CCA12-3991.

523. Gibson, A. T., "Magnetic bubble memory systems", Electron. Engineering, 49 (587), 36-7 (1977), CCA12-6647.

524. Hanson, M. M., Hewitt, F. G., Kaske, A. D., Lund, R. E. and Torok, E. J., "Bubbles and latrix elements", AIP Conf. Proc., no. 29, 626-8 (1976), CCA12-3990.

525. Hayes, P. J. and Walker, I. J., "A bubble domain memory cell", AIP Conf. Proc., no. 29, 632 (1976), CCA12-1499.

526. Hendel, R. J., Jamba, T. F., Jr., Keefe, G. E. and Rosier, L. L., "Magnetic bubble domain system having improved operating margins", Patent USA 3825885, Publ. July 1974.

527. Homma, N., Noro, Y. and Yoshizawa, S., "Magnetic single wall domain memory", Patent USA 3732551, Publ. May 1973.

528. Hunter, D. J., "A small magnetic bubble memory system", Proceedings of the Conference on Video and Data Recording, 1976, p. 57-62, CCA11-31603.

529. Juliussen, J. E., "Bubble memory organization with two port major/minor loop transfer", Patent USA 3838407, Publ. September 1974.

530. Juliussen, J. E., "Magnetic bubble memory interfacing", 11th IEEE Computer Society Conference, COMPCON 75, Digest of Papers, p. 87-90.

531. Juliussen, J. E., Lee, D. M. and Cox, G. M., "Bubbles appearing first as microprocessor mass storage", Electronics, 50 (16), 81-6 (1970).

532. Keefe, G. E. and Lin, Y. S., "High-density bubble domain memory", IBM Tech. Disclosure Bull., 13 (11), 3295 (1971), CCA6-20138.

533. Kish, D. E. and Smith, J. L., "Single wall domain memory organization", Patent USA 3697963, Publ. October 1972.

534. Kish, D. E. and Smith, J. L., "Transfer of magnetic domains in single-wall domain memories", Patent USA 3714639, Publ. January 1973.

535. Kish, D. E. and Smith, J. L., "Code translator employing sequential memories", Patent USA 3812480, Publ. May 1974.

536. Kiyasu, Z. and Tsuruhara, H., "Memory apparatus using cylindrical magnetic domain materials", Patent USA 3902166, Publ. August 1975.

537. Kohara, H., "Bubble domain circuit", Patent USA 3916396, Publ. October 1975.

538. Krupp, R. S. and Tomko, L. A., "Time coded signalling technique for writing control memories of time slot interchangers and the like", Patent USA 3743788, Publ. June 1973.

539. Kuhn, L., Lin, Y. S. and Sadagopan, V., "Optically controlled system for bubble domain memories", IBM Tech. Disclosure Bull., 16 (9), 3098-9 (1974), CCA9-18430.

540. Lacklison, D. E., "Method of magnetic data storage", Patent USA 3739360, Publ. June 1973.

541. LeCraw, R. C., Sherwood, R. C. and Wolfe, R., "Magnetic memory implementation", Patent USA 3506974, Publ. April 1970.

542. Lee, D. M. and Nader, R. A., "Bubble memory for military mass storage requirements", Proceedings of the National Electronic Packaging and Production Conference, NEPCON 76, p. 724-8.

543. Lienhard, H., Carter, W. S. and Wolf, I. W., "Magnetoresistive readout transducer for sensing magnetic domains in thin film memories", Patent USA 3883858, Publ. May 1975.

544. Lin, Y. S. and Yao, Y. I., "Bubble domain functional memory", IBM Tech. Disclosure Bull., 14 (7), 2144-5 (1971), EEA75-12979.

545. Marsh, A., "Applications for magnetic bubbles", Presented at Colloquium Digest on Computer Memories, London, England, February 1973, CCA8-11637.

546. Metzdorf, W., "Information storage with magnetic bubbles", J. Magn. & Magn. Mater., 4 (1-4), 145-58 (1977), German, CCA12-6660.

547. Michaelis, P. C. and Richards, W. J., "Magnetic bubble mass memory", IEEE Trans. Magn., MAG-11 (1), 21-5 (1975).

548. Murakami, H., "Cylindrical domain associative memory apparatus", Patent USA 3760390, Publ. September 1973.

549. Murakami, H., "Memory organization using imperfect bubble chips", IEEE Trans. Magn., MAG-13 (5), 1631-4 (1977).

550. Myer, J. H., "Biasing apparatus for magnetic domain stores", Patent USA 3831156, Publ. August 1974.

551. Myer, J. H., "Magneto-optical cylindrical magnetic domain memory", Patent USA 3806903, Publ. April 1974.

552. Naden, R. A. and West, F. G., Jr., "Bubble memory minor loop redundancy scheme", Patent USA 3909810, Publ. September 1975.

553. O'Dell, T. H., "Binary memory devices", Patent USA 3798622, Publ. February 1974.

554. Otala, M. N. T., "Magnetic bubble store having optical centering apparatus", Patent USA 3899780, Publ. August 1975.

555. Otala, M. N. T., "Magnetic domain storage disk", Patent USA 3905040, Publ. September 1975.

556. Parzefall, F., "High density data storage with magnetic bubbles", Int. Elektron. Rundsch., 27 (12), 267-9 (1973), German, CCA9-9205.

557. Parzefall, F., "Magnetic bubbles, memories for the future", Elektronik, 23 (2), 39-42 (1974), German, CCA9-9203.

558. Peltz, V. T. and Della Torre, E., "The impact of bubble memories on computer systems", 8th Asilomar Conference on Circuits, Systems and Computers, 1974, p. 723-5, Publ. 1975, CCA10-17637.

559. Plessey Co., Ltd., "Magnetic domain memory device", Patent UK 1292399, Publ. October 1972, CCA8-7342.

560. Potgiesser, J. A. L., "Device for the magnetic domain 'bubble'
     storage of data", Patent USA 3793640, Publ. February 1974.
561. Radner, R. J. and Wuorinen, J. H., Jr., "Magnetic bubble serial
     data store", 1976 International Solid-State Circuits Conference,
     Digest of Technical Papers, p. 178-9, CCA11-28277.
562. Raev, V. K., Potapov, V. S. and Shotov, A. E., "On data storage
     in accumulators on cylindrical magnetic domains when the supply
     is on/off", Prib. & Sist. Upr., no. 7, 12-15 (1976), Russian,
     CCA12-1501.
563. Reichard, R. W., "Static non-destructive single wall domain
     memory with Hall voltage readout", Patent USA 3701126, Publ.
     October 1972.
564. Salzer, J. M., "Bubble memories - where do we stand?", Computer,
     $\underline{9}$ (3), 36-41 (1976), CCA11-22652.
565. Sandscheper, G., "Storage by means of bubbles (cavities or
     closure domains). The search for faster and cheaper mass stor-
     age", ADL-Nachr., $\underline{18}$ (81), 20-2 (1973), German, CCA9-1229.
566. Shirakura, T. and Yoshihiro, S., "A conductor-Permalloy loop
     bubble domain memory", IEEE Trans. Magn., MAG-9 (3), 493-5
     (1973), CCA9-3413.
567. Singh, S. K. and Hubbell, W. C., "Operation of a bubble memory
     chip with a triangular drive field", AIP Conf. Proc., no. 29,
     46-7 (1976), CCA11-22646.
568. Takahashi, K., "Circulating access memory device", Patent USA
     3916397, Publ. October 1975.
569. Tomlinson, J. L. and Wieder, H. H., "Magnetic bubble domain
     memories in epitaxial garnet films", Radio & Electron. Eng.,
     $\underline{45}$ (12), 725-37 (1975), CCA11-9671.
570. Uher, L., "Magnetic bubble storage", Sdelovaci Tech., $\underline{24}$ (6),
     221-2 (1976), Czech, CCA12-1504.
571. Urai, H., "Cylindrical magnetic domain storage device having
     wave-like magnetic wall", Patent USA 3916395, Publ. October 1975.
572. Voegeli, O., Calhoun, B. A. and Rosier, L. L., "The use of
     bubble lattices for information storage", AIP Conf. Proc., no.
     24, 617-19 (1974), CCA11-6371.
573. Wong, C. K. and Yue, P. C., "Data organization in magnetic
     bubble lattice files", IBM J. Res. & Develop., $\underline{20}$ (6), 576-81
     (1976), CCA12-15499.
574. Yoshimi, K., "Magnetic bubble memory having by-pass for defec-
     tive loops", Patent USA 3921156, Publ. November 1975.
575. Ypma, J. E., "Magnetic bubble domain memories", IEEE Proceed-
     ings of National Aerospace Electron Conference, 1974, p. 50-4.

             c.   Logic

576. Almasi, G. S., "Cylindircal magnetic domain display system",
     Patent USA 3815107, Publ. June 1974.

577. Beausoleil, W. F. and Phelps, B. E., "Shift register storage
     unit with multi-dimensional ordering", Patent USA 3766534,
     Publ. October 1973.
578. Beausoleil, W. F. and Pugh, E. W., "Pulse sequence bubble
     domain logic", IBM Tech. Disclosure Bull., 15 (7), 2093-5
     (1972), CCA8-11631.
579. Bobeck, A. H., Scovil, H. E. D. and Shockley, W., "Magnetic
     logic arrangement", Patent USA 3541522, Publ. November 1970.
580. Bogholtz, W. E., Bosch, L. J., Kiseda, J. R., Simon, A. M.
     and Taylor, D. M., "Orthoferrite logic utilizing surface chan-
     nels", IBM Tech. Disclosure Bull., 13 (9), 2736-7 (1971),
     CCA6-12662.
581. Bogholtz, W. E. and Kiseda, J. R., "Exclusive OR bubble cir-
     cuit", IBM Tech. Disclosure Bull., 13 (7), 2053-4 (1970),
     CCA6-7990.
582. Bonyhard, P. I., "Single wall domain coding circuit", Patent
     USA 3786446, Publ. January 1974.
583. Braginski, A. I. and Kennedy, P. G., "Magnetic domain counter",
     Patent USA 3895363, Publ. July 1975.
584. Brown, D. T., "Dynamically double ordered shift register memory",
     Patent USA 3797002, Publ. February 1974.
585. Cannistra, A. T., Kiseda, J. R. and Shahan, V. T., "Expandable
     multiple input bubble domain OR/AND gate", IBM Tech. Disclosure
     Bull., 15 (1), 43-4 (1972), EEA75-35671.
586. Carlson, H. N., Perneski, A. J., Rago, L. F., Rothasuer, G. T
     and Wagner, W. D. P., "Field access bubble-to-bubble logic
     operations", IEEE Trans. Magn., MAG-8 (3), 367 (1972), CCA8-13674.
587. Caron, L., "Magnetic domain logic control arrangement", Patent
     USA 3763477, Publ. October 1973.
588. Chang, C. T. M., "Device characterisation of a complete set of
     passive bubble logic functional elements", AIP Conf. Proc., no.
     29, 32-3 (1976), CCA11-22549.
589. Chang, H., Chen, T. C. and Tung, C., "Symmetric switching func-
     tions using magnetic bubble domains", Patent USA 3919701, Publ.
     November 1975.
590. Chang, H. and Cooper, A., "Bubble domain analog-to-digital con-
     verter", IBM Tech. Disclosure Bull., 14 (7), 2218-19 (1971),
     EEA75-12712.
591. Chang, H., Keefe, G. E., Lin, Y. S. and Rosier, L. L., "Cylin-
     drical magnetic domain decoder", Patent USA 3689902, Publ.
     September 1972.
592. Chow, W. F., "Magnetic domain logic circuit", Patent USA
     3619636, Publ. November 1971.
593. Chow, W. F., "Magnetic domain logic circuit", Patent USA
     3638208, Publ. January 1972.
594. Chow, W. F., "Single wall magnetic domain logic arrangement",
     Patent USA 3711842, Publ. January 1973.
595. Clover, R. B. and Waites, R. F., "Magnetic bubble switches",
     Patent USA 3876994, Publ. April 1975.

596. Cohen, M. S., Hsieh, W. J. and Almasi, G. S., "Analytical model
     of a bubble switch", IEEE Trans. Magn., MAG-13 (4), 1035-41
     (1977).
597. Copeland, J. A., III, "Magnetic domain logic arrangement",
     Patent USA 3641518, Publ. February 1972.
598. Copeland, J. A., III, "Magnetic domain logic arrangement",
     Patent USA 3653010, Publ. March 1972.
599. Copeland, J. A., III, "Domain logic arrangement", Patent USA
     3676871, Publ. July 1972.
600. Copeland, J. A., III, "Single wall domain lateral displacement",
     Patent USA 3699548, Publ. October 1972.
601. Cordi, V. A. and Nickel, T. Y., "Implementation of the least
     recently used (LRU) algorithm using magnetic bubble domains",
     Patent USA 3737881, Publ. June 1973.
602. Danylchuk, I., "Operational characteristics of 10^3-bit garnet
     Y-bar shift register", J. Appl. Phys., 42 (4), 1358-9 (1971),
     CCA6-15132.
603. Danylchuk, I., Geusic, J. E. and Nelson, T. J., "Single wall
     domain logic arrangement", Patent USA 3813661, Publ. May 1974.
604. Danylchuk, I. and Michaelis, P. C., "Magnetic domain multiple
     input AND circuit", Patent USA 3651496, Publ. March 1972.
605. Feuersanger, A. E., "Bonded shift-registers for magnetic bubble
     memory modules", AIP Conf. Proc., no. 18, 105-9 (1973),
     CCA9-20756.
606. Gergis, I. S., Tocci, L. R. and Archer, J. L., "The operating
     characteristics of a compact bubble transfer replicate switch",
     IEEE Trans. Magn., MAG-13 (1), 898-900 (1977), CCA12-15380.
607. Grubb, H. R. and Liebschutz, L. C., "Magnetic domain decoder/
     encoder device", Patent USA 3786455, Publ. January 1974.
608. Hayashi, N., Chang, H., Romankiw, L. T. and Krongelb, S., "An
     analysis of a clear-view angelfish bubble-domain shift register",
     IEEE Trans. Magn., MAG-8 (1), 16-22 (1972), EEA75-12976.
609. Hayashi, N., Romankiw, L. T., Chang, H. and Krongelb, S., "Fab-
     rication and operation of indented angelfish bubble-domain
     shift register", IEEE Trans. Magn., MAG-8 (3), 370-2 (1972),
     CCA8-13608.
610. Heinz, D. M., "Magnetic bubble domain system", Patent USA
     3735145, Publ. May 1973.
611. Homma, N., Yoshizawa, S. and Noro, Y., "Magnetic bubble
     decoder", Patent USA 3757314, Publ. September 1973.
612. Joel, A. E., Jr., "Intermedium magnetic domain logic control
     arrangement", Patent USA 3811118, Publ. May 1974.
613. Keefe, G. E., "Multiphase magnetic bubble domain decoder",
     Patent USA 3858188, Publ. December 1974.
614. Keefe, G. E., Lin, Y. S. and Yao, Y. L., "Bubble domain decoder
     with built-in memory", IBM Tech. Disclosure Bull., 14 (6),
     1915-16 (1971), EEA75-9784.
615. Kinoshita, K., Sasao, T. and Matsuda, J., "On magnetic bubble
     logic circuits", IEEE Trans. Comput., C-25 (3), 247-53 (1976).

616. Kinsner, W. and Della Torre, E., "In-memory code convertors
     using magnetic bubbles", 8th Asilomar Conference on Circuits,
     Systems and Computers, 1974, p. 733-8, Publ. 1975, CCA10-17638.
617. Kluge, W., "Content-addressable memories based on magnetic-
     domain logic", Proc. Inst. Electr. Eng., $\underline{120}$ (11), 1308-14
     (1973), CCA9-6233.
618. Kluge, W. E., "Magnetic bubble domain switching device", Patent
     USA 3770978, Publ. November 1973.
619. Lee, S. Y. and Chang, H., "Magnetic bubble logic", IEEE Trans.
     Magn., MAG-$\underline{10}$ (4), 1059-66 (1974).
620. Lee, S. Y. and Chang, H., "Associative-search bubble devices for
     content-addressable memories", 11th IEEE Computer Society Con-
     ference, COMPCON 75, Digest of Papers, p. 91-2.
621. Lin, Y. S., "Bubble domain logic circuits", IBM Tech. Disclosure
     Bull., $\underline{13}$ (10), 3019-20 (1971), CCA6-12665.
622. Lin, Y. S., "Bubble domain logic device", IBM Tech. Disclosure
     Bull., $\underline{13}$ (10), 3068 (1971), CCA6-12666.
623. Lin, Y. S. and Yao, Y. L., "Threshold logic using magnetic bubble
     domains", Patent USA 3780312, Publ. December 1972.
624. Marsh, A., "Circular magnetic domain devices", Patent USA
     3731288, Publ. May 1973.
625. Matsuda, J. and Kinoshita, K., "Analysis of magnetic bubble logic
     systems", Syst. Comput. Controls, $\underline{5}$ (1), 35-41 (1974).
626. Minnaja, N., "Optical associative memory using complementary
     magnetic bubble shift registers", Patent USA 3887906, Publ. June
     1975, CCA11-2223.
627. Minnick, R. C., Bailey, P. T., Sandfort, R. M. and Semon, W. L.,
     "Magnetic bubble logic", 1972 WESCON Technical Papers, Vol. 16,
     Paper 8/4, 13 pp., CCA8-7345.
628. Minnick, R. C., Bailey, P. T., Sandfort, R. T. and Semon, W. L.,
     "Magnetic bubble computer", Patent USA 3798607, Publ. March 1974.
629. Minnick, R. C., Bailey, P. T., Sandfort, R. T. and Semon, W. L.,
     "Non-conservative bubble logic circuits", Patent USA 3866191,
     Publ. February 1975.
630. Minnick, R. C. and Sandfort, R. M., "Magnetic bubble two-rail
     logic gates", Patent USA 3909622, Publ. September 1975,
     CCA11-6337.
631. Nelson, T. J., "Practical magnetic bubble AND-OR gate", IEEE
     Trans. Magn., MAG-$\underline{11}$ (3), 953-4 (1975).
632. Nuyts, W., Valeminck, R. and De Brouckere, L., "A magnetic
     bubble passive time-slot interchanging gate", IEEE Trans. Magn.,
     MAG-$\underline{13}$ (3), 969-72 (1977), CCA12-15366.
633. Quadri, F., "Magnetic domain logic decoder", Patent USA 3760386,
     Publ. September 1973.
634. Rosier, L. L., "Magnetic bubble domain pump shift register",
     Patent USA 3913079, Publ. October 1975.
635. Sadagopan, V., Hatzakis, M., Ahn, K. Y., Plaskett, T. S. and
     Rosier, L. L., "High-density bubble domain shift register", AIP
     Conf. Proc., $\underline{5}$ (Pt. I), 215-19 (1971), EEA75-19535.

636. Sakalay, F. E., "Magnetic domain code repeater", Patent USA
     3824567, Publ. July 1974.
637. Sandfort, R. M. and Burke, E. R., "Logic functions for magnetic
     bubble devices", IEEE Trans. Magn., MAG-7 (3), 358-61 (1971).
638. Shroeder, D. H., Sweet, J. N. and Langenhorst, J. W., "Opera-
     tion of a 100-bit, 2.6 μm bubble shift register", IEEE Trans.
     Magn., MAG-13 (4), 1041-5 (1977).
639. Semon, W. L. and Rudolph, L. D., "Fast bubble logic gates",
     Patent USA 3863234, Publ. January 1975.
640. Show, W. F., "Input gate arrangement for domain wall device",
     Patent USA 3646530, Publ. February 1972.
641. Smith, J. L., "Magnetic domain logic arrangement", Patent USA
     3599190, Publ. August 1971.
642. Tanasoiu, C., "Orthoferrites and their application as logic
     elements in computer technique", Stud. Cercet Fiz., 23 (9),
     1107-22 (1971), Rumanian, EEA75-7103.
643. Tung, C., Chen, T. C. and Chang, H., "Bubble ladder for infor-
     mation processing", IEEE Trans. Magn., MAG-11 (5), 1163-4
     (1975), CCA11-4708.
644. Walker, E. L., "Read/write decoder", IBM Tech. Disclosure
     Bull., 13 (11), 3472-3 (1971), CCA6-20141.
645. Yamauchi, F., Yoshimi, K., Fujiwara, S. and Furuoya, T., "Bubble
     switch and circuit utilizing YY overlay", IEEE Trans. Magn.,
     MAG-8 (3), 372-4 (1972), CCA8-13609.

d.   Domain Tip

646. Anon., "MADOS - the DOT store from BASF", Buerotechnik, 24 (2),
     42-3 (1976), German, CCA11-22651.
647. Battarel, C. and Hanaut, M., "Domain tip memories", Proceedings
     of the Conference on Video and Data Recording, 1976, p. 63-78,
     CCA11-31604.
648. Deichelmann, H., "Magnetic domain tip memories - construction
     and applications", J. Magn. & Magn. Mater., 4 (1-4), 174-9
     (1977), German, CCA12-6663.
649. Jauvtis, H. I. and Spain, R. J., "Research in ferromagnetics
     Domain tip propagation devices", Report AD-741780, Cambridge
     Memories Inc., Newtonville, Mass., 1972, 92 pp., CCA8-12660.
650. Sanders, I. L. and Collins, A. J., "Domain tip propagation
     devices", Proceedings of the Conference on Video and Data
     Recording, 1976, p. 79-88, CCA11-31605.
651. Spain, R. J., "Domain tip propagation shift register", Patent
     USA 3438016, Publ. April 1969, CCA5-2689.
652. Spain, R. J., Jauvtis, H. I. and Duben, F. T., "Dot memory
     systems", AFIPS National Computer Conference, Proceedings,
     Vol. 43, 1974, p. 841-6.

e.   Others

653. Anon., "Current techniques utilising magnetic bubbles", Electron.
     & Microelectron. Ind., no. 220, 24-5 (1976), French, CCA11-22654.
654. Chang, H. and Daughton, J. M., "Bubble domain sensor arrays
     for magnetic discs", IBM Tech. Disclosure Bull., 14 (7), 2121-2
     (1971), EEA75-16245.
655. Chen, Y. S., Bonyhard, P. I. and Smith, J. L., "An experimental
     magnetic bubble time-slot interchanger", AIP Conf. Proc., no.
     24, 641-2 (1974), CCA11-6372.
656. De Bot, L. A. P. M., "Magneto-optical transducer using bubble
     domains", Patent USA 3824570, Publ. July 1974.
657. Goldstein, R. M. and Shoji, M., "Functional bubble domain cir-
     cuits employing bubble-bubble interactions", AIP Conf. Proc.,
     5 (Pt. I), 210-14 (1971), EEA75-19574.
658. Kita, Y., Inose, F. and Kasai, M., "Two-dimensional shift array
     of magnetic bubbles and its application to pattern processing",
     IEEE Trans. Magn., MAG-8 (3), 367-9 (1972), CCA8-13675.
659. Saito, N., Toyooka, T., Yoshizawa, S. and Sagita, Y., "A high
     speed drive coil for a large capacity bubble memory", IEEE
     Trans. Magn., MAG-12 (6), 639-41 (1976), CCA12-6653.
660. Slonczewski, J. C., "Bloch-line information writer for magnetic
     bubble store", IBM Tech. Disclosure Bull., 18 (3), 942-4
     (1975), CCA11-4806.

V.   SEMICONDUCTOR STORAGE
      1.   Charge-Coupled Devices (CCD)

661. Ablassmeier, U. and Doering, E., "CCD arrays of high storage
     density in Al-Si-gate-technology", Siemens Forsch. - & Ent-
     wicklungsber., 4 (4), 226-30 (1975), German, EEA78-37653.
662. Allen, R. A., Handy, R. J. and Sandor, J. E., "Charge coupled
     devices in digital LSI", 1976 International Electron Devices
     Meeting, Technical Digest, p. 21-6, CCA12-10620.
663. Amelio, G. F., "Charge-coupled devices for memory applications",
     AFIPS National Computer Conference, Proc., Vol. 44, 1975,
     p. 515-22.
664. Anon., "Charge coupled device memory comes in at ten for a
     penny", Datamation, 21 (2), 85-8 (1975).
665. Anon., "Prototype CCD system stores 128K bits, could find
     spaceborn applications", Digital Des., 6 (8), 12 (1976),
     CCA12-8889.
666. Anon., "The 65-k RAM looms large, but CCDs and bubbles are
     'iffy'", Electronics, 50 (2), 94-6 (1977), CCA12-8893.
667. Baertsch, R. D., Engeler, W. E. and Tiemann, J. J., "A new sur-
     face charge analog store", 1973 International Electron Devices
     Meeting, Technical Digest, p. 134-7, EEA77-14547.

668. Bankowski, W. F., Kimar, V. R. and Tartamella, J. D., "Charge-coupled devices optical sensor storage device", IBM Tech. Disclosure Bull., $\underline{16}$ (9), 2915-16 (1974), CCA9-16077.

669. Barsan, R. M., "Signal transfer and degradation in surface-channel charge-coupled devices", Proc. Inst. Electr. Eng., $\underline{124}$ (2), 103-8 (1977), EEA80-9028.

670. Beaumont, J., "Performance evaluation of the CCD450 digital memory", Report AD-019810/1SL, Air Force Inst. of Tech., Wright-Patterson, AFB, Ohio School of Engineering, 1975, 83 pp.

671. Belt, R. A., "A CCD memory for radar signal processing", 1975 International Conference on the Application of Charge-Coupled Devices, p. 413-21, EEA79-39081.

672. Belt, R. A., "Serpentine charge-coupled device memory circuit", Report AD-D000816/9SL, Department of the Air Force, Washington, D. C., 1975, 10 pp.

673. Benjamin, R., "Minimal-transfer CCD structures", Electron. Lett., $\underline{12}$ (25), 661-2 (1976), CCA12-4036.

674. Bhandarkar, D. P., "Digital CCD memory trade-offs: A systems viewpoint", Microelectronics, $\underline{7}$ (2), 30-5 (1975), EEA79-20486.

675. Bhandarkar, D. P., "Cost performance aspects of CCD fast auxiliary memory", 1975 International Conference on the Application of Charge-Coupled Devices, p. 435-42, EEA79-37702.

676. Brofterio, S. and Raffa, G., "Study and realization of a semi-conductor memory module in an experiment to reduce TV signal band", Alta Freq., $\underline{45}$ (3), 181-4 (1976), Italian, EEA79-34595.

677. Browne, V. A. and Perkins, K. D., "Nonoverlapping gate charge-coupling technology for serial memory and signal processing applications", IEEE J. Solid-State Circuits, SC-$\underline{11}$ (1), 203-7 (1976).

678. Butler, W. J., Engeler, W. E., Goldberg, H. S., Puckette, C. M., IV and Lobenstein, H., "Charge-transfer analog memories for radar and ECM systems", IEEE J. Solid-State Circuits, SC-$\underline{11}$ (1), 93-100 (1976), EEA79-17854.

679. Carnes, J. E., Kosonocky, W. F., Chambers, J. M. and Sauer, D. J., "Charge-coupled devices for computer memories", Proceedings of the AFIPS National Computer Conference, Vol. 43, 1974, p. 827-37.

680. Cases, M., Chang, F. Y. and Rubin, B. J., "High-density serial-parallel-serial CCD memory", IBM Tech. Disclosure Bull., $\underline{19}$ (7), 2519-20 (1976), EEA80-25013.

681. Chan, C. H., Rosenbaum, S. D., Caves, J. T., Poon, S. C. and Wallace, R. W., "10 MHz 16K CCD condensed SPS memory requiring only two clocks", International Electron Devices Meeting, Technical Digest, 1975, p. 309-12, CCA11-31630.

682. Chapple, K., "Semiconductor memories: the CCD answer", Electron, no. 88, 22, 24 (1976), EEA79-20533, CCA11-15065.

683. Chou, S., "Design of a 16384-bit serial charge-coupled memory device", IEEE J. Solid-State Circuits, SC-$\underline{11}$ (1), 10-18 (1976), EEA79-16878.

684. Collins, D. R., Barton, J. B., Buss, D. C., Kmitz, A. R. and
     Schroder, J. E., "CCD memory options", 1973 International Solid-
     State Circuits Conference, Digest of Technical Papers, p. 136-7,
     EEA76-23042.
685. Crouch, H. R., Cronett, J. B., Jr. and Eward, R. S., "CCDs in
     memory systems move into sight", Comput. Des., 15 (9), 75-80
     (1976), CCA12-1505.
686. Curtis, D. A., "The future of mass storage technology", Report
     L740901, Arthur D. Little, Inc., Cambridge, Mass., 1974, p. 4-14.
687. Fagan, J. L., White, M. H. and Lampe, D. L., "A high-density,
     read/write, nonvolatile charge-addressed memory", 1976 Inter-
     national Solid-State Circuits Conference, Digest of Technical
     Papers, p. 184-5.
688. Goser, K. and Knauer, K., "Nonvolatile CCD memory with MNOS
     storage capacitors", IEEE J. Solid-State Circuits, SC-9 (3),
     148-50 (1974), EEA77-23121.
689. Gottler, E., Gruter, O. and Schneider, P., "CCD memory circuits
     of high bit density", 1st European Solid State Circuits Con-
     ference - ESSCIRC, Extended Abstracts, 1975, p. 76-7, EEA79-28156.
690. Gutlove, N., Amelio, G., Green, A. and Wakeman, R., "Charge
     coupled device memories for synthetic aperture radar", Presented
     at Proceedings of the National Aerospace Electronics Conference,
     Dayton, Ohio, May 1973, EEA77-3632.
691. Hicks, B. W. and Andrews, C. J., "Reduction of latency in serial
     semiconductor memories", Aust. Comput. J., 8 (2), 47-50 (1976).
692. Ho, I. T. and Riseman, J., "Reach-through mode operation of
     single-electrode double-threshold charge-coupled memory cell",
     IBM Tech. Disclosure Bull., 15 (2), 412-3 (1972), EEA75-39938.
693. House, D., "CCD vs. RAM for bulk storage applications", 12th
     IEEE Computer Society International Conference, COMPCON 76,
     Digest of Papers, p. 58-61.
694. House, D. and MacKenzie, K., "Using charge coupled devices can
     reduce bulk memory costs", EDN, 22 (1), 70-3 (1977), EEA80-24994.
695. House, D., Oliphant, J. and Papenberg, B., "Design and applica-
     tion of a 16K charge coupled storage device", 1975 WESCON
     Technical Papers, Vol. 19, Paper 19/3, 32 pp., EEA79-20068.
696. Ibrahim, A. and Sellars, L., "4096-bit charge coupled device
     serial memory array", 1973 International Electron Devices
     Meeting, Technical Digest, p. 141-3, CCA9-11973.
697. Jamieson, J. M., "The CCD memory gains a foothold", Mini-Micro
     Syst., 10 (4), 18, 20, 22 (1977), EEA80-28391.
698. Knauer, K. and Goser, K., "Non-volatile CCD-memory with MNOS
     capacitors", International Conference on Technology and Appli-
     cations of Charge Coupled Devices, 1974, p. 274-80, EEA78-11740.
699. Kohyama, S., Hatano, H., Tanaka, T. and Kubota, N., "A new
     multiplexed electrode-per-bit structure for CCD memory", Inter-
     national Electron Devices Meeting, Technical Digest, 1976,
     p. 11-14.

700. Kornstein, H., "CCD - a replacement for drum stores", Micro-
     electron. & Reliab., 15 (4), 343-5 (1976), CCA11-31617.
701. Kosonocky, W. F. and Carnes, J. E., "Charge-coupled digital
     circuits", IEEE J. Solid-State Circuits, SC-6 (5), 314 (1971).
702. Kosonocky, W. F. and Carnes, J. E., "Advanced investigation
     of two-phase charge-coupled devices", Report N73-33722/2, RCA,
     Princeton, N. J., 1973, 77 pp.
703. Kosonocky, W. F. and Carnes, J. E., "Two-phase charge-coupled
     devices with overlapping polysilicon and aluminum gates", RCA
     Rev., 34 (1), 164-202 (1973), EEA76-26104.
704. Krambeck, R. H., Retajczyk, T. F., Jr., Silversmith, D. J. and
     Strain, R. J., "4160-bit C4D serial memory", IEEE J. Solid-State
     Circuits, SC-9 (6), 434-43 (1974).
705. Lind, L. F. and Matheson, W. S., "Minimal transfer CCD struc-
     tures", Electron. Lett., 12 (13), 343-4 (1976), EEA79-33045.
706. Linnenbrink, T. E., Monahan, M. J. and Rea, J. L., "A CCD-based
     transient data recorder", 1975 International Conference on the
     Application of Charge-Coupled Devices, p. 443-53, EEA79-39264.
707. Margraff, T. C., "Design and performance of a CCD memory system",
     Report AD-A008657/9SL, Air Force Inst. of Tech., Wright-Patter-
     son, AFB, Ohio School of Engineering, 1974, 82 pp.
708. Metzger, J., "CCD's add flexibility to memory system design",
     Electron. Prod., 18 (2), 29-33 (1975).
709. Mohsen, A. M., Bower, R. W., Wilder, E. M. and Erb, D. M., "A
     64-K bit block addressed charge-coupled memory", IEEE J. Solid-
     State Circuits, SC-11 (1), 49-58 (1976), EEA79-16883.
710. Mohsen, A. M., Tompsett, M. F., Fuls, E. N. and Zimany, E. J.,
     Jr., "A 16-K bit block addressed charge-coupled memory device",
     IEEE J. Solid-State Circuits, SC-11 (1), 40-8 (1976), EEA79-16882.
711. Nishizawa, J. and Nonaka, T., "Charge-coupled semiconductor
     memory device", Patent USA 3829885, Publ. August 1974.
712. Pao, H., Gunsagar, K., Guidry, M. and Amelio, G. F., "Application
     of CCD in memory systems", Proceedings of the IEEE 1974 National
     Aerospace and Electronics Conference, p. 62-70.
713. Patrin, N. A., "Fast access-low power charge-transfer device
     for high-performance memory", IBM Tech. Disclosure Bull., 18
     (2), 343-4 (1975), EEA78-37631.
714. Rosenbaum, S. D. and Caves, J. T., "CCD memory arrays with fast
     access by on-chip decoding", 1974 International Solid-State
     Circuits Conference, Digest of Technical Papers, p. 210-1.
715. Rosenbaum, S. D. and Caves, J. T., "8192-bit block addressable
     CCD memory", IEEE J. Solid-State Circuits, SC-10 (5), 273-80
     (1975), CCA10-25847.
716. Rosenbaum, S. D., Caves, J. T., Poon, S. and Chan, C. H., "A
     16 Kilobit high density CCD memory", 1975 International Confer-
     ence on the Application of Charge-Coupled Devices, p. 423-8,
     EEA79-37700.

717. Rosenbaum, S. D., Chan, C. H., Caves, J. T., Poon, S. C. and
     Wallace, R. W., "A 16384-bit high-density CCD memory", IEEE
     J. Solid-State Circuits, SC-11 (1), 33-40 (1976), EEA79-16881.
718. Ross, M., "What can CCD do for you?", Microelectron. & Reliab.,
     14 (4), 363 (1975), CCA11-9681.
719. Schneider, P. and Witte, J., "CCD memories in a working memory
     system", Siemens Forsch. - & Entwicklungsber., 4 (4), 231-7
     (1975), EEA78-41625.
720. Sequin, C. H., "Two-dimensional charge-transfer arrays", IEEE
     J. Solid-State Circuits, SC-9 (3), 134-42 (1974), CCA9-16072.
721. Sequin, C. H., Sealer, D. A., Bertram, W. J., Jr., Tompsett,
     M. F., Buckley, R. R., Shankoff, T. A. and McNamara, W. J.,
     "A charge-coupled area image sensor and frame store", IEEE
     Trans. Electron Devices, ED-20 (3), 244-52 (1973), CCA8-11644.
722. Siemens, K., Robinson, C. R. and Mastronardi, J., "A fast access
     bulk memory system using CCD's/A recorder buffer memory using
     CCD's", 1975 International Conference on the Application of
     Charge-Coupled Devices, p. 429-34, EEA79-37701.
723. Tchon, W. E., Elmer, B. R., Denboer, A. J., Negishi, S., Hira-
     bayashi, K., Nojima, I. and Kohyama, S., "4096-bit serial
     decoded multiphase serial-parallel-serial CCD memory", IEEE J.
     Solid-State Circuits, SC-11 (1), 25-33 (1976), EEA79-16880.
724. Tchon, W. E. and Huang, J. S. T., "Staggered oxide C4D structure
     with clocked source repeater", 1973 International Electron
     Devices Meeting, Technical Digest, p. 27-8, EEA77-14003.
725. Terman, L. M., "Small area stored charge memory cells", IBM
     Tech. Disclosure Bull., 15 (4), 1227-9 (1972), CCA8-5273.
726. Varshney, R. C., Guidry, M. R., Amelio, G. F. and Early, J. M.,
     "A byte organized NMOS/CCD memory with dynamic refresh logic",
     IEEE J. Solid-State Circuits, SC-11 (1), 18-24 (1976),
     EEA79-16879.
727. Waaben, S., "Charge-transfer electronics-tandem matrix semi-
     conductor memory selection", 1971 International Solid-State
     Circuits Conference, Digest of Technical Papers, p. 84-5,
     EEA74-22797.
728. Western Electric Co., "Charge coupled device which prevents
     trapping of the charge", Patent UK 1376639, Publ. December
     1974, EEA78-15372.
729. Western Electric Co., "Charge coupled devices", Patent UK
     1376640, Publ. December 1974, EEA78-15373.
730. Western Electric Co., "Reversible two-phase charge coupled
     storage device", Patent UK 1379141, Publ. January 1975,
     EEA78-19923.
731. White, M. H., Lampe, D. R. and Fagan, J. L., "CCD and MNOS
     devices for programmable analog signal processing and digital
     nonvolatile memory", 1973 International Electron Devices
     Meeting, Technical Digest, p. 130-3, CCA9-11971.

732. White, M. H., Lampe, D. R., Fagan, J. L., Kub, F. C. and Barth, D. A., "A nonvolatile charge-addressed memory (NOVCAM) cell", IEEE J. Solid-State Circuits, SC-10 (5), 281-8 (1975).

733. Windle, D. J., Williams, E. W., Vanstone, G. F., Bounden, J. E. and Mavor, J., "A serial parallel serial charge coupled array", New Electron., 7, 8 (1974), EEA77-26047.

734. Zaininger, K. H., "Charge-coupled devices for memory applications: a review", International Conference on Technology and Applications of Charge Coupled Devices, 1974, p. 178, EEA78-11739.

735. Zimmerman, T. A. and Miller, C. S., "Charge coupled devices in signal processing systems", Report AD-782574/8, TRW Systems Group, Redondo Beach, California, 1974, 145 pp.

2. Beam-Addressed Memories (BEAMOS)

736. Arntz, F. O., Rockstad, H. K. and Smith, D. O., "Electron beam-accessed data storage in MOS capacitors", 1976 IEEE International Solid-State Circuits Conference, Digest of Technical Papers, p. 176-7, EEA79-46320.

737. Chen, A. C. M., Dunham, A. M. and Wang, J. M., "Electron beam readout sensitivity of GeTeX amorphous thin films", 5th International Conference on Amorphous and Liquid Semiconductors, Proceedings, 1974, p. 701-6.

738. Cohen, M. S. and Moore, J. S., "Physics of the MOS electron-beam memory", J. Appl. Phys., 45 (12), 5335-48 (1974), PA78-25766.

739. Ellis, G. W., Possin, G. E. and Wilson, R. H., "Diode detection of information stored in electron-beam-addressed MOS structure", Appl. Phys. Lett., 24 (9), 419-21 (1974), CCA9-18396.

740. Hughes, W. C., Lemmond, C. Q., Parks, H. G., Ellis, G. W., Possin, G. E. and Wilson, R. H., "BEAMOS - a new electronic digital memory", AFIPS National Computer Conference, Proceedings, Vol. 44, 1975, p. 541-8.

741. Hughes, W. C., Lemmond, C. Q., Parks, H. G., Ellis, G. W., Possin, G. E. and Wilson, R. H., "Semiconductor nonvolatile electron beam accessed mass memory", Proc. IEEE, 63 (8), 1230-40 (1975).

742. Kelly, J., "The development of an experimental electron-beam-addressed memory module", Computer, 8 (2), 32-42 (1975), CCA10-15203.

743. Kelly, J., "Recent advances in electron beam-addressed memories", In: Advances in Electronics and Electron Physics, Marton, L., (Ed.), Academic Press, New York, p. 43-138 (1977).

744. Kelly, J., Moore, J. S. and Thornton, P. R., "Development of an experimental electron beam addressed memory module", IEEE Proceedings of National Aerospace Electronic Conference, 1974, p. 55-61.

745. Kelly, J., Moore, J. S. and Thornton, P. R., "Progress on a
     fast access electron beam addressed memory", 8th IEEE Computer
     Society International Conference, COMPCON 74, Digest of
     Papers, p. 26-8.
746. Lemmond, C. Q., Hughes, W. C., Kirkpatrick, C. G., Buschmann,
     E. C. and Grup, H. W., "Design, fabrication, and evaluation of
     an electron beam addressable high information density memory
     tube", Report AD-A026217/0SL, General Electric Corporate
     Research and Development, Schenectady, New York, 1976, 41 pp.
747. Newberry, S. P., "Electron beam memories - an evaluation",
     Electrochemical Society Spring Meeting, Extended Abstracts,
     1976, p. 444-5, CCA11-31625.
748. Parks, H. G., Hughes, W. C., Possin, G. E., Lemmond, C. Q.,
     Kirkpatrick, C. G., Ellis, G. W. and Wilson, R. H., "BEAMOS -
     an electron beam memory device using matrix lens optics",
     Electrochemical Society Spring Meeting, Extended Abstracts,
     1976, p. 499-500, CCA11-28305.
749. Possin, G. E., "Bit packing density of BEAMOS targets", IEEE
     Trans. Electron Devices, ED-$\underline{22}$ (11), 1068 (1975), CCA11-4812.
750. Possin, G. E., Hughes, W. C., Parks, H. G., Ellis, G. W.,
     Wilson, R. H., Lemmond, C. Q. and Kirkpatrick, C. G., "BEAMOS -
     a new electron beam digital memory device", 1975 International
     Electron Devices Meeting, Technical Digest, p. 305-8,
     CCA11-31629.
751. Speliotis, D. E., "Bridging the memory access gap", AFIPS
     National Computer Conference, Proceedings, Vol. 44, 1975,
     p. 501-8.

V.   MAGNETIC STORAGE
      1.   General

752. Anon., "Cartridge memory system has stationary disc and
     rotating head", Comput. Des., $\underline{11}$ (11), 130 (1972), CCA8-5258.
753. Backler, J., "Low-cost data storage: multiple options - hard
     choices", Digital Des., $\underline{6}$ (6), 60, 70 (1976), CCA12-3982.
754. Dorrell, C. E. and Meekleburg, A. W., "Mass storage device",
     IBM Tech. Disclosure Bull., $\underline{15}$ (9), 2943-5 (1973), CCA8-13657.
755. Doty, K. L. and Quanstrom, J. L., "Bit addressable storage
     access system", IBM Tech. Disclosure Bull., $\underline{18}$ (9), 2766-70
     (1976), CCA11-12314.
756. Godydke, H., Schumny, H. and Schroeder, H. J., "Computer aided
     evaluation of mean peak voltages for the calibration of mag-
     netic data storage media", Proceedings of the 6th Triennial
     World Congress of the International Federation of Automatic
     Control, Pt. IV, 1975, p. 56.3/1-8, CCA12-8887.
757. Greiner, J., "Comparison of metallic and oxidic storage layers",
     Int. Elektron. Rundsch., $\underline{24}$ (10), 251-5 (1970), German,
     CCA6-2061.

758. Grimes, D. W., "Skew detector and corrector for a disk, drum,
     or tape storage", IBM Tech. Disclosure Bull., 18 (7), 164-7
     (1975), CCA10-25834.
759. IBM Corp., "Storage system has separated read lines", Patent
     UK 1401262, Publ. July 1975, CCA11-9688.
760. King, G., "Cassette, cartridge and diskette drives", Digital
     Des., 7 (6), 50-66, 88-90 (1977).
761. Koldasov, G. D., "Distribution of magnetization in flat film
     memory elements", Izv. VUZ Priborostr., 17 (2), 49-53 (1974),
     Russian, CCA9-16054.
762. Manildi, A. B., "Inexpensive digital storage media", Electron.
     Des., 24 (25), 86-8 (1976), CCA12-10684.
763. Mazda, F. F., "The components of computers. VII. Magnetic film",
     Electron. Components, 14 (1), 34-8 (1973), CCA8-7332.
764. Nemeth, G., "Optimal writing working point of moving magnetic
     memory", Hiradastechnika, 27 (11), 334-6 (1976), Hungarian,
     CCA12-6639.
765. Parhami, B. and Avizienis, A., "Application of arithmetic error
     codes for checking of mass memories", International Symposium
     on Fault-Tolerant Computing, 1973, Digest of Papers, p. 47-51,
     CCA9-2717.
766. Petschauer, R. J., "Error logging in semiconductor storage
     units", Patent USA 3906200, Publ. September 1975, CCA11-9689.
767. Randet, D., "Electrochemical and magnetic bubble type magnetic
     memories", Rev. Polytech., no. 9, 761, 763, 765 (1976), French,
     CCA11-31602.
768. Sahakian, A., "Computer/cassette interface takes tone from
     clock", Electronics, 50 (1), 115 (1977), CCA12-3984.
769. Taylor, C. F., Blades, J. D. and Hewitt, F. G., "Residual in-
     formation in magnetic memory devices", J. Appl. Phys., 42 (4),
     1755-62 (1971), CCA6-17643.
770. Timofeev, B. B., Sukhomlinov, M. M., Stepko, D. P., Voloshin,
     P. A., Prokofev, P. A., Elpidina, G. M. and Sokolov, E. I.,
     "Construction of computational schemes and structures which use
     register tracks", Prib. & Sist. Upr., no. 9, 55-8 (1976),
     Russian, CCA12-6665.
771. VDC, "Data recording industry II, 1975-1980", Report, Venture
     Development Corporation, 1 Washington Street, Wellesley, Mass.,
     1977.
772. VDC, "Floppy disc markets II, 1977-1981", Report Venture
     Development Corporation, 1 Washington Street, Wellesley, Mass.,
     1977.
773. Wildman, M., "Terabit memory systems: a design history", Proc.
     IEEE, 63 (8), 1160-5 (1975).
774. Willaschek, K., "Reading of information in magnetic memory
     devices", Nachrichtentech. Elektron., 24 (7), 252-6 (1974),
     German, CCA9-23248.

2. Disks

775. Adamec, S. and Burda, K., "Magnetic discs and their use", Mech.
     Autom. Adm., 15 (2), 73-7 (1975), Czech, CCA10-15164.
776. Ahearn, G. R., Dishon, Y. and Snively, R. N., "Designs innova-
     tions of the IBM 3830 and 2835 storage control units", IBM J.
     Res. & Develop., 15 (1), 11-18 (1972), CCA7-14696.
777. Anon., "High density recording provides high performance head-
     per-track disc memories", Comput. Des., 10 (10), 100-1 (1971),
     CCA7-6283.
778. Anon., "The disc pack jungle", EDP Europa Rep., 2 (22), 1-6
     (1971), CCA7-6285.
779. Badische Anilin & Soda-Fabrik AG, "Magnetic disc assembly",
     Patent UK 1294824, Publ. November 1972, CCA8-9521.
780. Baranyay, P., "Orientation experiences with the magnetic disk
     memory tape CDC 9433", Funommechanika, 10 (8), 251-3 (1971),
     Hungarian, CCA6-24452.
781. Calfee, R. W. and Hatley, E. T., "Image storage and display
     using magnetic disc technology", 1973 IEEE International Con-
     vention and Exposition, Vol. V, Paper 16-2, 8 pp., CCA9-1227.
782. Charlton, R. J., Creelman, W. J. and Lucey, E. D., "Disk pack
     labeling", IBM Tech. Disclosure Bull., 13 (1), 119 (1970),
     CCA5-22421.
783. Chi, C. S., "Track mispositioning considerations in high density
     disk recording", IEEE Trans. Magn., MAG-12 (6), 731-3 (1976),
     CCA12-6642.
784. EG & G Inc., "Random access memory system", Patent UK 1281263,
     Publ. July 1972, CCA8-2636.
785. Engel, J. M., "Simple model describes error-causing defects in
     recording disks", IEEE Trans. Magn., MAG-11 (3), 926-8 (1975),
     CCA10-22913.
786. Franchini, R. C. and Wartner, D. L., "A method of high density
     recording on flexible magnetic discs", Comput. Des., 15 (10),
     106-9 (1976), CCA12-3981.
787. Freeman, A. E., Keefe, A. T. and Parker, L. E., "The design of
     disc stores for use in mobile operation", Proceedings of the
     Conference on Video and Data Recording, 1973, p. 157-64,
     CCA8-20876.
788. Gassoway, J. S., "Magnetic memory disc", Patent UK 1263890,
     Publ. February 1972, CCA7-22391.
789. Geffon, A. P., "A 6k BPI disk storage system using MOD-11
     interface", IEEE Trans. Magn., MAG-13 (5), 1205-7 (1977).
790. Ghose, S. and Kluth, A., "Magnetic disk pack assembly", IBM
     Tech. Disclosure Bull., 13 (1), 133 (1970), CCA5-22422.
791. Girdner, W. I. and Overton, W. H., "Reading and writing on
     the fast disc", Hewlett-Packard J., 23 (9), 12-14 (1972),
     CCA7-20078.

792. Goel, A. L., Nanda, P. S. and Mukhopadhyay, S., "Experimental design for the simulation study of a disk storage device", Bull. Op. Res. Soc. Am., 19 (Suppl. 2), B287 (1971), CCA7-4728.
793. Gotlieb, C. C. and MacEwen, G. H., "Performance of movable-head disk storage devices", J. Assn. Computing Mchy., 20, 604-23 (1973).
794. Hamada, M., Ishida, S. and Ogawa, S., "Mechanical strength of coating films for magnetic recording disks", Fujitsu Sci. & Tech. J., 12 (2), 191-212 (1976), CCA11-28273.
795. Hanson, C. C. and Jahnke, R. C., "Track location correction mechanism for magnetic disks", IBM Tech. Disclosure Bull., 19 (3), 1039-42 (1976), CCA12-3986.
796. Harker, J. M. and Chang, H., "Magnetic disks for bulk storage - past and future", AFIPS Joint Computer Conference, Vol. 40, 1972, p. 945-55, CCA7-20080.
797. Harris, J. P., Rohde, R. S. and Arter, N. K., "IBM 3850 mass storage system: design aspects", Proc. IEEE, 63 (8), 1171-6 (1975).
798. Honeywell Information Systems Inc., "Data storage system", Patent UK 1281919, Publ. July 1972, CCA8-2637.
799. IBM Corp., "Magnetic disc memory", Patent UK 1277177, Publ. June 1972, CCA7-24497.
800. IBM Corp., "Magnetic disc store", Patent UK 1283179, Publ. July 1972, CCA7-26482.
801. IBM Corp., "Recording of digital data in magnetic disc memory", Patent UK 1404541, Publ. September 1975, CCA11-4791.
802. Isozaki, I., Matsumoto, H., Miyazaki, O., Ito, M., Nakamura, S., Yamanura, S., Kageyama, S., Mitsuhashi, S. and Nakamura, K., "Fixed-head disk memory unit for high reliability applications", NEC Res. & Dev., no. 44, 57-66 (1977), CCA12-15495.
803. Ito, Y., Kaneko, R., Yasuda, K., Ueda, S., Takanami, S. and Fujiwara, Y., "Large capacity magnetic disc pack storage for DIPS system", Electr. Commun. Lab. Tech. J., 23 (12), 2669-93 (1974), Japanese, CCA10-15159.
804. Jack, R. W., "Head per track disk file mass memories", Memories, Terminals, and Peripherals, Proceedings of the 1970 IEEE International Computer Group Conference, p. 357-9, CCA6-2054.
805. Johnson, C., "IBM 3850 mass storage system", AFIPS National Computer Conference, Proceedings, Vol. 44, 1975, p. 509-14.
806. Johnson, C. T., "IBM 3850: A mass storage system with disk characteristics", Proc. IEEE, 63 (8), 1166-70 (1975).
807. Kaneko, R., Yasuda, K., Ono, K. and Sato, I., "Large capacity magnetic disc pack for DIPS system", Electr. Commun. Lab. Tech. J., 23 (12), 2695-707 (1974), Japanese, CCA10-15160.
808. Kellogg, R. D., Pendy, W. J. and Wheeler, S. E., "Magnetic disk cartridge", IBM Tech. Disclosure Bull., 18 (10), 3397-9 (1976), CCA11-17398.
809. Kirby, R., "Disc storage for business and industrial systems", Comput. Wkly., 20 (479), 16, 27 (1976), CCA11-15190.

810. Lugger, J. and Melenk, H., "A software package for dynamic and long term storage allocation of a disk file", Elektron. Rechenanlagen, 14 (4), 171-6 (1972), German, CCA7-24493.

811. Mallmann, G., "Disc storage unit - a reasonably priced medium for eighty years", Online, 14 (1-2), 59-60, 65 (1976), German, CCA11-15049.

812. Maruyama, K. and Smith, S. E., "Optimal reorganization of distributed space disk files", Commun. ACM, 19 (11), 634-42 (1976).

813. Matsukura, J., "Magnetic disc memories", J. Inst. Electron. & Commun. Eng. Jap., 54 (4), 555-8 (1971), Japanese, CCA6-17635.

814. Mauduit, C., "Standardization of magnetic disc packs", Courrier Normalisation, 38 (221), 576-8 (1971), French, CCA7-4729.

815. May, G. H., "Beam addressable file with flexible disks", IBM Tech. Disclosure Bull., 16 (9), 3026-7 (1974), CCA9-18384.

816. Memorex Corp., "Disc storage unit", Patent UK 1294150, Publ. October 1972, CCA8-9519.

817. Miller, T. H., "Error activated state memory for disc file", IBM Tech. Disclosure Bull., 14 (7), 2004-5 (1971), CCA7-8312.

818. Ogawa, S., Handa, C., Ishida, S. and Soma, H., "Stress demagnetization in magnetic coated disks", IEEE Trans. Magn., MAG-13 (5), 1394-6 (1977).

819. Peters, T. R., "Controlled pause and restart of magnetic disc memories and the like", Patent USA 3678464, Publ. July 1972, CCA7-26483.

820. Pohm, A. and Zingg, R. J., "Variable diameter disc pack with cooperating head", Patent USA 3703713, Publ. November 1972, CCA8-13658.

821. Shiraki, T., Kawata, K., Kajimura, H., Chiba, H. and Nakata, M., "Flying heads for magnetic disc memories", Natl. Tech. Rep., 19 (6), 592-9 (1973), Japanese, CCA9-16058.

822. Sidhu, P. S., "Group-coded recording reliably doubles diskette capacity", Comput. Des., 15 (12), 84-6, 88 (1976).

823. Sindelar, F. L. and Hoffman, L. J., "A two-level disc protection system", 2nd USA-Japan Computer Conference Proceedings, 1975, p. 139-42, CCA11-15052.

824. Soboll, H., "The new semiconductor store and the new magnetic disc unit for the large computer CYBER 76", Elektron. Rechenanlagen, 19 (1), 43-4 (1977), German, CCA12-10688.

825. Speliotis, D. E. and Chi, C. S., "The remanent magnetisation in particulate disk coatings", IEEE Trans. Magn., MAG-12 (6), 734-6 (1976), CCA12-6643.

826. Stevenson, T. J., "Disc file optimization", IEEE Trans. Magn., MAG-11 (5), 1237-9 (1975), CCA11-4786.

827. Takano, H., Kawata, K., Wada, J., Sibuya, C., Nakada, K., Hirahata, A. and Ichiyanagi, T., "Small size magnetic disc memory JK-811", Natl. Tech. Rep., 19 (6), 571-7 (1973), Japanese, CCA9-16055.

828. Talke, F. E. and Su, J. L., "Mechanism of wear in magnetic recording disk files", Tribol Int., 8 (1), 15-20 (1975).

829. Taylor, C. J., "Disk pack testing - a new approach", Radio &
     Electron. Eng., 44 (3), 119-24 (1974), CCA9-18385.
830. Tseng, R. C., "Rarefaction effects of gas-lubricated bearings
     in a magnetic recording disk file", ASME Paper No. 75-LUB-F,
     1975, 6 pp.
831. Waters, S. J., "Estimating magnetic disc seeks", Comput. J.,
     18 (1), 12-17 (1975).
832. Weiss, R. D., "Magnetic orientation in disc media", AIP Conf.
     Proc., no. 29, 630-2 (1976), CCA12-1493.
833. Wieselman, I. L., "Moveable head, fixed disc systems", Memories,
     Terminals, and Peripherals, Proceedings of the 1970 IEEE Inter-
     national Computer Group Conference, p. 360-2, CCA6-2055.
834. Wilhelm, N. C., "General model for the performance of disk
     systems", J. Assoc. Comput. Mach., 24 (1), 14-31 (1977).
835. Wright, T. W., "Servo control technique enhances performances
     of disc storage units", Comput. Des., 16 (3), 99-103 (1977).
836. Yang, H. S., "Design directions for rotating mass storage",
     Memories, Terminals, and Peripherals, Proceedings of the 1970
     IEEE International Computer Group Conference, p. 363-6,
     CCA6-2056.

3.   Floppy Disks

837. Anon., "Flexible magnetic disk patent activity", Digital Des.,
     7 (8), 62-4 (1977).
838. Caswell, S. A., "Floppy disk drives and systems", Mod. Data,
     no. 8, 32-41 (1975).
839. Conway, J., "Flexible discs vs. magnetic tape - are they com-
     plementary or competitive", EDN, 22 (3), 36-46 (1977).
840. Dorie, L. A., "Floppy disks: the low-cost storage medium of
     the 70's", Data Manage., 14 (1), 6-8 (1976), CCA11-15191.
841. Eidson, M. E. and Parker, L. A., "Synchronous adapter reduces
     complexity of floppy disc controller", Comput. Des., 16 (4),
     102-6 (1977).
842. Floyd, W., "When it comes to floppys, there is much to know
     before you buy", EDN, 21 (1), 18-24 (1976).
843. Franchini, R. C. and Wartner, D. L., "Method of high density
     recording on flexible magnetic discs", Comput. Des., 15 (10),
     106-9 (1976).
844. Franson, P., "Floppy discs spin into systems", Electronics,
     48 (2), 59-60 (1975).
845. Greenberg, H. J., Stephens, R. L. and Talke, F. E., "Dimen-
     sional stability of floppy disks", IEEE Trans. Magn., MAG-13
     (5), 1397-9 (1977).
846. Harris, R. G., Sustman, J. E. and McDonald, J. F., "Controller
     for a flexible disk", Proceedings of the National Computer
     Conference, AFIPS, Vol. 43, 1974, p. 545-52.
847. Kalstrom, D. J., "Simple encoding schemes double capacity of
     a flexible disc", Comput. Des., 15 (9), 98, 100, 102 (1976).

848. Kaye, D. N., "Floppy discs, cartridges and fixed-hard discs rise in use at savings to you", Electron. Des., 23 (22), 54-7 (1975), CCA11-9661.
849. Meister, W., "The floppy disc-data store for the future", Radio Elektron. Schau, 52 (2), 18-21 (1976), German, CCA11-15051.
850. Pohm, A. V. and Zingg, R. J., "Proposal for a 10^{12} bit flexible disk pack memory", IEEE Trans. Magn., MAG-8 (3), 574-6 (1972), CCA8-17690.
851. Sherry, K. J., "Universal floppy disk controller", Proceedings of International Symposium and Course on Mini- and Micro-computers, 1976, p. 79-82.

4.   Tapes/Cassettes

852. Anon., "Buffered magnetic tape transport provides design flexibility", Comput. Des., 10 (11), 122-4 (1971), CCA7-6284.
853. Anon., "Development and application of magnetic tape memories", Wiad. Elektrotech., 42 (13-14), 318-20 (1974), Polish, CCA9-23250.
854. Anon., "The DATEV data processing system (DES): real time data tracks in magnetic tape cassettes", Buerotechnik, 24 (11), 48-9 (1976), German, CCA12-8912.
855. Azuma, N., "Incremental magnetic tape memory", KDD Tech. J., no. 67, 25-33 (1971), Japanese, CCA6-12717.
856. Baker, B. R., "A dropout model for a digital tape recorder", IEEE Trans. Magn., MAG-13 (5), 1196-8 (1977).
857. Bartlett, K. G., "Consider MSI for tape controllers", Electron. Des., 23 (16), 50-4 (1975).
858. Bernstein, K., "Matching unit for coupling the magnetic tape store MB 1250 with the data remote transfer system DFE 550", Mess. Steuern Regeln Mit Automatisierungsprax, 19 (8), 181-4 (1976), German, CCA11-31592.
859. Bogunovic, N., "The application of magnetic tape cassette memories in computer systems", Elektrotehnika, Zagreb, 18 (6), 355-60 (1975), Croatian, CCA11-15048.
860. Bony, J. and Laussel, P., "Device for signal positioning on a magnetic tape and for automatic control of processing", Onde Electr., 55 (4), 237-40 (1975), French, CCA10-17634.
861. Bratieres, J., "Standardization of magnetic tapes and cassettes", Courrier Noramlisation, 38 (221), 569-71 (1971), French, CCA7-1955.
862. Chao, A., Hagemeier, W. and Williams, J., "Optical spectrum analysis and imaging of magnetic tape patterns using the Kerr effect", Proceedings of the Society of Photo-Optical Instrumentation Engineers, Vol. 52, Publ. 1975, p. 82-7, CCA11-22645.
863. Chaudhary, B. D., Mahabala, H. N., Jha, K. M. L. and Mehta, G. K., "Interface for digital recording on incremental magnetic tape", J. Inst. Electron Telecommun. Eng., 21 (2), 45-8 (1975).
864. Cuddihy, E. F., "Hygroscopic properties of magnetic recording tape", IEEE Trans. Magn., MAG-12 (2), 125-35 (1976).

865. Cudmore, P. H., "Magnetic tape head function switching system",
     Report N75-25134/8SL, National Aeronautics and Space Administra-
     tion, Goddard Space Flight Center, Greenbelt, Md., 1975, 17 pp.
866. Davis, S., "Digital cassette and cartridge recorders today",
     Comput. Des., 12 (1), 59-71 (1973).
867. Davis, S., "Update on magnetic tape memories", Comput. Des.,
     13 (8), 127-40 (1974).
868. Dillon, W. C. and Mesecar, R. S., "Magnetic tape translator",
     Report AD-D001435/7SL, Department of the Navy, Washington,
     D. C., 1974, 12 pp.
869. Drayton, W., "How to interpret the ECMA-34", Syst. Int., 4 (8),
     29 (1976), CCA12-15496.
870. Frazier, G. R., "Magnetic tape velocity variation measurement
     using a read while write head", IEEE Trans. Magn., MAG-12 (1),
     44-5 (1976).
871. Geller, S. B., "A universal dropout tester for magnetic storage
     media", Report NBS TN 739, Nat. Bur. Stand., Washington, D. C.,
     1972, 27 pp., CCA8-13659.
872. Glockmann, H. P., "Data storage by PCM on magnetic tape cas-
     settes", Mess. & Pruef., no. 9, 395-400 (1975), German,
     CCA11-6368.
873. Gonzalez, F. C. and Alcaraz, J. S., "Macropacking and micro-
     packing in particles on magnetic tapes", An. Fis., 69 (4-6),
     157-65 (1973), Spanish, CCA9-16052.
874. Goron, A. I., "Variational method of determining the residual
     magnetization of thin magnetic tapes", Telecommun. Radio Eng.,
     29-30 (2), 124-8 (1975).
875. Heuer, K., "Magnetic tape store ZMB 51", Jena Rev., 16 (Spe.),
     63-5 (1971), CCA6-12716.
876. Heuer, K., "Magnetic tape stores ZMB61 and ZMB101", Jena Rev.,
     17 (2), 64-5 (1972), CCA7-20079.
877. Hicks, J., "Cassettes - a 'standard' code", Systems, 4 (1),
     28-9 (1976), CCA11-28274.
878. Higuchi, S., Hisagen, Y., Ozu, T. and Ohata, T., "Magnetic
     recording tape with double layer coatings", IEEE Trans. Magn.,
     MAG-13 (5), 1668-72 (1977).
879. Hirota, E., Kawamata, T., Mihara, T. and Terada, Y., "Sulfur
     modified chromium dioxide (magnetic tape)", IEEE Trans. Magn.,
     MAG-8 (3), 430 (1972), CCA8-15432.
880. Hori, Y., Hasuike, A., Hagashi, T. and Nagase, Y., "A study on
     foil bearing - an application to a magnetic memory device", J.
     Fac. Eng. Univ. Tokyo A, no. 12, 16-17 (1974), Japanese,
     CCA10-15163.
881. Ikeda, Y., "Magnetic tape memories", J. Inst. Electron. &
     Commun. Eng. Jap., 54 (4), 548-52 (1971), Japanese, CCA6-17633.
882. Javid, B., "New techniques squeeze more data onto magnetic
     tape", EDN, 20 (11), 63-4 (1975).

883. Johnston, H. T., Jr. and Cobb, R. O., "Preparation of read-only magnetic tape", IBM Tech. Disclosure Bull., 14 (2), 377 (1971), CCA6-24453.

884. Kawasaki, M. and Higuchi, S., "Alloy powders for magnetic recording tape", IEEE Trans. Magn., MAG-8 (3), 430-2 (1972), CCA8-15434.

885. Koenig, R. A., "Scroll mass storage drive", 8th IEEE Computer Society International Conference, COMPCON 74, p. 93-5.

886. Kohler, H., "Program controlled measurement data processing by a hybrid-magnetic tape storage unit", Elek. Ausrustung, 11 (5), 25-8 (1970), German, CCA6-2197.

887. Kosorok, J. R. and Ratcliffe, C. A., "Memory data storage via DEC tape", DECUS Fall 1974 Symposium (abstracts), p. 94, CCA11-9664.

888. Lapin, V. S., "The problem of automatic correction of grouped errors on magnetic tape", Probl. Inf. Transm., 4 (1), 22-6 (1968), Russian, CCA6-2053.

889. Lomnicky, H., "The technique of threading magnetic tapes automatically in tape decks for data processing", Feinwerktech. & Messtech., 84 (1), 34-7 (1976), German, CCA11-15195.

890. Martin, V. C., "Full erasure of interblock gap between records on magnetic tape", IBM Tech. Disclosure Bull., 15 (2), 641-2 (1972), CCA7-26480.

891. Meiners, E. P., Lenz, L. L., Dalby, A. E. and Hornsby, J. M., "Recommended standards for digital tape formats", Geophysics, 37 (1), 36-44 (1972), CCA7-20077.

892. Mohus, I. and Osmundsen, R., "CATSY (cassette tape system) developed by Sintef", Elektrotek. Tidsskr., 85 (12), 15-18 (1972), Norwegian, CCA7-22388.

893. Montgomery, T. Y., "Designing magnetic tape files for variable length records", Mange. Inf., 3 (6), 271-6 (1974).

894. Moritomo, T., Sasaoka, H., Takesue, M., Hosokawa, S. and Shimozuma, K., "DIPS high speed magnetic tape subsystem", Electr. Commun. Lab. Tech. J., 23 (12), 2709-18 (1974), Japanese, CCA10-15161.

895. Muller, J., "Multi-purpose data storage unit utilising a magnetic tape cassette", Elektron. J., 9 (3), 16, 54-6 (1974), German, CCA9-16053.

896. Muslay, R., "Storing large volume of data on magnetic tapes", Inf. Elektron., 6 (3), 232-5 (1971), Hungarian, CCA7-1959.

897. Nierhaus, R. and Taylor, B. G., "Video-tape nuclear physics digital data recording", Proceedings of the Conference on Video and Data Recording, 1973, p. 165-72, CCA8-20877.

898. Novak, I., "Memodyne systems for the recording of digital data in 1/8" cassette", Mess. & Zachlen Regist., no. 8, 15-17 (1975), German, CCA11-4789.

899. Pastoriza, J. J., "Tradeoffs among binary codes in magnetic tape cassettes", Comput. Des., 15 (1), 102-4 (1976).

900. Pawlikowska-Kozlowska, J. and Swiderska, L., "Operations on
     magnetic tapes", Algorytmy, Spec. Issue, 77-93 (1971), Polish,
     CCA7-2387.
901. Probyn, B. A. and Schotten, H. G. K., "Demagnetisation in
     digital tape recordings", 1971 Digests of the International
     Magnetics (Intermag.) Conference, 1 pp., CCA6-17639.
902. Prusinkiewicz, P. and Budkowski, S., "Double track error-cor-
     rection code for magnetic tape", IEEE Trans. Comput., C-$\underline{25}$ (6),
     642-5 (1976).
903. Purdy, S., "Understanding computers. Backing stores-magnetic
     tape", Data Manage., $\underline{3}$ (5), 4 (1971), CCA6-15215.
904. Robinson, G. R., "A high redundancy method for recording digital
     data on commercially available cassettes", J. Phys. E, $\underline{9}$ (2),
     110-11 (1976), CCA11-9663.
905. Rot, A., "Magnetic tape as information storage - 500 cycles per
     mm", Nachr. Elektron., $\underline{30}$ (12), 294 (1976), German, CCA12-3987.
906. Schaefer, W. C. and Young, F. C., "Tape address detector sub-
     system", IBM Tech. Disclosure Bull., $\underline{13}$ (10), 2855-6 (1971),
     CCA6-12715.
907. Schalk, K. and Schmidt, G., "7FS1 reader for magnetic tape
     cassettes of the datareg system", Siemens Rev., $\underline{42}$ (10), 438-40
     (1975).
908. Shimamoto, S., "Experiments on magnetic instability of super-
     conducting tape", Electr. Eng., $\underline{95}$ (1), 9-14 (1975).
909. Shroff, B., "Tightening tape for improved cartridge-drive opera-
     tion", Digital Des., $\underline{6}$ (12), 28, 30, 32 (1976), CCA12-15493.
910. Sloane, N. J. A., "Simple description of an error-correcting
     code for high-density magnetic tape", Bell Syst. Tech. J., $\underline{55}$
     (2), 157-65 (1976).
911. Stange, R., "The compact tape cassette in data processing",
     Radio Elektron. Schau, $\underline{51}$ (11), 644-7 (1975), German, CCA11-6369.
912. Starostin, Y. V., Krasnykh, V. I., Lyubenko, E. A. and Bocharov,
     V. I., "Magnetic properties of Ni-Fe-Mo tape in relation to
     technological factors", Met. Sci. Heat Treat., $\underline{16}$ (11-12),
     957-9 (1974).
913. Straubel, R., "The physical limits of the storage characteris-
     tics of magnetic tapes", Funommechanika, $\underline{10}$ (8), 228-30 (1971),
     Hungarian, CCA6-24451.
914. Tanaka, T. and Mizugaki, A., "Magnetic tape facility in DEX-21
     electronic switching system", Rev. Electr. Commun. Lab., $\underline{22}$
     (11-12), 1019-25 (1974).
915. Tsunekawa, S., Hanada, H., Maejima, H. and Toyozumi, S., "Recent
     technical view of magnetic tape station", Hitachi Hyoron, $\underline{55}$
     (10), 25-30 (1973), Japanese, CCA9-11953.
916. Vasilevskii, Y. A., Zelenina, L. I., Nemirovskaya, N. M. and
     Subbotin, S. S., "Temperature dependence of magnetic tape prop-
     erties", Tekh. Kino & Telev., no. 2, 19-22 (1975), Russian,
     CCA10-15165.

917. Winkler, K. and Olbrich, O., "Design features of the Model 3530
     and 3531 magnetic tape decks", Feinwerktech. & Messtech., 84
     (1), 28-34 (1976), German, CCA11-15194.
918. Zsom, G., "Magnetic tape cassette data acquisition equipment",
     Meres & Autom., 24 (2), 71-5 (1976), Hungarian, CCA11-15196.

     5.   Drums

919. Baudet, G., Boulenger, J. and Ferrie, J., "Analysis of a drum
     with bulk arrivals", International Computer Symposium, Proceed-
     ings, 1975, p. 213-24.
920. Denning, P. J., "A note on paging drum efficiency", Comput.
     Surv., 4 (1), 1-3 (1972), CCA7-14695.
921. Gal, J. and Reszler, A., "Magnetic disc and magnetic drum
     external storages", Inf. Elektron., 7 (3), 201-6 (1972), Hun-
     garian, CCA8-2632.
922. Hayashi, T. and Yamazaki, A., "Small record capacity magnetic
     drum memory subsystem for DIPS system", Electr. Commun. Lab.
     Tech. J., 23 (12), 2719-23 (1974), Japanese, CCA10-15162.
923. Hayashi, Y., Osawa, A., Hamada, N. and Kurane, K., "High re-
     cording density magnetic drum memory for control computer",
     Hitachi Hyoron, 55 (10), 19-23 (1973), Japanese, CCA9-11952.
924. Kameyama, T. and Kawada, T., "DEX-All magnetic drum memory",
     Rev. Electr. Commun. Lab., 22 (9-10), 810-17 (1974).
925. Kaneko, R., Matuda, R. and Hara, S., "Designing a magnetic
     drum floating head mechanism", Rev. Electr. Commun. Lab., 20
     (7-8), 621-42 (1972), CCA8-2634.
926. Kaye, D. N., "Focus on disc and drum memories", Electron. Des.,
     20 (10), C16-24 (1972), CCA7-20076.
927. Maslov, S. P., "Ternary magnetic drum memory", Vychisl. Tekh.
     & Vopr. Kibera, no. 10, 124-37 (1974), Russian, CCA9-21986.
928. Tatuta, T., "Magnetic drum memories", J. Inst. Electron. &
     Commun. Eng. Jap., 54 (4), 552-4 (1971), Japanese, CCA6-17634.
929. Tominaga, H. and Fujiwara, Y., "DEX-21 electronic switching
     system magnetic drum controller design", Rev. Electr. Commun.
     Lab., 20 (7-8), 613-20 (1972), CCA8-2633.

     6.   Recording

930. Abecia, V. L., Yarza, A. R. and Royo, F. R., "Recording and
     recovery of digital information on commercial magnetic tape
     cassettes and tape recorders", Rev. Inf. & Autom., 9 (27),
     7-15 (1976), Spanish, CCA11-22644.
931. Anon., "A 30-Kbpi data recording system. A 'self-timing' re-
     corder eliminates the need for complex tape deskew hardware
     and packs 70 gigabits on one 7200-ft, 28-track reel of 1" (2.54
     cm) tape", Digital Des., 6 (6), 72-5 (1976), CCA12-3983.

932. Belyayev, R. M. and Shibanov, L. S., "Some problems in increasing
     the density of digital magnetic recording", Telecommun. Radio
     Eng., 29-30(5), 85-9 (1975).
933. Chi, C. S. and Speliotis, D. E., "Isolated pulse and two-pulse
     interactions in digital magnetic recording", IEEE Trans. Magn.,
     MAG-11 (5), 1179-81 (1975).
934. Davidson, M., Haase, S. F., Machamer, J. L. and Wallman, L. H.,
     "High density magnetic recording using digital block codes of
     low disparity", IEEE Trans. Magn., MAG-12 (5), 584-6 (1976),
     CCA11-31591.
935. Flanders, P. J., "Remanence changes of magnetic tapes and disks
     stressed in zero and reverse fields", IEEE Trans. Magn., MAG-13
     (5), 1673-80 (1977).
936. Flanders, P. J., "Changes in the magnetization of a recording
     tape produced by heating", IEEE Trans. Magn., MAG-13 (5),
     1681-7 (1977).
937. Fluitman, J., "Effect of nonzero write field rise time in digital
     magnetic recording", IEEE Trans. Magn., MAG-12 (3), 218-23 (1976).
938. Haynes, M. K., "Experimental determination of the loss and phase
     transfer function of a magnetic recording channel", IEEE Trans.
     Magn., MAG-13 (5), 1284-6 (1977).
939. Hoagland, A. S., "Magnetic recording for data storage", 8th IEEE
     Computer Society International Conference, COMPCON 74, Digest
     of Papers, p. 293-6.
940. Hoagland, A. S., "Magnetic recording storage", IEEE Trans.
     Comput., C-25 (12), 1283-8 (1976), CCA12-1492.
941. Honeywell Information System Inc., "Magnetic recording apparatus",
     Patent UK 1294397, Publ. October 1972, CCA8-9520.
942. Hsieh, E. J. and Winslow, J. W., "Radiation hardness of a new
     magnetic thin-film recording system", IEEE Trans. Nucl. Sci.,
     NS-18 (1), 224-7 (1971), CCA6-15221.
943. Huber, W. D., "Maximization of lineal recording density", IEEE
     Trans. Magn., MAG-13 (5), 1208-10 (1977).
944. IBM Corp., "Magnetic recording/replay device", Patent UK
     1233683, Publ. May 1971, CCA6-20135.
945. IBM Corp., "Digital data tape recorder", Patent UK 1394696,
     Publ. May 1975, CCA10-25837.
946. Inagaki, N., Hattori, S., Ishii, Y. and Katsuraki, H., "Ferrite
     thin films for high recording density", IEEE Trans. Magn.,
     MAG-11 (5), 1191-3 (1975), CCA11-4784.
947. Iwasaki, S. I. and Nakamura, Y., "An analysis for the magneti-
     zation mode for high density magnetic recording", IEEE Trans.
     Magn., MAG-13 (5), 1272-7 (1977).
948. Jacoby, G. V., "A new look-ahead code for increased data
     density", IEEE Trans. Magn., MAG-13 (5), 1202-4 (1977).
949. Janssens, N., "Static models of magnetic hysteresis", IEEE
     Trans. Magn., MAG-13 (5), 1379-81 (1977).
950. Johnson, C. E., Jr., "Digital recording in low cost transports",
     Digital Des., 7 (6), 38-48 (1977).

951. Kaganowicz, G., Hockings, E. F. and Robinson, J. W., "Influence
     of zinc on cobalt substituted magnetic recording media", IEEE
     Trans. Magn., MAG-11 (5), 1194-6 (1975).
952. Krey, K. H. and Liberman, A. G., "Analyses in magnetic recording
     of VHF signals", Report AD-A013525/1SL, Harry Diamond Labs.,
     Adelphi, Md., 1975, 57 pp.
953. Lewkowicz, J. and Stephens, T. R., "Tri-bit servo detection",
     IBM Tech. Disclosure Bull., 19 (3), 810-3 (1976), CCA12-3985.
954. Mallinson, J. C., "A unified view of high density digital re-
     cording theory", IEEE Trans. Magn., MAG-11 (5), 1166-9 (1975),
     CCA11-4783.
955. Mallinson, J. C. and Bertram, H. N., "Write processes in high
     density recording", IEEE Trans. Magn., MAG-9 (3), 329-31 (1973),
     CCA9-3392.
956. Middleton, B. K. and Wisely, P. L., "The development and applica-
     tion of a simple model of digital magnetic recording to thick
     oxide media", Proceedings of the Conference on Video and Data
     Recording, 1976, p. 33-42, CCA11-31593.
957. Minukhin, V. B., "Phase distortions of signals in magnetic
     recording equipment", Telecommun. Radio Eng., 29-30(1),
     114-20 (1975).
958. Nemeth, G., "Examination of the writing process in moving
     magnet stores", Meres & Autom., 24 (8), 305-7 (1976), Hungarian,
     CCA12-6645.
959. Ogawa, K., Ogawa, S. and Suenaga, T., "A method for predicting
     missing-error counts at high recording densities", IEEE Trans.
     Magn., MAG-13 (5), 1193-5 (1977).
960. Ortenburger, I. B., Cole, R. W. and Potter, R. I., "Improve-
     ments to a self-consistent model for the magnetic recording
     properties of non-particulate media", IEEE Trans. Magn.,
     MAG-13 (5), 1278-83 (1977).
961. Patel, A. M., "Zero-modulation encoding in magnetic recording",
     IBM J. Res. & Dev., 19 (4), 366-78 (1975), CCA11-6366.
962. Patel, A. M., "New method for magnetic encoding combines ad-
     vantages of older techniques", Comput. Des., 15 (8), 85-91 (1976).
963. Phillips, W. B. and McDonough, H. P., "Maximizing the areal
     density of magnetic recording", 8th IEEE Computer Society Inter-
     national Conference, COMPCON 74, p. 101-3.
964. Portigal, D. L., "Magnetic recording simulation program having
     an improved fit to actual hysteresis loops", IEEE Trans. Magn.,
     MAG-11 (3), 934-41 (1975).
965. Schneider, R. C., "Improved pulse-slimming method for magnetic
     recording", IEEE Trans. Magn., MAG-11 (5), 1240-1 (1975).
966. Schulze, G. H., "An emerging standard format for high density
     digital magnetic recording systems", Proceedings of the Con-
     ference on Video and Data Recording, 1976, p. 290, CCA12-1497.
967. Speliotis, D. E. and Chi, C. S., "Design of advanced digital
     magnetic recording systems", IEEE Trans. Magn., MAG-11 (5),
     1234-6 (1975), CCA11-4785.

968. Speliotis, D. E. and Chi, C. S., "Plated films for digital mag-
     netic recording", Plat. & Surf. Finish., $\underline{63}$ (1), 31-4 (1976),
     CCA11-15055.
969. Speliotis, D. E. and Chi, C. S., "The effect of thickness in
     high bit density recording on particulate and plated disks",
     IEEE Trans. Magn., MAG-$\underline{13}$ (5), 1287-9 (1977).
970. Tahara, Y., Takagi, H. and Ikeda, Y., "Optimum design of
     channel filters for digital magnetic recording", Fujitsu Sci.
     & Tech. J., $\underline{12}$ (4), 119-27 (1976), CCA12-8886.
971. Tjaden, D. L. A. and Tercic, E. J., "Theoretical and experi-
     mental investigations of digital magnetic recording on thin
     media", Philips Res. Rep., $\underline{30}$ (2-3), 120-61 (1975).
972. Velichkin, V. A., "Comparison of different methods for the mag-
     netic recording of binary data with regard to noise immunity
     and density", Telecommun. & Radio Eng. Pt. 2, $\underline{30}$ (12), 118-21
     (1975), CCA12-1494.
973. Wilson, D. M., "Effect of switching field distributions and
     coercivity on magnetic recording properties", IEEE Trans. Magn.,
     MAG-$\underline{11}$ (5), 1200-2 (1975).

7.   Heads

974. Anon., "An integrated magnetoresistive read, inductive write
     high density recording head", AIP Conf. Proc., no. 24, 548-9
     (1974), CCA11-4782.
975. Bate, G., Echelmeier, R. H., Everett, L. H., Kehr, W. B., Schaps,
     J. E., Taylor, G. and Thornley, R. F. M., "A multi-track head
     for low frequency applications", IEEE Trans. Magn., MAG-$\underline{8}$ (3),
     542-4 (1972), CCA8-17689.
976. Brownlow, L. W. and King, C. C., "Write field analysis for inte-
     grated heads of the finite pole-tipe configuration", IEEE
     Trans. Magn., MAG-$\underline{8}$ (3), 539-41 (1972), CCA8-17688.
977. Bulavenko, O. M., "Magnetic head for contactless magnetic
     recording", Sov. Autom. Control, $\underline{8}$ (1), 63-6 (1975).
978. Charbonnier, P. P., "Flight of flexible disk over recording
     heads", IEEE Trans. Magn., MAG-$\underline{12}$ (6), 728-30 (1976), CCA12-6641.
979. Chynoweth, W. and Kayser, W., "Thin film inductive recording
     heads", AIP Conf. Proc., no. 24, 534-40 (1974), CCA11-4780.
980. Desserre, J., Helle, M. and Lazzari, J. P., "The track width
     on integrated heads", Proceedings of the Conference on Video
     and Data Recording, 1976, p. 227-34, CCA12-1495.
981. Desserre, J., Helle, M. and Lazzari, J. P., "New possibilities
     offered by integrated heads to reduce crossfeed (Write to Read)",
     IEEE Trans. Magn., MAG-$\underline{13}$ (5), 1469-71 (1977).
982. Feng, J. S. Y., Romankiw, L. T. and Thompson, D. A., "Magnetic
     self-bias in the Barber pole MR structure", IEEE Trans. Magn.,
     MAG-$\underline{13}$ (5), 1466-8 (1977).

983. Goron, A. I., "Effect of partial saturation of the pole pieces of a recording head on its field", Telecommun. Radio Eng., 28-29 (6), 123-5 (1974).

984. Hahn, F. W., Jr., "Design considerations for a helical-scan recording head contour", Proceedings of the Conference on Video and Data Recording, 1976, p. 261-6, CCA12-1496.

985. Hattori, S. and Kashiwabara, M., "Magnetic recording head with fixed coil", Rev. Electr. Commun. Lab., 23 (3-4), 387-93 (1975).

986. Hughes, G. F. and Bloomberg, D. S., "Recording head side read/write effects", IEEE Trans. Magn., MAG-13 (5), 1457-9 (1977).

987. Kawakami, K., Kaneko, E. and Ono, K., "On magnetic path saturation in thin film heads", AIP Conf. Proc., no. 24, 547 (1974), CCA11-4781.

988. Kehr, W. D., Thornley, R. F. M. and Meldrum, C. B., "Wear of non-stoichiometric polycrystalline nickel-zinc ferrite", IEEE Trans. Magn., MAG-13 (5), 1478-9 (1977).

989. Kugimiya, K. and Hirota, E., "A hot pressed coprecipitated ferrite and its application to magnetic heads", IEEE Trans. Magn., MAG-13 (5), 1472-4 (1977).

990. Lagrange, A., Nicolas, J. and Hildebrandt, M., "Preparation and properties of hot-pressed Ni-Zn ferrites for magnetic head application", IEEE Trans. Magn., MAG-8 (3), 494-7 (1972), CCA8-17683.

991. Lindholm, D. A., "Magnetic fields of finite track width heads", IEEE Trans. Magn., MAG-13 (5), 1460-2 (1977).

992. Lindholm, D. A., "Image fields for two-dimensional recording heads", IEEE Trans. Magn., MAG-13 (5), 1463-5 (1977).

993. Luoma, R. W., "Coil lead termination system for magnetic head assemblies", IBM Tech. Disclosure Bull., 19 (4), 1377-9 (1976), CCA12-6644.

994. Marconi Co. Ltd., "Magnetic record reading head", Patent UK 1235857, Publ. June 1971, CCA6-20136.

995. Miyazaki, T., Sawada, R. and Ishijima, Y., "New magnetic alloys for magnetic recording heads", IEEE Trans. Magn., MAG-8 (3), 501-2 (1972), CCA8-17684.

996. Monson, J. E., "Field analysis for nonlinear magnetic heads", IEEE Trans. Magn., MAG-8 (3), 533-6 (1972), CCA8-17685.

997. Moxley, A. E., "Computer support of magnetic tape head manufacture", 1971 IEEE International Convention Digest, p. 576-7, CCA6-17636.

998. Nakagawa, S., Kanai, K. and Kobayashi, F., "Investigation of head saturation using a large scale ferrite model head", IEEE Trans. Magn., MAG-8 (3), 538 (1972), CCA8-17687.

999. Onozuka, C., Wada, T. and Naruse, Y., "The development and evaluation of multi-track magnetic heads for digital use", Nat. Tech. Rep., 17 (1), 71-80 (1971), Japanese, CCA6-15217.

1000. Pierce, J. T., "Method for fabricating magnetic read-write head array and product", Patent USA 3601871, Publ. August 1971, CCA6-24454.

1001. Romankiw, L. T. and Simon, P., "Batch fabrication of thin film
      magnetic recording beads: A literature review and process
      description for vertical single turn heads", IEEE Trans. Magn.,
      MAG-11 (1), 50-5 (1975).
1002. Roscamp, T. A. and Frank, P. D., "Thin-film magnetic heads
      excel in packing and moving data", Electronics, 50 (5), 97-103
      (1977), CCA12-8885.
1003. Sansom, D. J., "Recording head design calculations", IEEE
      Trans. Magn., MAG-12 (3), 230-3 (1976).
1004. Senno, H., Yanagiuichi, Y., Satomi, M., Hirota, E. and Haya-
      kawa, S., "Newly developed Fe-Si-Al alloy heads by squeeze
      casting", IEEE Trans. Magn., MAG-13 (5), 1475-7 (1977).
1005. Shibaya, H. and Fukuda, I., "The effect of the B_s of recording
      head cores on the magnetization of high coercivity media", IEEE
      Trans. Magn., MAG-13 (3), 1005-8 (1977).
1006. Suzuki, T. and Iwasaki, S. I., "An analysis of magnetic re-
      cording head fields using a vector potential", IEEE Trans.
      Magn., MAG-8 (3), 536-8 (1972), CCA8-17686.
1007. Talke, F. E. and Tseng, R. C., "Submicron transducer spacings
      in rotating head/tape interface", IEEE Trans. Magn., MAG-12
      (6), 725-7 (1976), CCA12-6640.
1008. van Herk, A., "Side fringing fields and write and read cross-
      talk of narrow magnetic recording heads", IEEE Trans. Magn.,
      MAG-13 (4), 1021-8 (1977).
1009. van Symons, C., "Logic sparing of fixed heads on drums and
      discs", Comput. Des., 11 (9), 91-4 (1972), CCA8-2628.
1010. Zapala, R. J., "Flexible recording head mounting assembly",
      Report AD-D002337/4SL, Department of the Air Force, Washing-
      ton, D. C., 1976, 9 pp.

        8.   Others

1011. Bardidge, V. V., Larin, E. M. and Perekatov, V. I., "The
      physical analysis of laminated ferrite wafer individual stor-
      age element", IEEE Trans. Magn., MAG-6 (3), 677-82 (1970),
      CCA6-10766.
1012. Boll, R., Koster, H. J. and Weiss, G., "Plated tape for mag-
      netic memories", Z. Angew. Phys., 30 (1), 120-4 (1970),
      German, CCA6-2065.
1013. Burroughs Corp., "Magnetic memory system", Patent UK 1229152,
      Publ. April 1971, CCA6-20133.
1014. Burroughs Corp., "Magnetic memory system", Patent UK 1229153,
      Publ. April 1971, CCA6-20134.
1015. Kozima, K., Gamow, Y., Uesaka, T., Nambu, H., Taniguchi, K.
      and Ozaki, H., "16-KB coupled film memory", Electron. &
      Commun. Jap., 54 (8), 136-42 (1971), CCA8-2643.

1016. Oshima, S., Kobayashi, T., Kamibayashi, T., Okada, A., Koma-
      zawa, Y. and Komuro, K., "Improving fine striped memory mag-
      netic properties", IEEE Trans. Magn., MAG-$\underline{6}$ (3), 725-8 (1970),
      CCA6-10771.
1017. Shahbender, R., Herkart, P., Karstad, K., Kurlansik, H. and
      Onyshkevych, I., "Sonic film memory", IEEE Trans. Magn.,
      MAG-$\underline{5}$ (3), 427 (1969), CCA5-6490.
1018. Subscription Television Inc., "Magnetic memory", Patent UK
      1296026, Publ. November 1972, CCA8-9531.
1019. Torok, E. J., "Film only a few atoms thick promises very large
      mass memories", Electronics, $\underline{45}$ (23), 106-12 (1972), CCA8-2641.

      VI. OPTICAL STORAGE
         1.   General

1020. Anderson, L. K., "Application of holographic optical techniques
      to bulk memory", IEEE Trans. Magn., MAG-$\underline{7}$, 601-5 (1971).
1021. Bartolini, R. A., Weakleim, H. A. and Williams, B. F., "Review
      and analysis of optical recording media", Opt. Eng., $\underline{15}$ (2),
      99-108 (1976).
1022. Boltz, C. L., "From Stonehenge to laser (storage devices)",
      Phys. Technol., $\underline{6}$ (1), 3-10 (1975), CCA10-19929.
1023. Chomat, M., "Semipermanent optical memories with holograms",
      Slaboproudy Obzor, $\underline{32}$ (3), 120-5 (1971), Czech, CCA6-10785.
1024. Gaylord, T. K., "Is optical holography the answer?", Opt.
      Spectra, $\underline{6}$ (11), 25-6, 35-7 (1972), CCA8-9548.
1025. Gillis, A. K., Hoffmann, G. E. and Nelson, R. H., "Holographic
      memories - fantasy or reality?", Proceedings of the AFIPS
      National Computer Conference, Vol. 44, 1975, p. 535-9.
1026. Graf, P. and Kiemle, H., "Holographic storage of digital data",
      Electron. Equip. News, $\underline{15}$ (11), 42-4 (1974), CCA9-18424.
1027. Hill, B., Schmidt, U. J. and Schmitt, H. J., "Optical memor-
      ies", J. Appl. Sci. Eng. Sect. A, $\underline{1}$ (1), 39-44 (1975).
1028. Hoff, F., "On some materials for optical information recording",
      Slaboproudy Obz., $\underline{37}$ (2), 55-60 (1976), Czech, CCA11-15074.
1029. Huignard, J. P., Micheron, F. and Spitz, E., "Optical systems
      and photosensitive materials for information storage", In:
      Optical Properties of Solids - New Developments, Seraphin,
      B. O., (Ed.), North-Holland, Amsterdam, Netherlands, p. 847-925
      (1975), CCA11-17416.
1030. Jeppson, K., "Optic memories - the memories of the future?",
      Eltek. Aktuell Elektron., $\underline{17}$ (2), 52-3 (1974), Swedish,
      CCA9-16090.
1031. Killat, U. and Terrell, D. R., "Performance and limitations of
      photothermoplastic devices", Opt. Acta, $\underline{24}$ (4), 441-52 (1977),
      CCA12-15542.
1032. Knight, G. R., "Holographic memories", Opt. Eng., $\underline{14}$ (5),
      453-9 (1975).

1033. Kock, W. E., "Optical computing, an example of change", Proc.
      IEEE, $\underline{65}$ (1), 6-9 (1977), CCA12-4037.
1034. Kurtz, R. L. and Owen, R. B., "Holographic recording materials
      - a review", Opt. Eng., $\underline{14}$ (5), 393-401 (1975).
1035. Lohman, R. D., "Optical memories", IEEE Region 6 Conference
      (West USA) on Optoelectronics and Laser Technology, Proceed-
      ings, 1974, p. 1-4.
1036. Minnaja, N. and Nobile, M., "Holographic memories", Rend.
      Riunione Assoc. Elettrotec. Ital., $\underline{49}$ (3), B2/1-7 (1974),
      Italian, CCA11-6389.
1037. Rajchman, J. A., "Promise of optical memories", J. Appl.
      Phys., $\underline{41}$, 1376-83 (1970).
1038. Waterworth, P., "Holographic memories", New Electron., $\underline{7}$ (7),
      37, 41 (1974), CCA9-18433.
1039. Waterworth, P., "Holographic memories", New Electron., $\underline{9}$ (21),
      91-2 (1976), CCA12-1526.

2.   Recording Media
a.   Magneto-optic

1040. Aagard, R. L., "Signal-to-noise ratio for magnetooptic readout
      from MnBi films", IEEE Trans. Magn., MAG-$\underline{9}$ (4), 705-7 (1973),
      CCA9-14366.
1041. Ahrenkiel, R. K. and Coburn, T. J., "Hot-pressed $CoCr_2S_4$: a
      magneto-optical memory material", Appl. Phys. Lett., $\underline{22}$ (7),
      340-1 (1973), CCA8-13712.
1042. Akaev, A., Golubkova, M. N. and Maiorov, S. A., "Contribution
      to the theory of reconstruction of the image of a page of
      numeric information from a magnetic hologram", Sov. J. Quantum
      Electron., $\underline{6}$ (3), 366-8 (1976).
1043. Akaev, A., Golubkova, M. N. and Smirnov, N. A., "Analysis of
      images of pages of digital information, reconstructed from a
      magnetic hologram", Izv. VUZ Priborostr., $\underline{19}$ (4), 56-62 (1976),
      Russian, CCA11-22676.
1044. Brown, B. R., "Optical data storage potential of six materials",
      Appl. Opt., $\underline{13}$ (4), 761-6 (1974), CCA9-18416.
1045. Chen, D., "Magnetic materials for optical recording", Appl.
      Opt., $\underline{13}$ (4), 767-78 (1974), CCA9-16082.
1046. Chen, T. and McKinley, T., "Observation of magnetic domains in
      MnSb and MnBi single crystals by the Kerr effect and a replica
      method", IEEE Trans. Magn., MAG-$\underline{13}$ (5), 1580-2 (1977).
1047. Chida, K., Tsujiyama, B., Katsui, A. and Egashira, K., "Magneto-
      optical memory experiments on a rotating Mn-Cu-Bi disk medium",
      IEEE Trans. Magn., MAG-$\underline{13}$ (3), 982-8 (1977), CCA12-15494.
1048. DuBois, G., Misra, P., Szenlesi, O., Wile, G. and Zaky, S.,
      "Toward a magneto-optic memory", Telesis, $\underline{2}$ (4), 20-6 (1972),
      CCA8-7367.

1049. Esho, S., Noguchi, S., Ono, Y. and Nagao, M., "Optimum thick-
      ness of MnBi films for magnetooptical memory", Appl. Opt.,
      13 (4), 779-83 (1974), CCA9-16083.
1050. Feldmann, P., Guillot, M. and LeGall, H., "Temperature and
      field dependence of the magnetic and magnetooptical properties
      of single crystal dysprosium iron garnet", IEEE Trans. Magn.,
      MAG-13 (5), 1574-6 (1977).
1051. Haskal, H., Bernal, E., Chen, G. and Chen, D., "Subnanosecond
      laser recording on MnBi thin films", Appl. Opt., 13 (4), 866-8
      (1974), CCA9-16088.
1052. Heitmann, H., Krumme, J. P. and Witter, K., "Magneto-optic
      memory materials", Opt. Acta, 24 (4), 483-94 (1977), CCA12-15534.
1053. Heyworth, R. C. F. and Hoffmann, G. R., "Thermomagnetic writing
      on thin-film MnBi using an electron beam", J. Appl. Phys., 45
      (6), 2789-91 (1974), CCA9-21990.
1054. Hill, B., Sander, I. and Much, G., "Magneto-optic memories",
      Opt. Acta, 24 (4), 495-504 (1977), CCA12-15535.
1055. Huth, B. G., "Calculations of stable domain radii produced by
      thermomagnetic writing", IBM J. Res. & Dev., 18 (2), 100-9
      (1974), CCA9-18388.
1056. Katsui, A., Shibukawa, A., Terui, H. and Egashira, K., "New
      magneto-optic thin films for storage application", J. Appl.
      Phys., 47 (11), 5069-71 (1976), CCA12-4003.
1057. Kokubu, A. and Tanaka, T., "Lengthwise modulated laser record-
      ing on MnBi films", IEEE Trans. Magn., MAG-13 (5), 1290-2 (1977).
1058. Krishman, R., Vien, T. K., Canit, J. C. and Visnovsky, S.,
      "Magnetooptical properties of $Y_3Fe_{5-x}Sc_xO_{12}$ garnets", IEEE
      Trans. Magn., MAG-13 (5), 1577-9 (1977).
1059. Krumme, J. P. and Hansen, P., "New magneto-optic memory con-
      cept based on compensation wall domains", Appl. Phys. Lett.,
      23 (10), 576-8 (1973), CCA9-6230.
1060. Krumme, J. P., Heitmann, H., Mateika, D. and Witter, K., "MOPS,
      a magneto-optic-photoconductor sandwich for optical informa-
      tion storage", J. Appl. Phys., 48 (1), 366-8 (1977), CCA12-10696.
1061. Krumme, J. P. and Schmitt, H. J., "Ferrimagnetic garnet films
      for magnetooptic information storage", IEEE Trans. Magn.,
      MAG-11 (5), 1097-102 (1975), CCA11-4822.
1062. Kuroda, C., "Magneto-optical effects and their applications",
      J. Inst. Electron. Commun. Eng. Japan, 53 (7), 914-24 (1970),
      Japanese, CCA6-3780.
1063. Langlet, R., Pivot, J. P. and Carre, B., "High speed optical
      writing and reading on magnetic discs", Revue Tech. Thomson-
      CSF, 4 (2), 257-81 (1972), French, CCA7-22390.
1064. Lee, T. C., "Writing conditions of holographic storage in MnBi",
      Digest of Technical Papers of the 1971 IEEE/OSA Conference on
    · Laser Engineering and Applications, p. 62-3, CCA6-15196.
1065. Lewicki, G. and Guisinger, J. E., "Laser recording on MnBi
      films", IEEE Trans. Magn., MAG-9 (4), 700-4 (1973), CCA9-14365.

1066. Loshkareva, N. N., Plotnikov, A. F., Rodionov, A. N., Samokh-
      valov, A. A. and Chebotaev, N. M., "EuO films as a storage
      medium for a reversible optical memory", Sov. Phys. Lebedev
      Inst. Rep., no. 9, 6-10 (1974), CCA11-15075.
1067. MacDonald, R. E. and Beck, J. W., "Magneto-optical recording",
      J. Appl. Phys., 40, 1429-35 (1969).
1068. Marchenko, S. N. and Khromov, A. V., "Apparatus for data re-
      cording on magnetic film with a laser beam", Sov. J. Opt.
      Technol., 40 (8), 493-4 (1973), CCA9-20766.
1069. Matsumoto, M. and Koyama, Y., "Preparation of MnAl thin films
      usable as a storage medium in optical memory system", J. Fac.
      Eng. Shinshu Univ., no. 32, 271-8 (1972), CCA8-13715.
1070. Matsushita, S., Sunago, K. and Sakuri, Y., "Thermomagnetic
      writing in Ho-Co films", Jap. J. Appl. Phys., 14 (11), 1851-2
      (1975), CCA11-4808.
1071. Matsushita, S., Yamada, Y., Sugano, K. and Sakura, Y., "Thermo-
      magnetic contact printing in amorphous Tb-Fe films", IEEE
      Trans. Magn., MAG-13 (5), 1382-4 (1977).
1072. Mayer, L., "Curie point writing on magnetic films", J. Appl.
      Phys., 29, 1003 (1958).
1073. Msserschmitt-Bolkov-Blohm GmbH, "Magneto-optic-memory element",
      Patent UK 1282533, Publ. July 1972, CCA8-2697.
1074. Mezrich, R. S., "Reconstruction effects in magnetic holography",
      IEEE Trans. Magn., MAG-6 (3), 537-41 (1970), CCA6-10783.
1075. Plotnikov, A. F., Rodionov, A. N., Samokhvatov, A. A. and
      Seleznev, V. N., "Holographic storage of information in EuO
      films", Sov. J. Quantum Electron., 6 (9), 1136-7 (1976),
      CCA12-15538.
1076. Rull, H. and Kempter, K., "Recording of one-dimensional holo-
      grams in MnBi films", Opt. Commun., 16 (1), 83-5 (1976),
      CCA11-9694.
1077. Shakhgedanov, V. N. and Shimko, A. A., "Investigation of the
      use of MnBi films in holographic memory systems", Sov. J.
      Quantum Electron., 3 (4), 331-3 (1974), CCA9-23260.
1078. Siemens AG, "Magneto-optical memory", Patent UK 1288479,
      Publ. September 1972, CCA8-7371.
1079. Tanaka, M. and Nishimura, Y., "Characteristics of recording and
      readout of magnetic holograms on MnBiTe films", Electron.
      Commun. Jpn., 58 (2), 145-52 (1975).
1080. Ukita, H. and Iwasaki, S., "Analysis of thermomagnetic record-
      ing of optical memory using Co-system thin films", Electron.
      & Commun. Jap., 57 (11), 116-23 (1974), CCA11-12339.
1081. Volz, H., "Application of magnetic materials for image storage",
      J. Signalaufzeichnungsmater., 3 (4), 257-87 (1975), German,
      CCA11-4788.
1082. Zook, J. D., "Beam addressed optical memory", Patent USA
      3696346, Publ. October 1972, CCA8-7373.

b.   Ferroelectric

1083. Abdullaev, G. B., Belenkii, G. L., Larionkina, L. S., Nani,
      R. K. and Salaev, E. Y., "Optical memory effects in $CdIn_2S_4$
      crystals", Sov. Phys.-Semicond., 7 (4), 561 (1973), CCA9-11979.
1084. Aldrich, R. E. and Caruso, P. J., "Recording in an electro-
      optic medium", Patent USA 3564560, Publ. February 1971,
      CCA6-10786.
1085. Alphonse, G. A., Alig, R. C., Staebler, D. L. and Phillips,
      W., "Time-dependent characteristics of photoinduced space-
      charge field and phase holograms in lithium niobate and other
      photorefractive media", RCA Rev., 36 (2), 213-29 (1975).
1086. Amodei, J. J. and Staebler, D. L., "Holographic storage in
      electro-optic crystals", 1973 IEEE/OSA Conference on Laser
      Engineering and Applications, Digest of Technical Papers,
      p. 90-1, CCA9-11983.
1087. Anderson, L. K., "Optical applications of ferroelectrics",
      Ferroelectrics, 7 (1-4), 55-63 (1974).
1088. Antsygin, V. D., Belinicher, V. L., Kanaev, I. F., Malinovskii,
      V. K. and Sturman, B. I., "Spatial and transient characteris-
      tics of optical storage in non-alloyed $LiNbO_3$ crystals",
      Avtometriya, no. 4, 7-13 (1976), Russian, CCA12-1524.
1089. Asam, A. R., "Ferroelectric-photoelectric storage unit", Patent
      USA 3693171, Publ. September 1972, CCA8-7369.
1090. Belinicher, V. I. and Malinovskii, V. K., "Resonances in a
      solid body and optical data storage", Avtometriya, no. 5,
      31-4 (1976), Russian, CCA12-6675.
1091. Feldtkeller, E., "Arrangement for reading an electro-optical
      memory", Patent USA 3693172, Publ. September 1972, CCA8-7370.
1092. Gaplevskaya, S. P., Zavertannaya, L. S. and Rvachev, A. L.,
      "Some features of cadmium sulphide photoelectric memory",
      Ukr. Fiz. Zh., 19 (2), 292-5 (1974), Russian, CCA9-11980.
1093. Hadni, A. and Thomas, R., "Irreversible and localised reversal
      of spontaneous polarisation in triglycine sulphate with the
      aid of a laser beam, application to ferroelectric memories",
      Opt. Commun., 6 (4), 314-16 (1972), French, CCA8-11633.
1094. Hadni, A. and Thomas, R., "Localized irreversible thermal
      switching in ferroelectric TGS by an argon laser (memory device
      application)", Ferroelectrics, 6 (3-4), 241-5 (1974), CCA9-21998.
1095. Hou, S. L. and Oliver, D. S., "Pockels Readout Optical Memory
      using $Bi_{12}SiO_{20}$", Appl. Phys. Lett., 18 (8), 325-8 (1971),
      CCA6-15278.
1096. Huignard, J. P., Herriau, J. P. and Micheron, F., "Two different
      holographic recording process in iron doped $LiNbO_3$ and applica-
      tions", Electrochemical Society Spring Meeting, Extended
      Abstracts, 1976, p. 1064-70, CCA12-6679.
1097. Huignard, J. P., Herriau, J. P. and Micheron, F., "Holographic
      data storage and processing in photosensitive electro-optic
      crystals", Proceedings of the Conference on Video and Data
      Recording, 1976, p. 45-7, CCA11-31634.

1098. Huignard, J. P., Herriau, J. P. and Micheron, F., "Image storage display and processing in photosensitive electrooptic crystals: LiNbO3:Fe", Conference on Laser and Electrooptical Systems, Digest of Technical Papers, 1976, p. 68, 70, CCA12-8906.

1099. Huignard, J. P., Herriau, J. P. and Micheron, F., "Photosensitive electro-optical materials for holographic data storage", Rev. Tech. Thomson-CSF, 8 (4), 671-97 (1976), French, CCA12-15544.

1100. Huignard, J. P. and Micheron, F., "Control of sensitivity to hologram storage in LiNbO3. Using an accessory photovoltaic crystal", Opt. Commun., 16(1), 80-2 (1976), French.

1101. Jacobs, J. T., Keester, K. L. and Silverman, B. D., "Thermal capacitive-ferroelectric storage device", IBM Tech. Disclosure Bull., 15 (4), 1294-5 (1972), CCA8-5284.

1102. Janta, J., "Photorefractive effect and its applications", Cesk. Cas. Fis. A, 26 (2), 191-5 (1976), Czech, CCA11-22675.

1103. Kim, D. M., Shah, R. R., Rabson, T. A. and Tittel, F. K., "A time dependent analysis of holographic optical data storage properties of LiNbO3 crystals", Electrochemical Society Spring Meeting, Extended Abstracts, 1976, p. 1073-4, CCA11-31633.

1104. Kratzig, E. and Kurz, H., "Photorefractive and photovoltaic effects in doped LiNbO3", Opt. Acta, 24 (4), 475-82 (1977), CCA12-15533.

1105. Kurz, H., "Photorefractive recording dynamics and multiple storage of volume holograms in photorefractive LiNbO3", Opt. Acta, 24 (4), 463-73 (1977), CCA12-15532.

1106. Land, C. E., "Variable birefringence, light scattering, and surface-deformation effects in PLZT ceramics", Ferroelectrics, 7 (1-4), 45-51 (1974).

1107. Li, L. B., "Detector for optical memory", IBM Tech. Disclosure Bull., 16 (6), 1831-2 (1973), CCA9-14367.

1108. Micheron, F., Mayeux, C. and Trotier, J. C., "Electrical control in photoferroelectric materials for optical storage", Appl. Opt., 13 (4), 784-7 (1974), CCA9-18417.

1109. Micheron, F., Morell, A. and Hermosin, A., "Optical recording in PLZT ceramics", Rev. Tech. Thomson-CSF, 8 (4), 699-719 (1976), French, CCA12-15537.

1110. Micheron, F., Rouchon, J. M. and Vergnolle, M., "Optical recording of digital data in PLZT ceramics", Appl. Phys. Lett., 24 (12), 605-7 (1974), CCA9-21997.

1111. Micheron, F., Rouchon, J. M. and Vergolle, M., "Ferroelectric image memory", Ferroelectrics, 10 (1-4), 15-18 (1976).

1112. Nogami, G. and Imasaki, M., "Holographic recording and erasing properties in LiNbO3", Ferroelectrics, 11 (1-2), 403-5 (1976).

1113. Roberts, H. N., "PLZT input transducers for coherent optical memories and processors", Digest of Technical Papers of the 1971 IEEE/OSA Conference on Laser Engineering and Applications, p. 65, CCA6-15210.

1114. Shah, R. R., Kim, D. M., Rabson, T. A. and Tittel, F. K., "Characterization of iron-doped lithium niobate for holographic storage applications", J. Appl. Phys., $\underline{47}$ (12), 5421-31 (1976).

1115. Smith, A. W., "Electron-laser beam addressable memory", IBM Tech. Disclosure Bull., $\underline{15}$ (3), 957-8 (1972), CCA8-2695.

1116. Thaxter, J. B. and Kestigian, M., "Unique properties of SBN and their use in a layered optical memory", Appl. Opt., $\underline{13}$ (4), 913-24 (1974), CCA9-18423.

1117. von der Linde, D., Glass, A. M. and Rodgers, K. F., "Multiphonon photorefractive processes for optical storage in LiNbO$_3$", Appl. Phys. Lett., $\underline{25}$ (3), 155-7 (1974), CCA9-23259.

1118. Waterworth, P., "PLZT modulator arrays for 100 Mbit/sec holographic recording systems", Presented at 2nd European Electro-Optics Conference, Abstracts only received, Montreux, Switzerland, April 1974, CCA9-16098.

1119. Wu, S. Y., "A new ferroelectric memory device, metal-ferroelectric-semiconductor transistor", IEEE Trans. Electron Devices, ED-$\underline{21}$ (8), 499-504 (1974), CCA9-21995.

c. Thermoplastic

1120. Butter, C. D. and Lee, T. C., "Thermoplastic holographic recording of binary patterns of PLZT line composer", IEEE Trans. Comput., C-$\underline{24}$ (4), 402-6 (1975).

1121. Cindrich, I., Currie, G. D., Leonard, C. and Salmer, R., "Thermoplastic modulator brassboard", Report AD-A025450/8SL, Environmental Research Institute of Michigan, Ann Arbor, 1976, 108 pp.

1122. DuBow, J. B. and Colburn, W. S., "Ghost images in thermoplastic-photoconductor holographic storage media", J. Opt. Soc. Am., $\underline{63}$ (10), 1301 (1973), CCA9-1238.

1123. Eschler, H., "Holographic mass stores", Funk-Tech., no. 14, 497-500 (1974), German, CCA9-25103.

1124. Friesem, A. A. and Stadler, K., "Experimental aspects of photoconductor-thermoplastic devices for information storage", Sci. Electr., $\underline{22}$ (2), 72-83 (1976), CCA12-1527.

1125. Gale, M. T., Knop, K. and Russell, J. P., "Color-encoded focused-image holograms for micropublishing", J. Microgr., $\underline{8}$ (5), 225-9 (1975).

1126. Lee, T. C., "Holographic recording on thermoplastic films", Appl. Opt., $\underline{13}$ (4), 888-95 (1974), CCA9-18420.

1127. Rogers, J. W., "Analytical and experimental evaluation of techniques for the fabrication of thermoplastic hologram storage devices", Report N76-18407/6SL, Mississippi State University, 1975, 24 pp.

1128. Saito, T., Oshima, S., Honda, T. and Tsujiuchi, J., "Improved technique for holographic recording on a thermoplastic photoconductor", Opt. Commun., $\underline{16}$ (1), 90-5 (1976).

d.   Photochromic

1129. Berezin, P. D., Dyatlov, M. K., Kompanets, I. N. and Narzullaev,
      K. N., "Holographic data storage on photochromic plates using
      a helium-cadmium laser", Avtometriya, no. 3, 107-9 (1976),
      Russian, CCA11-28312.
1130. Blume, H., Bader, T. and Luty, F., "Bi-directional holographic
      information storage based on the optical reorientation of F/A
      centers in KCl:Na", Opt. Commun., 12 (2), 147-51 (1974).
1131. Burt, J. V., "A photodichroic readout device using circularly
      polarized light", Report AD-D000484/6SL, Department of the
      Navy, Washington, D. C., 1974, 15 pp.
1132. Georgiev, M. and Todorov, T., "Optical storage by photochromic
      F-F type conversions in alkali halides", J. Signalaufzeichungs-
      mater., 4 (3), 187-91 (1976).
1133. Lehmann, M., "Photodichroic material for high density informa-
      tion storage", Electro-Optical System Design Conference, 1974,
      p. 52-7.
1134. Schneider, I., "Three-dimensional optical storage element using
      M centres in a KCl crystal", Appl. Opt., 10 (4), 980-1 (1971),
      CCA6-12759.
1135. Stadnik, B., Khomat, M. and Khoff, F., "Investigation of new
      materials for holographic storage", Avtometriya, no. 1, 18-22
      (1974), Russian, CCA9-14356.
1136. Stadnik, B. and Tronner, Z., "Optical information storage in
      KCl and KBr crystals based on conversion of F and X centres",
      Opt. Commun., 6 (2), 199-201 (1972), CCA8-9549.
1137. Tanaka, K., Hamakawa, Y., Wakino, K. and Murata, M., "Photo-
      chromic effect in Fe-doped PLZT ceramics", J. Am. Ceram. Soc.,
      59 (11-12), 465-9 (1976).
1138. Tokes, S. B., "New colour-centre conversion in NaCl for photo-
      graphic and holographic storage", Electro-Optical International
      Conference, Proceedings of the Technical Programme, 1974,
      p. 115-17.

e.   Amorphous Semiconductors

1139. Feinleib, J. and Shaw, R. F., "Optical mass memory employing
      amorphous thin films", Patent USA 3696344, Publ. October 1972,
      CCA8-7355.
1140. Okoshi, T. and Masamura, T., "Phenomenological model of phase-
      grating formation in amorphous semiconductors (optical record-
      ing applications)", Electron. & Commun. Jap., 58 (3), 96-100
      (1975), CCA12-4026.
1141. Ovshinsky, S. R., "Amorphous materials as optical information
      media", J. Appl. Photogr. Eng., 3 (1), 35-9 (1977).
1142. Smith, A. W., "Injection laser writing on chalcogenide films",
      Appl. Opt., 13 (4), 795-8 (1974), CCA9-16084.

1143. Tanaka, M. and Minami, T., "Glass semiconductors for optical memory devices", Kagaku, 31 (3), 234-6 (1976), Japanese.

       f.   Others

1144. Anon., "MOSFET-controlled array readies data for optical storage", Digital Des., 7 (1), 55 (1977), CCA12-15526.
1145. Bogenberger, R., Helmcke, C., Kritikos, A., Kroy, W. and Mehnert, W. E., "Electrochemical data storage with electron beam accessing", Patent USA 3691533, Publ. September 1972, CCA8-7372.
1146. Booth, B. L., "Photopolymer laser recording materials", J. Appl. Photogr. Eng., 3 (1), 24-30 (1977).
1147. Boyd, S. H., "Recording media for use in COM systems", Proceedings of the Technical Program, Electrical Optical Systems Design Conference 1971 West, p. 275-84, CCA8-2700.
1148. Bulankov, N. I., Zhuravov, V. D., Plotnikov, A. F., Seleznev, V. N., Tokarchuk, D. N. and Ferchev, G. P., "Optically controlled memory element based on a metal-nitride-oxide-semiconductor structure with a gallium arsenide substrate", Sov. J. Quantum Electron., 6 (9), 1137-9 (1976), CCA12-15518.
1149. Chen, D., "Reversible optical storage media", Laser 75 Opto-Electronics Conference Proceedings, p. 252-4, Publ. 1976, CCA12-4033.
1150. Glosser, R., "MIS optical memory with phase-sensitive readout", CRC Crit. Rev. Solid State Sci., 5 (3), 337-40 (1975), CCA11-9672.
1151. Guro, G. M. and Popov, Y. M., "Optical memory based on a metal-insulator-semiconductor-insulator-metal structure", Sov. J. Quantum Electron., 4 (1), 24-7 (1974), CCA9-25105.
1152. Igo, T., Zembutsu, S., Toyoshima, Y. and Noguchi, Y., "Holographic memory characteristics in chalcogenide glass films", Rev. Electr. Commun. Lab., 23 (5-6), 559-63 (1975).
1153. Keokebakker, J., "New storage medium may challenge mag tapes and discs", Can. Datasyst., 8 (9), 67, 69-70 (1976), CCA12-8903.
1154. Maslowski, S., "High density data storage on ultraviolet sensitive tape", Appl. Opt., 13 (4), 857-60 (1974), CCA9-16086.
1155. Ragnarsson, S. I., "Holograms recorded in extremely thick photographic emulsions", Opt. Commun., 14 (1), 39-41 (1975).
1156. Zech, R. G., "Applications of electrophotography to optical data storage", Conference Record of the IAS Annual Meeting, 1975, p. 301-8, CCA11-28315.
1157. Zech, R. G., Dwyer, J. C., Fichter, H. and Lewis, M., "Heat-processed photoresist for holographic data storage applications", Appl. Opt., 12 (12), 2822-7 (1973), CCA9-6242.

3.   Storage Techniques
a.   Read-Write

1158. Alphonse, G. A., "Optical storage in lithium niobate", Ann. NY Acad. Sci., $\underline{267}$, 373-83 (1976).

1159. Alphonse, G. A. and Phillips, W., "Read-write holographic memory with iron-doped lithium niobate", Ferroelectrics, $\underline{11}$ (1-2), 397-401 (1976).

1160. Alphonse, G. A. and Phillips, W., "Iron-doped lithium niobate as a read-write holographic storage medium", RCA Rev., $\underline{37}$ (2), 184-5 (1976), CCA11-28314.

1161. Bailey, G. A., "Optical mass memories", Ann. NY Acad. Sci., $\underline{267}$, 411-16 (1976).

1162. Cashman, M. W., "A read/write optical memory system", Datamation, $\underline{19}$ (3), 67-9 (1973), CCA8-13713.

1163. D'Auria, L., Huignard, J. P., Slezak, C. and Spitz, E., "Experimental holographic read-write memory using a 3-D storage", Appl. Opt., $\underline{13}$ (4), 808-18 (1974), CCA9-18419.

1164. Fleming, D. L. and Johansen, T. R., "Stripe domain light deflector", Report AD-A012410/7SL, Sperry Univac, St. Paul, Minn., 1975, 41 pp.

1165. Hill, B., Krumme, J. P., Much, G., Pepperl, R., Schmidt, J., Schmidt, K. P., Witter, K. and Heitmann, H., "Polycube optical memory: a 6.5 x 10^7 bit read-write and random access optical store", Appl. Opt., $\underline{14}$ (11), 2607-13 (1975), CCA11-4821.

1166. Hill, B., Krumme, J. P., Much, G., Riekmann, D. and Schmidt, J., "Fully operational write-read-write and random-access optical store", J. Appl. Phys., $\underline{47}$ (8), 3697-701 (1976), CCA11-28275.

1167. Hill, B., Schmidt, U. J. and Schmitt, H. J., "Optical memories", European Conference on Electrotechnics, EUROCON '74 Digest, Extended Abstracts, Paper D3-1, 2 pp., CCA9-22001.

1168. Huignard, J. P., "Holographic storage in thick photosensitive materials", Presented at 2nd European Electro-Optics Conference, (Abstracts only received), 1974, 1 pp., CCA9-16096.

1169. Kaufman, A. B., "An expandable ferroelectric random access memory", IEEE Trans. Comput.,C-$\underline{22}$ (2), 154-8 (1973), CCA8-7363.

1170. Kenney, G. C., "Special purpose applications of the optical videodisc system", IEEE Trans. Consum. Electron., CE-$\underline{22}$ (4), 327-38 (1976).

1171. Knoebel, H. W., Krone, H. V., Kirkwood, B. D., Burt, J. V. and Harris, D. G., "High density optical memory", Report AD-A021673/9SL, Illinois Univ. at Urbana-Champaign, Coordinated Science Lab., 1975, 105 pp.

1172. Kozma, A., "Holograms for archival storage", 1973 IEEE International Convention and Exposition, Vol. V, Paper 9-2, 3 pp., CCA9-829.

1173. Lang, M. and Eschler, H., "Gigabyte capacities for holographic memories", Opt. Laser Technol., $\underline{6}$ (5), 219-24 (1974).

1174. Mathieu, M., "Random access system adapted for the optical
      videodisc: Its impact on information retrieval", J. SMPTE,
      $\underline{86}$(2), 80-3 (1977).
1175. Verber, C. M., Schwerzel, R. E., Perry, P. J. and Craig, R. A.,
      "Holographic recording materials development", Report N76-
      23544/9SL, Battelle Columbus Labs., Ohio, 1976, 99 pp.
1176. Welch, J. D., "A landmark recognition and tracking experiment
      for flight on the shuttle/advanced technology laboratory
      (ATL)", Report N76-14580/4SL, General Electric Co., Philadel-
      phia, Pa., Space Division, 1975, 88 pp.

               b.   Read-Only

1177. Bruel, A. and Cazaux, J. C., "Conception by computer of planes
      for holographic mass memory", Onde Electr., $\underline{52}$ (10), 439-45
      (1972), French, CCA8-5285.
1178. Goldmann, G., "Holographic data storage", Feinwerktech. &
      Micronic, $\underline{78}$ (2), 57-63 (1974), German, CCA9-16091.
1179. Knight, G. R., "Holographic memories", Opt. Eng., $\underline{14}$ (5),
      453-9 (1975).
1180. Lee, M. D., "Some features of the interface between the
      electro-optic and electronic parts of a large random access
      holographic memory", Presented at 2nd European Electro-Optics
      Conference (Abstracts only received), 1974, 1 pp., CCA9-16099.
1181. Lee, M. D., "Random access photodiode array system with non-
      destructive readout for use in computer applications", Electro-
      Optic International Conference, Proceedings of the Technical
      Programme, 1974, p. 143-9.
1182. Minnaja, N. and Nobile, M., "Associative optical memories",
      Rend. Riunione Assoc. Elettrotec. Ital., $\underline{49}$ (3), B3/1-4 (1974),
      Italian, CCA11-6390.
1183. Phillips, W., Burke, W. J. and Staebler, D. L., "Materials for
      volume phase holography", Report AD-A035676/6SL, RCA Labs.,
      Princeton, N. J., 1976, 56 pp.
1184. Williams, D. N., "The improved read-only optical memory sys-
      tem", Report AD-D000562/9SL, Department of the Navy, Washing-
      ton, D. C., 1974, 16 pp.
1185. Wilson, M. J. D. and Aleksander, I., "RAM, ROM, PROM circuits
      for simple image processing", Microelectronics, $\underline{6}$ (4), 30-2
      (1975).

               c.   Read-Write-Erase

1186. Hayashi, Y., Nagai, K. and Tarui, Y., "Electrically erasable
      non-volatile optical memory for visible and infrared light
      pattern", Oyo Buturi, $\underline{43}$ (Suppl.), 362-6 (1974), EEA78-2881.
1187. Hayashi, Y., Nagai, K. and Tarui, Y., "Electrically erasable
      nonvolatile optical semiconductor memory", Bull. Electrotech.
      Lab., $\underline{26}$ (4-5), 16-22 (1976), Japanese, EEA79-46904.

1188. Strehlow, W. H., "Optical memory system for reading, writing and erasing information", Patent USA 3902788, Publ. September 1975, CCA11-4826.
1189. Tseng, S. C. C. and Pennington, K. S., "Electro-optical memory with write, read and erase characteristics", IBM Tech. Disclosure Bull., 15 (4), 1327-8 (1972), CCA8-7364.

4.   Holographic Recording
      a.   General

1190. Babenko, M. K., Kolomiyets, O. D. and Nishchenets, V. M., "Recording of binary data in a holographic memory", Sov. Autom. Control, 8 (2), 1-5 (1975), CCA11-9696.
1191. Bardos, A. M., Nelson, R. B. and Shuman, C. A., "High data rate holographic recording", Electro-Optical System Design Conference, Proceedings of Technical Program, 1974, p. 126-33.
1192. Becker, C. H., "Non-photographic digital laser storage", Laser 75 Opto-Electronics Conference Proceedings, p. 260-2, Publ. 1976, CCA12-4035.
1193. Brun, K., "Holographic data stores", Funkschau, 47 (25), 83-6 (1975), German, CCA11-9691.
1194. Carson, A. N., "Large capacity, high speed holographic memory", SPIE Seminar Proceedings, Vol. 45, 1974, p. 157-60.
1195. Cobb, R. O. and Moore, A. C., "Associatively addressed holographic storage", IBM Tech. Disclosure Bull., 13 (5), 1070-1 (1970), CCA6-3778.
1196. David, R. M., Parten, C. R., Hagler, M. O. and Kristiansen, M., "A holographic data storage system design project", Int. J. Electr. Eng. Educ., 10 (4), 296-305 (1972), CCA9-18428.
1197. Gueret, P., Mohr, T. O. and Wolf, P., "Single flux-quantum memory cells", IEEE Trans. Magn., MAG-13 (1), 52-5 (1977), CCA12-10695.
1198. Hessel, K. R., Stalker, K. T. and McCarthy, A. E., "Data recording using one-dimensional holography", Opt. Commun., 15 (2), 213-17 (1975).
1199. Hill, B. and Schdmit, K. P., "The realisation and technology of interfaces between optics and electronics in holographic memories", Laser 75 Opto-Electronics Conference Proceedings, p. 255-9, Publ. 1976, CCA12-4034.
1200. Huignard, J. P. and Spitz, E., "Holographic mass memories using volume holograms", Laser 75 Opto-Electronics Conference Proceedings, p. 248-51, Publ. 1976, CCA12-4032.
1201. Ingebrigsten, K. and Stern, E., "Holographic storage of acoustic surface waves with Schottky diode arrays", Proceedings of Ultrasonic Symposium, 1975, p. 212-16.
1202. Jenkins, R. W., "Holographic data processing", Instrum. & Contr. Syst., 44 (3), 124-7 (1971), CCA6-12762.
1203. Kiemle, H., "Holographic storage of digital data", Ingenieur, 84 (51-52), E83-8 (1972), German, CCA8-9547.

1204. Kiemle, H., "Considerations on holographic memories in the giga-
      byte region", Appl. Opt., 13 (4), 803-7 (1974), CCA9-18418.
1205. Knight, G. R., "Page-oriented associative holographic memory",
      Appl. Opt., 13 (4), 904-12 (1974), CCA9-18422.
1206. Kostov, E. G., Malinovskii, V. K., Nesterkhin, Y. E. and Pota-
      pov. A. N., "Features of the physical realisation of an opera-
      tive optical memory", Avtometriya, no. 4, 3-6 (1976), Russian,
      CCA12-1523.
1207. Kovtonyuk, N. F., Kostryukov, V. V., Morozov, V. A., Nikitin,
      V. V. and Samoilov, V. D., "Optical working memory based on
      injection lasers", Sov. J. Quantum Electron., 4 (1), 41-3
      (1974), CCA9-25106.
1208. Lee, W. H., "Optical data storage using PPM and PWM", Appl.
      Opt., 14 (9), 2217-24 (1975), CCA11-4819.
1209. Lohmann, A. W., "Optical data technology", Laser 75 Opto-
      Electronics Conference Proceedings, p. 240, Publ. 1976,
      CCA12-4030.
1210. Maslowski, S., "Data transmission and recording by laser",
      Toute Electron., no. 352, 23-6 (1971), French, CCA6-15279.
1211. Mikaelyan, A. L. and Bobrinev, V. I., "Holographic memory
      systems", Telecommun. Radio Eng., 28-29 (5), 47-57 (1974).
1212. Pohl, D., "Stacked optical memories", Appl. Opt., 13 (2),
      341-6 (1974), CCA9-14357.
1213. Popov, Y. M., "Optoelectronics", Proceedings of the European
      Semiconductor Device Research Conference on Solid State
      Devices, 1971, p. 67-76, CCA8-2701.
1214. Russell, J. T. and Walker, R. A., "Optical digital recording",
      Opt. Eng., 15 (1), 20-3 (1976), CCA11-17415.
1215. Schmitt, H. J., "Data handling and recording by laser", Opt.
      Acta, 24 (4), 407-12 (1977), CCA12-15528.
1216. Takeda, Y., "Information storage density of hologram memory",
      Electron. & Commun. Jap., 56 (4), 80-6 (1973), CCA9-18426.
1217. Toffer, H., Hildebrand, B. P. and Albrecht, R. W., "Ultrasonic
      holography through metal barriers", 5th International Sym-
      posium on Acoustic Holography and Imaging, Proceedings, 1973,
      p. 133-57, Publ. 1974.
1218. Verbovetskii, A. A. and Fedorov, V. B., "Small-size phase
      holograms for storing binary information", Opt. & Spectrosc.,
      32 (5), 530-1 (1972), CCA8-9550.
1219. Witte, J., "Holographic working memories", Elektron. Rechenan-
      lagen, 18 (5), 214-19 (1976), German, CCA12-1490.

b.   Characteristics

1220. Akaev, A., Kowalevskii, L. V., Maiorov, S. A., Maltsev, L. N.
      and Meskin, I. V., "Suppression of mutual interference between
      holograms in high-capacity memories", Opt. & Spectrosc., 40
      (2), 209-12 (1976), CCA11-28313.

1221. Akahori, H., "Phase coding methods for computer-generated hologram memories. Comparison of deterministic phase coding with random phase coding", Bull. Electrotech. Lab., 37 (11), 973-84 (1973), Japanese, CCA9-20772.

1222. Akahori, H., "Deterministic phase coding for digitally generated hologram", Electron. & Commun. Jap., 58 (3), 101-7 (1975), CCA12-4027.

1223. Burke, W. J. and Sheng, P., "Crosstalk noise from multiple thick-phase holograms", J. Appl. Phys., 48 (2), 681-5 (1977), CCA12-15527.

1224. Esikov, V. B. and Pafomov, D. N., "Power consumption of a ferroacoustic information accumulator", Izv. VUZ Priborostr., 18 (7), 57-61 (1975), Russian, CCA11-2221.

1225. Gifford, D. L., "Perceived holographic image distortions", SPIE Seminar Proceedings, Vol. 45, 1974, p. 207-14.

1226. Haskal, H. M., "Polarization and efficiency in magnetic holography", IEEE Trans. Magn., MAG-6 (3), 542-5 (1970), CCA6-10784.

1227. Hill, R., "Point efficiency and signal-to-background ratio in exponential holograms for optical memories", Appl. Opt., 11 (12), 2937-44 (1972), CCA8-5282.

1228. Inagaki, T., Furukawa, Y., Goto, Y., Akimuta, T. and Nishimura, Y., "Hologram writer using a plasma display panel", Appl. Opt., 13 (4), 819-24 (1974), CCA9-16085.

1229. Killat, U., "Holographic microfiche of picture-like information", Opt. Acta, 24 (4), 453-72 (1977), CCA12-15543.

1230. Kryuchin, A. A. and Petrov, V. V., "Influence of the information carrier nonlinearity on the density of information recorded in optical memories", Sov. J. Quantum Electron, 7 (1), 109-11 (1977).

1231. Mityakov, V. G. and Fedorov, V. B., "Enhancement of the effective density of data storage in optoelectronic memories", Sov. J. Quantum Electron, 6 (11), 1284-6 (1976).

1232. Morozov, V. N. and Popov, Y. M., "High-capacity holographic memories with synthesized aperture", Sov. J. Quantum Electron, 6 (11), 1265-71 (1976).

1233. Nomura, H. and Okoshi, T., "Capacity limitation of volume hologram memory", Electron. & Commun. Jap., 58 (3), 108-15 (1975), CCA12-4028.

1234. Nomura, H. and Okoshi, T., "Storage density limitation of a volume-type hologram memory: theory", Appl. Opt., 15 (2), 550-5 (1976), CCA11-15072.

1235. Pafamov, D. N., "Evaluation of the physical volume of the acoustic part of a ferroacoustic memory", Izv. VUZ Priborostr., no. 1, 57-60 (1973), Russian, CCA8-13656.

1236. Rull, H., "Storage of binary electrical signals in one-dimensional holograms with extremely narrow tracks", Opt. Commun., 11 (3), 242-4 (1974), CCA9-22000.

1237. Rull, H., "Holographic techniques for recording of electrical
      signals on moving storage media", Opt. Eng., 14 (3), 232-6
      (1975).
1238. Rull, H., "Holographic storage of information in narrow tracks",
      Laser 75 Opto-Electronics Conference Proceedings, p. 241-7,
      Publ. 1976, CCA12-4031.
1239. Stroke, G. W., Halioua, M., Thon, F. and Willasch, D. H.,
      "Image improvement and three-dimensional reconstruction using
      holographic image processing", Proc. IEEE, 65 (1), 39-62 (1977),
      CCA12-4038.
1240. Tsunoda, Y. and Takeda, Y., "High-density image-storage holo-
      grams by a random-phase sampling method", 1973 IEEE/OSA Con-
      ference on Laser Engineering and Applications, Digest of Tech-
      nical Papers, p. 89, CCA9-11982.
1241. von der Linde, D. and Glass, A. M., "Photorefractive effects
      for reversible holographic storage of information", Appl. Phys.,
      8 (2), 85-100 (1975).

              c.   Equipment

1242. Akos, G., Kiss, G. and Varga, P., "Effect of lens aberrations
      on the storage capacity of holographic memories", Opt. Commun.,
      20 (1), 63-7 (1977), CCA12-8904.
1243. Bardos, A., "Wideband holographic recorder", Appl. Opt., 13
      (4), 832-40 (1974), CCA9-14361.
1244. Castera, J. P. and Hepner, G., "Isolator in integrated optics
      using the Faraday and Cotton-Mouton effects", IEEE Trans.
      Magn., MAG-13 (5), 1583-5 (1977).
1245. Jantsch, O., Hundelshausen, U. V., Feight, I. and Hering, W.,
      "Detector matrix for a holographic memory", Siemens Forsch. -
      & Entwicklungsber., 2 (1), 34-8 (1973), CCA8-11655.
1246. Malina, V. and Chomar, M., "Preparation for a holographic mem-
      ory", Slaboproudy Obz., 37 (12), 572-7 (1976), Czech,
      CCA12-6678.
1247. Minami, M. and Yamada, K., "Artificial diffusers for the holo-
      graphic picture memory", Electro-Opt International Conference,
      Proceedings of the Technical Programme, 1974, p. 125-30.
1248. Nishi, K., Fujiwara, T., Yamaoka, T., Imai, S. and Maeda, M.,
      "Analysis of double-emitter phototransistor operation for
      optical memory", Electron. & Commun. Jap., 58 (1), 98-105
      (1975), CCA11-15073.
1249. Ogiwara, H. and Ishii, A., "Hologram correlation detection by
      frequency spectrum", Electron. Commun. Jap., 57 (4), 115-22
      (1974).
1250. Olivei, A., "Performance evaluation of holographic storage and
      access systems - I. Modeling of the reconstruction and image
      scanning apparatus", Optik, 42 (5), 463-85 (1975).

1251. Olivei, A., "Performance evaluation of holographic storage
      and access systems - II. Modeling of the transmitting and
      receiving apparatus", Optik, 43 (1), 1-24 (1975).
1252. Rajchman, J. A., "Array of electrically controlled light valves
      for use in an optical memory", Report TN879, RCA, Princeton,
      N. J., 1971, 2 pp, CCA6-15280.
1253. Som, S. C. and Lessard, R. A., "New technique for holographic
      multiplexing", Presented at the 1970 Annual Meeting of the
      Optical Society America (abstracts only received), Hollywood,
      Fla., September-October, CCA6-3782.
1254. Thire, J., "Semiconductor photodetection matrix for holographic
      memory reading", Onde Electr., 52 (10), 452-61 (1972), French,
      CCA8-5286.
1255. Yamaoka, T., Fujiwara, T. and Nishi, H., "Photodetector array
      for a holographic optical memory system", Fujitsu Sci. & Tech.
      J., 8 (3), 137-51 (1972), CCA8-5283.
1256. Yonezawa, S., "A deterministic phase shifter for holographic
      memory devices", Opt. Commun., 19 (3), 370-3 (1976), CCA12-6677.

d.   Applications

1257. Carlsen, W. J., "Holographic page synthesis for sequential input
      of data", Appl. Opt., 13 (4), 896-903 (1974), CCA9-18421.
1258. Carlson, F. P., "Generalized optical data processors", Nav.
      Res. Rev., 26 (1), 40-2 (1973), CCA9-13789.
1259. Cosentino, L. S. and Stewart, W. C., "Membrane page composer -
      further developments", RCA Rev., 35 (4), 539-66 (1974).
1260. Dyakov, V. A. and Tarasov, I. V., "Lasers and coherent optical
      systems in cybernetics and computer technology. - I. Optical
      processing of information - a new trend in cybernetics and
      computer techniques", Eng. Cybern., 12 (3), 85-99 (1974).
1261. Eschler, J., "Multifrequency acousto-optic page composers for
      holographic data storage", Opt. Commun., 13 (2), 148-53 (1975).
1262. Gibin, I. S., Gofman, M. A., Pen, E. F. and Tverdokhleb, P. E.,
      "Associative information retrieval in hologram-based core
      stores", Avtometriya, no. 5, 12-18 (1973), Russian, CCA9-6241.
1263. Goto, Y., Furukawa, Y. and Inagaki, T., "Construction of data
      pattern for hologram memory", Fujitsu Sci. Tech. J., 13 (1),
      61-72 (1977).
1264. Hallock, J. N., "Holographic position determination of space
      vehicles", Laser, 2 (2), 39-40 (1970), German, CCA6-3781.
1265. Ishii, A., Sakai, T. and Ogiwara, H., "Recording and retrieval
      system in PANDORA", Rev. Electr. Commun. Lab., 23 (3-4),
      191-204 (1975).
1266. Jeudy, M., "Binary processes in holographic memories and ap-
      plication to the calculation of capacity", Opt. & Laser
      Technol., 5 (6), 253-5 (1973), CCA9-11978.

1267. Labrunie, G., Robert, J. and Borel, J., "Nematic liquid crystal
      1024 bits page composer (holographic memories)", Appl. Opt.,
      13 (6), 1355-8 (1974), CCA9-20771.
1268. Leighty, R. D., "Coherent optics potential applications to
      mapping", Opt. Eng., 13 (5), 440-50 (1974).
1269. Muhlenfeld, E., "A holographic associative memory for infor-
      mation retrieval", International Optical Computing Conference,
      Digest of Papers, 1976, p. 111, CCA11-31636.
1270. Nelson, R. R., Vander Lugt, A. and Zech, R. G., "Holographic
      data storage and retrieval", Opt. Eng., 13 (5), 429-34 (1974).
1271. Schmidt, K. P., "A 16 x 18 bit CdS spatial modulator (page
      composer)", Opt. Acta, 24 (4), 433-40 (1977), CCA12-15531.
1272. Strehlow, W. H., Packard, J. R. and Dennison, R. L., "Holo-
      graphic data search and retrieval system", Electro-Opt System
      Des. Conf., 1974 West and International Laser Expo, Proceedings
      of Technical Program, p. 35-7.

D. MICROPROCESSORS/MICROCOMPUTERS

I. BIBLIOGRAPHIES

1. Grooms, D. W., "Microcomputers: Four bibliographies with abstracts", Reports PS-75/251/9WC, PS-76/0204/2SL, PS-76/0203/2SL, PS-76/0204/3SL, NTIS, Springfield, Va., 22151.
2. Jensen, C. W., "Microcomputer publication reference list", Report UCID-17123, California University, Livermore, 1976, 6 pp.
3. Nauful, E. S., "Microprocessor bibliography", Nauful, 2635 Merrywood Road, Charlotte, N. C., 1976.
4. Ward, A. R., "LSI microprocessors and microcomputers - a bibliography", Computer, $\underline{7}$ (7), 19-39 (1974); also: ibid $\underline{9}$ (1), 42-53 (1976).

II. BOOKS

5. Agerwala, T., Masson, G. and Westgate, R., "Designing with microprocessors: Tutorial", COMPCON 76, IEEE Catalog No. 76CH1178-3C.
6. Altman, L., (Ed.), "Microprocessors", Electronic Book Series, McGraw-Hill, Inc., New York, 1975.
7. Anon., "Study of the market and applications for microprocessors and microcomputers", Bahners Publishing Co., Inc., Boston, Mass., 1974.
8. Anon., "Microprocessors: An ERA assessment of LSI computer components", Ovum, Ltd., 22 Grays Inn Road, London WCI, 1973.
9. Anon., "Microprocessor handbook", Texas Instruments Learning Center, Dallas, Texas, 1975.
10. Anon., "Minicomputer and microcomputer markets", Frost & Sullivan, 106 Fulton Street, New York, 1975.
11. Barna, A. and Porat, D. I., "Introduction to microcomputers and microprocessors", Wiley-Interscience, New York, 1976.
12. Elphick, M. S., (Ed.), "Microprocessor basics", Hayden Book Co., Inc., Rochelle Park, N. J., 1977.
13. Hilburn, J. L., "Microcomputers/microprocessors: hardware, software and applications", Prentice-Hall, Englewood Cliffs, N. J., 1976.
14. Klingman, E. E., "Microprocessor systems design", Prentice-Hall, Inc., Englewood Cliffs, N. J., 1977.
15. Korn, G. A., "Microprocessors and small digital computer systems for engineers and scientist", McGraw-Hill Book Co., New York, 1977.
16. Leahy, W. F., "Microprocessor architecture and programming", Wiley-Interscience, New York, 1977.
17. Lin, W. C., (Ed.), "Microprocessors: fundamentals and applications", IEEE Press, New York, 1977.

18. McGlynn, D. R., "Microprocessors: technology, architecture and applications", Wiley-Interscience, New York, 1976.
19. McMurran, M. W., "Programming microprocessors", Tab Books, Blue Ridge Summit, Pa., 1977.
20. Martin, D. P., "Microcomputer design", Martin Research, Chicago, IL, 1974, (2 volumes).
21. Moore, A. W., et al, "Microprocessor applications manual", McGraw-Hill Book Co., New York, 1975.
22. Motorola Semiconductor Products, Inc., "Microprocessor applications manual", McGraw-Hill Book Co., New York, 1975.
23. Nashelsky, L., "Introduction to digital computer technology", Wiley-Interscience, New York, 1977, CCA12-12938.
24. Osborne, A., "An introduction to microcomputers, Vol. 1, basic concepts", Osborne Associates, Inc., Berkeley, CA, 1976.
25. Osborne, A., "An introduction to microcomputers, Vol. 2, some real products", Osborne Associates, Inc., Berkeley, CA, 1976.
26. Osborne, A., "6800 programming for logic design", Osborne Associates, Inc., Berkeley, CA, 1976.
27. Osborne, A., "8080 programming for logic design", Osborne Associates, Inc., Berkeley, CA, 1976.
28. Rony, P. R., Larsen, D. G. and Titus, J., "Bugbook 3: Microcomputer interfacing", E & L Instruments, Derby, Conn., 1975.
29. Sawin, D. H., "Microprocessors and microcomputer systems", Lexington Books, Lexington, Mass., 1977.
30. Sippl, C. J., "Microcomputer handbook", Petrocelli/Charter, New York, 1977.
31. Sippl, C. J. and Kidd, D. A., "Microcomputer dictionary and guide", Matrix Publishers, Inc., 207 Kenyon Road, Champaign, IL, 1976.
32. Soucek, B., "Microprocessors and microcomputers", Wiley-Interscience, New York, 1976.
33. Spencer, J. P., "Minicomputer systems", Infotech International, Maidenhead, Berks, England, 1977, CCA12-12937.
34. Torrero, E. A., (Ed.), "Microprocessors: new directions for designers", Hayden Book Co., Rochelle Park, N. J., 1975.
35. United Technical Publications, "Modern guide to digital logic: processors, memories and interfaces", Tab Books, Blue Ridge Summit, Pa., 1976.
36. Waite, M. and Pardee, M., "Microcomputer primer", Howard W. Sams & Co., Ind., Indianapolis, Indiana, 1976.
37. Ward, B., "Microprocessor-microprogramming handbook", Tab Books, Blue Ridge Summit, Pa., 1975.
38. Wester, J. G. and Simpson, W. D., "Software design for microprocessors", Texas Instruments Learning Center, Dallas, Texas, 1976.
39. Wilmink, J., Sami, M. G. and Zaks, R., (Eds.), "Microprocessing and microprogramming", North-Holland Publishing Co., New York, 1977.

III. REVIEW ARTICLES

40. Anon,, "Emergence of the microcomputer", EDP In-Depth Rep., $\underline{3}$,
    1-16 (1974).
41. Anon., "Microprocessor survey", Conference on Microcomputers:
    Fundamentals and Applications, 1974, p. 15-20, Publ. by:
    MINICONSULT, London, 1975.
42. Anon., "A twenty dollar chip worth a million (microprocessors)",
    Comput. Wkly., $\underline{21}$ (522), 13 (1976), CCA12-10699.
43. Bews, M., "Microprocessor survey", New Electron, $\underline{7}$ (22),
    pp. 40ff (1974).
44. Bishop, P. G., "Microprocessors: Computing in miniature", Phys.
    Technol., $\underline{7}$ (2), 47-53 (1976).
45. Boufaissal, J., "What is a microprocessor?", SAE Special Publi-
    cation No. 393, 1975, p. 51-63.
46. Callahan, B. L. J., "Microprocessors and microcomputers: The
    state-of-the-art", Mod. Data, $\underline{7}$ (5), 30-2 (1974).
47. Cushing, R., "Overview of microprocessor hardware", Proceedings,
    1st National Microprocessor Conference, Microprocessors:
    Economics, Technology, Applications, 1974.
48. Davis, S., "A fresh view of minicomputers and microcomputers",
    Comput. Des., $\underline{13}$ (5), 67-79 (1974).
49. Dieckmann, H. W., "Computer on a semiconductor crystal. The
    microprocessor, a programmable LSI circuit", Elektrotech. Z.
    (ETZ) B, $\underline{27}$, 378-80 (1975), German, EEA78-37568.
50. Falk, H., "Microprocessor: Jack-of-all trades", IEEE Spectrum.,
    $\underline{11}$ (11), 46-51 (1974).
51. Falk, H., "Microprocessors talk it up", IEEE Spectrum., $\underline{14}$ (3),
    48-53 (1977).
52. Goessler, R. and Schwerte, J., "On the way to practice with
    microprocessors", Elektronik, $\underline{25}$ (3), 74-85 (1976).
53. Goss, L., "The first of the third generation microcomputers",
    Solid State Technol., $\underline{20}$ (7), 42-5 (1977).
54. Groves, B., "Microprocessor - a few picoacres of silicon rule
    the future of digital control", Instrum. Control Syst., $\underline{48}$ (3),
    47-53 (1975).
55. Herzog, G. B., "The impact of LSI technology on computer systems",
    Proceedings, IFIP Congress, 1974, p. 34-8.
56. Hoff, M. E., "Microprocessors", 6th International Congress on
    Microelectronics, 1974, 6 pp., EEA78-24520.
57. Hubert, R., "Microcomputers. A situation report", Inf. Elektron.,
    $\underline{12}$ (2), 65-9 (1977), Hungarian, CCA12-12927.
58. Jones, M. R. and Pooch, U. W., "Evolution and design of micro-
    computer", Symposium on Trends and Applications 1976: Micro and
    Mini Systems, p. 1-9.
59. Kielbasinski, J. and Sobczyk, J., "Microcomputers", Informatyka,
    $\underline{10}$ (9), 10-12 (1974), Polish.

60. Koch, G. R., "State-of-the-art and trends in programming of microprocessors - 1, 2", Elektronik, $\underline{26}$ (1, 2), 63-6, 66-71 (1977), German.
61. Kohler, V., "Structural features of mini- and micro-computers", Nachrichtentech. Elektron., $\underline{27}$ (2), 51-3 (1977), German, CCA12-12932.
62. Lee, I., "LSI microcomputer and digital control", ASME Paper No. 75 - Aut-V, 1975, 8 pp.
63. Lee, I., "Overview of microprocessor technology and its impact on applications and education", Proceedings of International Symposium and Course on Mini- and Microcomputers and their Applications (MIMI 76), p. 3-10.
64. Liles, W. C. and Tenner, M. D., "Microcomputer survey", In: New Components and Subsystems for Digital Design: a Report, Technology Service, Santa Monica, CA, p. 179-85 (1975).
65. Marshall, T., "Microcomputers", Report R-556, Stanford Research Inst., Menlo Park, Calif., 1975, 28 pp.
66. Meier, H., "Microprocessors, an introduction", Mess. Pruef, $\underline{12}$ (3), 94, 96, 98 (1976), German.
67. Neunert, H., "Microcomputer system technology", Elektronik, $\underline{23}$ (10), 391-5 (1974), German.
68. Ogdin, J. L. and McPhillips, A. S., "Microprocessor survey", In: New Components and Subsystems for Digital Design: a Report, Technology Service, Santa Monica, CA, p. 153-78 (1975).
69. Posthumus, R., "The microprocessor; the centre of modern technique", Polytech. Tijdschr. Elektrotech. Elektron., $\underline{31}$ (11), 647-56 (1976), Dutch, CCA12-4061.
70. Rattner, J., "Building block microprocessors", 10th IEEE Computer Society International Conference, COMPCON 75, Digest of Papers, p. 79-82.
71. Reyling, G., Jr., "LSI building blocks for parallel digital processors", IEEE International Convention Technical Papers, 1973, Paper 21/3.
72. Russo, P. M., "Microprocessors at work - session overview", Proceedings, National Computer Conference, 1975, p. 21-2.
73. Simpson, H. K., "Getting smaller: microcomputers", Digital Des., $\underline{7}$ (11), 43-4, 46, 48, 52, 54, 56, 58 (1977).
74. Soucek, B., "Micro-, mini- and portable computer: a balance", Proceedings of International Symposium and Course on Mini- and Microcomputer and their Applications, MIMI 76, p. 31-48.
75. Sowinski, A., "Microprocessor - the only future of electronics", Elektronika, $\underline{18}$ (3), 89-92 (1977), Polish, CCA12-15559.
76. Stuart, J. P., "Introduction to microcomputers", In: Microcomputers: Fundamentals and Applications, MINICONSULT, London, p. 1-14 (1975).
77. Theis, D. J., "Microprocessor and microcomputer survey", Datamation, $\underline{20}$ (12), 90-1ff (1974).
78. Theis, D. J., "Microprocessors and microcomputers", Report AD-A014823/9SL, Aerospace Corp., El Segundo, Calif., 1975, 35 pp.

79. Tireford, H., "Microprocessors (advantages, disadvantages and applications)", Systems, 6 (10), 19-20 (1976), CCA12-15581.
80. Toong, H. D., "Implication of new technology to information storage, manipulation, and transfer", Proceedings, ASIS Annual Meeting, 1975.
81. Toong, H. M. D., "Microprocessors", Sci. Am., 237(3), 146-61 (1977).
82. Vacroux, A. G., "Microcomputers", Sci. Am., 232 (5), 32-40 (1975).
83. Waddington, D. E. O'N., "Microprocessors", Wireless World, 81 (1480), 550-5 (1975).
84. Wiatrowski, C. A., "Microprocessor restroom robot", Comput. Des., 16 (4), 98-100 (1977).
85. Wolf, H., "Microcomputers: a new revolution", EE/Syst. Eng. Today, 33 (1), 80-3 (1974).
86. Wright, D., "Microprocessor survey", In: Microcomputers: Fundamentals and Applications, MINICONSULT, London, 1974, Publ. 1975.
87. Wright, D. W., "Review of microprocessors", Minicomputers and Small Business Systems, Forum, 1976, p. 189-206.
88. Wuthrich, K., "Micro-computers", Elektroniker, 13 (4), EL1-7 (1974), German.
89. Yasaki, E. K., "The emerging microcomputer", Datamation, 20 (12), 81, 83, 86 (1974).

IV. ARCHITECTURE

90. Alford, C. O. and Sledge, R. B., "Microprocessor architecture for discrete manufacturing control - 1: History and problem definition", IEEE Trans. Manuf. Technol., MFT-5 (2), 43-9 (1976).
91. Alford, C. O. and Sledge, R. B., "Microprocessor architecture for discrete manufacturing control - 2: Identification of system requirements", IEEE Trans. Manuf. Technol., MFT-5 (3), 62-8 (1976).
92. Alford, C. O. and Sledge, R. B., "Microprocessor architecture for discrete manufacturing control - 3. System architecture", IEEE Trans. Manuf. Technol., MFT-5 (4), 84-9 (1976).
93. Allison, D. R., "Design philosophy for microcomputer architectures", Computer, 10 (2), 35-41 (1977).
94. Anderson, G. L. and Bartlett, K., "Hardware allocation of data system resources", Comput. Des., 13 (7), 89-97 (1974).
95. Armstrong, C. V. M., "Functional memory techniques applies to the microprogrammed control of an associative processor", 2nd Annual Symposium on Computer Architecture, Proceedings, 1975, p. 34-40.
96. Barrett, R. C., Bork, R. H., Fox, G., Hattendorf, D. W. and Vahey, M. D., "High speed microprocessor study", Report AD-A018702/1SL, Hughes Aircraft Co., Culver City, Calif., 1975, 73 pp.
97. Bass, J. E., "Distributed processing in microcomputers", 1975 WESCON Technical Papers, Paper 6/4.

98. Bass, J. E., "System-oriented microcomputers", 10th IEEE Computer Society International Conference, COMPCON 75, Digest of Papers, p. 87-90.

99. Broadman, T. L., Jr., "A microprocessor architecture for digital device implementation", Proceedings of the 1977 National Computer Conference, AFIPS, Vol. 46, p. 201-6.

100. Clegg, F. W., "Microcomputer architecture", Computer, 10 (2), 11-12 (1977).

101. Conn, R. W., "Microprocessor as a computer mainframe", Report UCRL-77443, California Univ., Lawrence Livermore Lab., 1975, 7 pp.

102. Cushman, R. H., "Understand the 8-bit μP you'll see a lot of it", EDN, 19 (2), 48-54 (1974).

103. Cushman, R. H., "Very complete chip set joins the great microprocessor race", EDN, 19 (22), 87-94 (1974).

104. Dagless, E. L., "Hardware of the microprocessor", In: Microcomputers: Fundamentals and Applications, MINICONSULT, London, p. 51-71 (1975).

105. Deuel, D. R. and Gault, J. W., "An integrated hardware/software microprocessor based development system", 11th IEEE Computer Society International Conference, COMPCON 75, Digest of Papers, p. 163-5.

106. Dumstorff, E. M. and others, "System input/output architecture for LSI microprocessor", IBM Tech. Disclosure Bull., 15 (12), 3845-6 (1972).

107. Ebertin, M. A., "System approach to microprocessor architectures", Colloque International sur les Circuits Integres Complexes, 1974, p. 437-50.

108. Elphick, M., "Analog boards for microcomputer: You can't always get what you want", Electron. Des., 25 (19), 26-32 (1977).

109. Faber, U. and McAllister, J. P., "A bit serial microprocessor", IEEE International Convention Technical Papers, 1974, Paper 17/3.

110. Faiman, M., Weaver, A. C. and Catlin, R. W., "MUMS - a reconfigurable microprocessor architecture", Computer, 10 (1), 11-17 (1977), CCA12-15438.

111. Falk, H., "Self-contained microcomputers ease system implementation", IEEE Spectrum., 11 (12), 53-5 (1974).

112. Fisher, E. R., "A modular approach to microprocessors", 9th IEEE Computer Society International Conference, COMPCON 74, Digest of Papers, p. 55-7.

113. Fisher, E. R., "A system approach for processor implementation", Proceedings, Symposium on Electronics, 1974, p. 161-76.

114. Ford, M. A., "Microcomputers: The next step in small computer systems", EE/Syst. Eng. Today, 33 (1), 46 (1974).

115. Foster, C. C., "View of computer architecture", Commun. ACM, 15 (7), 557-65 (1972).

116. Gallacher, J., "Miniaturization makes microcomputers a practical proposition", Contr. Instrum., 7 (3), 16-17 (1975).

117. Garrow, R. S., Hou, J. L. and Walker, H., "Microcomputer-development system achieves hardware-software harmony", Electronics, 48 (11), 95-102 (1975).
118. Gebler, P., "The new generation of single-chip microcomputers", Electron. Engineering, 49 (588), 59, 61 (1977), CCA12-10700.
119. Gonzalez, M. J., Jr., "Microprocessor architecture and systems", Computer, 9 (9), 49-51 (1976), CCA12-1447.
120. Greenfield, S. E., "Microcomputer architecture", 1975 Semiconductor Test Symposium, Digest of Papers, p. 47-51.
121. Haring, R. R., "Microprocessor CPU architecture", Proceedings, 1st National Microprocessor Conference. Microprocessors: Economics, Technology, Applications, 1974.
122. Holt, R. M., "Architecture and organization of the AMI 7200 microprocessor: A new breed of minicomputer", IEEE Region 6 Conference on Minicomputers and Their Applications Conference Record, 1973, p. 192-5.
123. Holt, R. M. and Lemas, M. R., "Current microcomputer architecture", Comput. Des., 13 (2), 65-73 (1974).
124. House, D. L., "Micro level architecture in minicomputer design", Comput. Des., 12 (10), 75-80 (1973).
125. Jordan, B. W., Jr. and Baatz, E. L., "C. Mup - Northwestern University's multimicrocomputer network", Proceedings, Computer Networks: Trends and Applications, 1974, p. 51-5.
126. Kartashev, S. I., "Microcomputer with a shift-register memory", IEEE Trans. Comput., C-25 (5), 470-84 (1976).
127. Keele, R. V. and Stephenson, G. R., "Microprocessor developments for Project 2175", Report AD 771382/9, 1973, 182 pp.
128. Kehl, T. H., "Basil architecture - an HLL minicomputer", 3rd Annual Symposium on Computer Architecture, Proceedings, 1976, p. 86-92.
129. Leventhal, L. A., "Microcomputer architecture: memory and input/output sections", Simulation, 28 (1), 22-8 (1977), CCA12-10659.
130. Lewin, M. H., "Anatomy of microprocessors", Proceedings of the IEEE International Symposium on Circuits and Systems, 1974, p. 220-3.
131. Lewsin, M. H., "Integrated microprocessors", IEEE Trans. Circuits Syst., CAS-22 (7), 577-85 (1975).
132. Lewis, D. R. and Siena, W. R., "How to build a microcomputer", Electron. Des., 21 (19), 60-5 (1973).
133. Liccardo, M. A., "Architecture of microcontroller system", Proceedings, National Computer Conference, 1975, p. 75-84.
134. Lipovski, G. J., "Varistructured fail-soft cellular computer", Proceedings of the 1st Annual Symposium on Computer Architecture, 1973, p. 161-5.
135. Lipovski, G. J., "On gray box descriptions of microprocessors", International Symposium on Computer Hardware Description Languages and their Applications, 1975, p. 184-6, CCA11-34554.

136. Lipovski, G. J., "On a varistructured array of microprocessors",
     IEEE Trans. Comput., C-26 (2), 125-38 (1977), CCA12-10652.
137. Lycklama, H., "UNIX on a micro-processor", Proceedings of the
     1977 National Computer Conference, AFIPS, Vol. 46, p. 237-42.
138. McPhillips, A. S., "Inside microprocessors", Mod. Data, 8,
     37ff (1975).
139. Malinovsky, B. N. and Kozlov, L. G., "Real-time information
     processing and special-purpose microprocessors", Euromicro
     Newsl., 3 (1), 67-72 (1977), CCA12-12926.
140. Malinovsky, B. N., Palagin, A. V. and Ivanov, V. A., "Archi-
     tecture and structure of modern minicomputers: Principal design
     problems", Digital Processes, 2 (1), 3-25 (1976).
141. Mazur, T., "Put together a complete microcomputer", Electron.
     Des., 24 (15), 66-77 (1976).
142. Mizuno, M., Suzuki, M., Shimakura, K. and Takemae, K., "An N-
     channel 16-bit variable architecture microprocessor", 6th
     International Congress on Microelectronics, 1974, 5 pp.
     EEA78-24521.
143. Morris, J. H., Patel, H. and Schwartz, M., "Architecture: Key
     to microprocessor performance", Mach. Des., 49 (3), 108-11 (1972).
144. Murphy, J. P. and others, "Enhancing an LSI computer to handle
     decimal data", Electronics, 46 (5), 77-83 (1973).
145. Parasuraman, B., "Pipelined architectures for microprocessors",
     9th IEEE Computer Society International Conference, COMPCON 74,
     Digest of Papers, p. 225-8.
146. Parasuraman, B., "High-performance microprocessor architectures",
     Proc. IEEE, 64 (6), 851-9 (1976).
147. Pathak, J., "Time-slicing offers an alternative to multipro-
     cessor systems", Comput. Des., 16 (7), 95-104 (1977).
148. Payne, J., "Microbuilding blocks can make bridge to user",
     Comput. Wkly., 21 (520), 20 (1976), CCA12-6594.
149. Pease, M. C., III, "The indirect binary n-cube microprocessor
     array", IEEE Trans. Comput., C-26 (5), 458-73 (1977), CCA12-12840.
150. Peddle, C. I., "Microprocessor-simulators: What and why", In:
     New Components and Subsystems for Digital Design: a Report,
     Technology Service, Santa Monica, CA, p. 19-24 (1975).
151. Peppiette, G., "Memory architecture and the influence of the
     microprocessor", Microelectron. & Reliab., 15 (4), 307-13
     (1976), CCA11-31588.
152. Peuto, B. L., "Current issues in the architecture of micropro-
     cessors", Computer, 10 (2), 20-5 (1977).
153. Phillips, D. and Goodman, A., "Slave microcomputer lightens
     main microprocessor load", Electronics, 50 (14), 109-12 (1977).
154. Radoy, C. H. and Lipovski, G. J., "A microprocessor architec-
     ture for effective use of program memory", 9th IEEE Computer
     Society International Conference, COMPCON 74, Digest of Papers,
     p. 11-14.

155. Radoy, C. H. and Lipovski, G. J., "Switched multiple instruc- tion, multiple data stream processing", 2nd Annual Symposium on Computer Architecture, Proceedings, 1975, p. 183-7.

156. Schoeffler, J. D., "Microprocessor architecture", IEEE Trans. Ind. Electron. Control Instrum., IECI-$\underline{22}$ (3), 256-72 (1975).

157. Seipp, W. H., "Boost μP bit-manipulation capability by using simple logic external to the CPU chip", Electron. Des., $\underline{24}$ (5), 50-8 (1976).

158. Stakem, P. H., "Using a calculator chip to extend a micropro- cessor's capabilities", Comput. Des., $\underline{14}$ (9), 98-9 (1975).

159. Threewitt, B., "Microprocessor rationale", Proceedings, National Computer Conference, 1975, p. 3-7.

160. Toong, H. D. and Hampson, B. E., "Time-shared multi-user approach to microprocessor system development", Proceedings of International Symposium on Mini and Micro Computers, MIMI 76, p. 16-23.

161. Wakerly, J. F., "Microprocessor input/output architecture", Computer, $\underline{10}$ (2), 26-33 (1977).

162. Weisbecker, J., "Simplified microcomputer architecture", Computer, $\underline{7}$ (3), 41-7 (1974).

163. Weissberger, A. J., "Distributed function microprocessor archi- tecture", Comput. Des., $\underline{13}$, 77-83 (1974).

164. Weissberger, A. J., "Analysis of multiple-microprocessor system architecture", Comput. Des., $\underline{16}$ (6), 151-63 (1977).

165. Weissberger, A. J. and Toal, T., "Tough mathematical tasks are child's play for Number Cruncher (special-purpose micropro- cessor)", Electronics, $\underline{50}$ (4), 102-7 (1977), EEA80-16798.

166. Wittie, L. D., "Efficient message routing in mega-micro- computer networks", 3rd Annual Symposium on Computer Architec- ture, Proceedings, 1976, p. 134-40.

167. Zaks, R., "Structure of microprocessors", Automatisme, $\underline{19}$ (8-9), 421-7 (1974), French.

168. Zellweger, A., "Computer architectures for advanced air traffic control applications", Proceedings of the International Con- ference on Parallel Processing, 1976, p. 132-9.

V.   TECHNOLOGIES
      1.   General

169. Ahlgren, D. R., "Restored processor function saves logic and improves performance in microcomputer system", Comput. Des., $\underline{14}$ (8), 88ff (1975).

170. Altman, L., "Memory tapes multiply, microprocessor families grow", Electronics, $\underline{49}$ (22), 76-8, 81-2 (1976), CCA12-1508.

171. Altman, L. and Capece, R. P., "One-chip controllers and 4-k static RAM's star", Electronics, $\underline{50}$ (22), 96-102 (1977).

172. Anon., "Microcomputers using integrated microprocessors", Communitronics, $\underline{1}$ (3), 3-7 (1974).

173. Aspinall, D., "Processor components: data path and control circuits", Microprocessors, 1 (6), 380-4 (1977).

174. Bell, B. and Ogden, D., "Single-chip microprocessor rules the roast", Electronics, 49 (25), 105-10 (1976).

175. Callahan, B. L. J., "The state of the art", Mod. Data, 7 (5), 30-2 (1974).

176. Catlin, R. W., "MUMS - a modular unified microprocessor system", Report R-76-809, Illinois Univ., Urbana, 1976, 5 pp.

177. Chroust, G., "Memory organization for two parallel readouts", Microprocessors, 1 (6), 369-70 (1977).

178. Clymer, J., "Use 4-bit slices", Electron. Des., 25 (10), 62-71 (1977).

179. Cushman, R. H., "Very complete chip set joins the great microprocessor race", EDN, 19 (22), 87-94 (1974).

180. DeMan, H., "Technology of microcomputers", Proceedings of International Symposium and Course on Mini- and Microcomputers and Their Applications, MIMI 76, p. 11-22.

181. Derman, S., "PLAs or μPs? At times they compete, and at other times they cooperate", Electron. Des., 24 (18), 24, 26 (1976), CCA12-1448.

182. Dessoulavy, R., "Technology of microprocessors and memories", 1974 Electronic Meeting on Microprocessors, p. 45-71, French, CCA11-9644.

183. Drew, R., "Semiconductors and IC's - Microprocessors and a new mix of monolithic memories blaze the way in solid state", EDN, 20 (13), 56-61 (1975).

184. Francis, B., "Microprocessors: the minicomputer/random logic alternative?", Electron. Eng., 47 (565), 45-8 (1975).

185. Franson, P., "IC's and semiconductors - new devices pack everything you always wanted on a chip - and more", EDN, 21 (22), 86-8 (1976), EEA80-24395.

186. Fuller, S. H., Lesser, V. R., Bell, C. G. and Kaman, C. H., "Effects of emerging technology and emulation requirements on microprogramming", IEEE Trans. Comput., C-25 (10), 1000-9 (1976).

187. Graff, M., "Boundary conditions for PROMs, PLA, and microprocessors", 1975 WESCON Technical Papers, Paper 26/3.

188. Hoff, M. E., Jr., "New LSI components", 6th IEEE Computer Society International Conference, Digest of Papers, 1972, p. 141-3.

189. Juliussen, J. E. and Mowle, F. J., "Multiprocessors with common main and control memories", IEEE Trans. Comput., C-22 (11), 999-1007 (1973).

190. Kornstein, H., "Memories and microcomputers", Microprocessors, 1 (4), 252-7 (1977).

191. Kurtz, R. L. and Reynouard, B., "A shared memory technique for different microprocessors", Proceedings of the 1976 International Conference on Parallel Processing, p. 221, CCA12-12943.

192. Leventhal, L. A., "Semiconductor technologies and semiconductor memories", Simulation, 27 (2), 65-77 (1976), CCA11-28286.

193. Leventhal, L. A., "Cut your processor's computation time",
     Electron. Des., 25 (17), 82-9 (1977).
194. Lewandowski, R., "Preparation: the key to success with micro-
     processors", Electronics, 48 (6), 101-6 (1975).
195. Lewin, M. H. "Integrated microprocessors", IEEE Trans. Circuits
     & Syst., CAS-22 (7), 577-85 (1975).
196. Lewis, D. R. and Siena, W. R., "Microprocessor or random
     logic?", Electron. Des., 21 (18), 106-10 (1973).
197. Lofthus, A. and Ogden, D., "16-bit processor performs like
     minicomputer", Electronics, 49 (11), 99-105 (1976).
198. McCaskill, R., "Wring out 4-bit µP slices", Electron. Des., 25
     (10), 74-7 (1977).
199. McDermott, J., "Improved solid-state memories and micropro-
     cessors altering the structure of computers", Electron. Des.,
     22 (22), 34-6 (1974).
200. Manning, P., "Making microprocessors by the module", Instrum.
     Contr. Syst., 47 (11), 57-9 (1974).
201. Mick, J., "Speed up digital processors", Electron. Des., 25
     (19), 78-82 (1977).
202. Ogdin, J. L., "Microprocessors: the inevitable technology",
     Mod. Data, 8 (1), 42, 44-7 (1975).
203. Parrish, E. A., Jr., Aylor, J. H. and MacDonald, W. E., "Modular
     microcomputer system", Annual Conference Proceedings - Indus-
     trial Applications of Microprocessors, 1977, p. 47-51.
204. Raphael, H., "How to expand a microcomputer's memory", Elec-
     tronics, 49 (26), 67-9 (1976).
205. Reyling, G. F., "Single-chip microprocessor employs a minicom-
     puter word length", Electronics, 47 (26), 87-93 (1974).
206. Roybal, P., "FACE - a field-alterable control element", 1974
     WESCON Technical Papers, Paper 15/3.
207. Schmid, H., "Monolithic processors", Comput. Des., 13 (10),
     87-95 (1974).
208. Sharp, D. E., "The world of µP: bits, chips and memories", Can.
     Controls & Instrum., 15 (9), 30 (1976), CCA12-8914.
209. Toong, H. D., "The new technology - microprocessors", Proceed-
     ings, ASIS Annual Meeting, 1975.
210. Verhofstadt, P. W. J., "Technology for microprocessor hardware",
     12th IEEE Computer Society International Conference, COMPCON
     76, Digest of Papers, p. 19-22.
211. Verhofstadt, P. W. J., "Evaluation of technology options for
     LSI processing elements", Proc. IEEE, 64 (6), 842-51 (1976).
212. Walker, R., "Microprocessors in perspective", Electron. Power,
     20(13), 528-30 (1974).
213. Washburn, J., "Microprocessor chip mania", 12th IEEE Computer
     Society Inernational Conference, COMPCON 76, Digest of Papers,
     p. 165-7.

2.   Bipolar

214. Anon., "First I^{2}L processor in four-bit design surpasses N-MOS",
     Electronics, <u>48</u> (2), 29-32 (1975).
215. Erickson, C., Hingarh, H., Moeckel, R. and Wilnai, D., "A 16-
     bit monolithic I^{2}L processor", 1977 International Solid-State
     Circuits Conference, Digest of the Technical Papers, p. 140-1.
216. Hart, C. M., Slob, A. and Wulms, H. E. J., "Bipolar LSI takes
     a new direction with integrated injection logic", Electronics,
     <u>47</u> (20), 111-18 (1974).
217. Hoff, M. E., Jr., "Designing central processors with bipolar
     microcomputer components", Proceedings, National Computer
     Conference, 1975, p. 55-62.
218. Horton, R. L., Englade, J. and McGee, G., "I^{2}L takes bipolar
     integration a significant step forward", Electronics, <u>48</u> (3),
     83-90 (1975).
219. Huse, H., "I^{2}L: The technology and its application in a high-
     speed 4-bit microprocessor", Elektronik, <u>25</u> (2), 79-82 (1976),
     German, CCA11-12488.
220. Lewis, D. R. and Siena, W. R., "Microprocessor or random
     logic", Electron. Des., <u>21</u> (18), 106-10 (1973).
221. Mazor, S., "A new bipolar microcomputer", Proceedings, Symposium
     on Electronics, 1974, p. 121-9.
222. Nemec, J., "One-chip bipolar microcontroller approaches bit-
     slice performance", Electronics, <u>50</u> (18), 91-6 (1977).
223. Rattner, J., Cornet, J. C. and Hoff, M. E., Jr., "Bipolar LSI
     computing elements usher in new era of digital design", Elec-
     tronics, <u>47</u> (18), 89-96 (1974).
224. Rattner, J., Cornet, J. C. and Hoff, M, E., "Bipolar LSI ele-
     ments yield flexible and high-performance microcomputer system",
     Funk-Tech., <u>31</u> (22), 726-30 (1976), German, CCA12-15442.
225. Smith, K., "High performance memories and micros among clutch
     of new I^{2}L", New Electron., <u>10</u> (4), 107, 108, 111 (1977),
     EEA80-28904.
226. Wyland, D. C., "Bipolar microprocessor design configurations",
     Proceedings, National Computer Conference, 1975, p. 63-6.
227. Yates, V., "Microprocessors. II", Rev. Esp. Electron., <u>23</u> (261,
     262), 62-6 (1976), Spanish, CCA12-1454.

3.   MOS

228. Altman, L., "Fast 8-bit microprocessor is versatile", Elec-
     tronics, <u>47</u> (13), 149, 151 (1974).
229. Anon., "NMOS LSI brings fast 16-bit architecture to versatile
     microcomputer system", JEE, no. 109, 32-3 (1976), CCA11-17515.
230. Bromme, I., Paulmichl, E., Pfrenger, E. and von Sichart, F., "A
     high-speed 4-bit microprocessor in n-channel silicon gate tech-
     nology", Siemens Forsch. - & Entwicklungsber., <u>5</u> (6), 319-23
     (1976), EEA80-3542.

231. Chung, D., "Four-chip microprocessor family reduces system parts counts", Electronics, $\underline{48}$ (5), 87-93 (1975), EEA78-14896.

232. Leventhal, L. A., "Microprocessor CPU: Some unique features", Simulation, $\underline{28}$ (5), 153-5 (1977).

233. Luce, R., "Latest developments in solid state technology and their application", Proceedings of National Electron Conference, Vol. 29, 1974, p. 29.

234. Lynch, F., "Keep the PACE up and running", Electron. Des., $\underline{24}$ (25), 64-70 (1976), CCA12-10704.

235. Mick, J. R. and Schopmeyer, R., "MOS support microprocessor teams with bit-slice prototype for easier microprogram debugging", Electronics, $\underline{50}$ (19), 127-30 (1977).

236. Peterson, J. J., "Design and development of MOS sealer integrated circuit", Report HDL-TR-050-1, Collins Radio Co., Newport Beach, Calif., 1974, 124 pp., EEA78-23776.

237. Shima, M. and Faggin, F., "In switch to n-MOS microprocessor gets a 2-µs cycle time", Electronics, $\underline{47}$ (8), 95-100 (1974), EEA77-22481.

238. Shima, M., Faggin, F. and Mazor, S., "N-channel 8-bit single chip microprocessor", 1974 IEEE International Solid-State Circuits Conference, Digest of Technical Papers, p. 56-7.

239. Uebe, F., "n-channel MOS memories - new possibilities for microprocessor memory design", Microelectron. & Reliab., $\underline{13}$ (5), 387-400 (1974), EEA78-5366.

240. Vittera, J., "Nonvolatile microprocessor memory uses 4K n-channel MOS RAM's", EDN, $\underline{20}$ (8), 53-6 (1975).

241. Weissberger, A. J., "MOS/LSI microprocessor selection", Electron. Des., $\underline{22}$ (12), 100-4 (1974), EEA78-15403.

242. Wickes, W. E., "Compatible MOS/LSI microprocessor device family", Comput. Des., $\underline{12}$ (7), 75-81 (1973).

243. Young, L., Bennett, T. and Lavell, J., "N-channel MOS technology yields new generation of microprocessors", Electronics, $\underline{47}$ (8), 88-95 (1974), EEA77-22480.

4.   CMOS

244. Altman, L., "Special report: C-MOS enlarges its territory", Electronics, $\underline{48}$ (10), 77-88 (1975).

245. Anon., "Evolution of COS/MOS into MSI and LSI components by R tarrant", Microelectron. & Reliab., $\underline{15}$ (1), 85-92 (1976).

246. Bangen, G., "CMOS random access memory replaces ROM", Electron. Engineering, $\underline{48}$ (584), 29 (1976), CCA12-1506.

247. Capell, A., Knoblock, D., Mather, L. and Lopp, L., "Process refinements bring C-MOS on sapphire into commercial use (microprocessors)", Electronics, $\underline{50}$ (11), 99-105 (1977), EEA80-24991.

248. Dang, L. G., Ashkins, P. B., Yee, R. and O'Brien, M., "A CMOS/SOS 16-bit parallel µCPU", 1977 International Solid-State Circuits Conference, Digest of the Technical Papers, p. 134-5.

249. Dingwall, A. G. F., Stricker, R. E. and Sinniger, J. O., "High
     speed bulk CMOS C^{2}L microprocessor", IEEE J. Solid-State Cir-
     cuits, SC-12 (5), 457-62 (1977).
250. Grieshaber, B., "Consider the 6100 CMOS microprocessor", Elec-
     tron. Des., 24 (23), 50-3, 55-7 (1976), CCA12-15421.
251. Oliphant, J., "A non-volatile memory for 4040s", New Electron.,
     8 (23), 44, 46 (1975), CCA11-9684.
252. Thomas, A. T., "Architecture and applications of a 12-bit CMOS
     microprocessor", Proc. IEEE, 64 (6), 873-81 (1976), CCA11-22817.
253. Whitmore, J., "Monolithic 10-bit CMOS multiplying DAC", Analog
     Dialogue, 9 (3), 10-11 (1975), EEA79-46406.
254. Winder, R. O., "The COS/MOS microprocessor", 1974 WESCON Tech-
     nical Papers, Paper 15/4.

        5.   ROM/PROM

255. Anon., "A versatile microcomputer (prototyping) system", Elec-
     tron. Ind., 3 (3), 21 (1977), CCA12-12925.
256. Blacksher, R., "PROM decoder replaces chip-enabling logic",
     Electronics, 49 (18), 100-2 (1976).
257. Blume, H., Budde, D., Raphael, H. and Stamm, D., "Single-chip
     8-bit microcomputer fills gap between calculator types and
     powerful multichip processors", Electronics, 49 (24), 99-105
     (1976), EEA80-6249.
258. Brooks, K. R., "MPU-controlled PROM programmer", New Electron.,
     9 (23), 39-40 (1976), CCA12-3899.
259. Feeney, H. V., Jr., "Micro computer applications of electrically
     alterable ROMs", 1972 WESCON Technical Papers, Paper 4/4.
260. Haas, D., "Single chip microcomputer with integrated EPROM",
     Elektronik, 25 (12), 54-7 (1976), German, EEA80-6255.
261. Haynes, G., "Optimising fixed content memory systems for micro-
     processors", Electron, no. 104, 53, 55-6 (1976), CCA12-4014.
262. Johnson, D. W., "Go from flow chart to hardware", Electron.
     Des., 24 (18), 90-5 (1976).
263. Martin-Jones, A. J. and Payne, D. A., "Using microprocessors
     without ROM", Microprocessors, 1 (4), 237-40 (1977).
264. Peatman, J. B., Dack, D. G. and Warren, D. A., "ROMs in micro-
     processors can test themselves", Electronics, 47 (23), 153,
     155 (1974).
265. Tarui, T., Namimoto, K. and Takahashi, Y., "Twelve-bit micro-
     processor nears minicomputer's performance level", Electronics,
     47 (6), 111-16 (1974).

        6.   Others

266. Burns, J. R., "Silicon-on-sapphire integrated circuits", Inter-
     national Microelectronics Symposium, 1972, Paper 2A-1, 4 pp.

267. Elmasry, M. I. and Card, H. C., "Variable-program microprocessor
     systems using nonvolatile ERLIM arrays", Canadian Communica-
     tions and Power Conference, 1976, p. 164-7, CCA12-6610.
268. Feinberg, B., "The microcomputer - genesis of a new computer
     generation", Manuf. Eng. & Manage., no. 6, 36-7 (1974).
269. Herbert, P., "SOS and its place in LSI technology", Proceedings,
     Symposium on Electronics, 1974, p. 139-48.
270. Sherwood, W., "PLATO - PLA translator/optimizer", Proceedings
     of the Symposium on Design Automation and Microprocessors, 1977,
     p. 28-35.

VI. USER'S GUIDE
1.   General

271. Allison, D. R., "Small scale computing", Computer, $\underline{10}$ (3), 8-9
     (1977).
272. Altman, L., "Single-chip microprocessors open up to a new world
     of applications", Electronics, $\underline{47}$ (8), 81-7 (1974).
273. Anon., "Microcomputers throw the industry off balance", Business
     Week, 56-7 (March 16, 1974).
274. Anon., "Updating circuit symbols, the graphic language of elec-
     tronics", Electronics, $\underline{48}$ (7), 90-4 (1975).
275. Anon., "Primer on microprocessors - I", Electron. Prod., $\underline{17}$ (8),
     24-32 (1975).
276. Anon., "Primer on microprocessors - II.", Electron. Prod., $\underline{17}$
     (9), 37-45 (1975).
277. Anon., "Microprocessor DVMs, with new features, to hit the market
     shortly", Electron. Des., $\underline{23}$ (16), 32-4 (1975).
278. Anon., "Microcomputers - Mind boggling potential", Infosystems,
     $\underline{22}$, 54 (1975).
279. Anon., "Microprocessor terms, definitions and abbreviations",
     OEM Des., 86, 88 (1976), EEA80-9502.
280. Anon., "One-chip micros spur new race", Electronics, $\underline{50}$ (20),
     69, 71 (1977).
281. Barton, M. and Dagless, E., "Graphical approach to microprocessor
     comparison", Microprocessors, $\underline{1}$ (6), 371-9 (1977).
282. Beddoes, J. M., "Digital integrated circuit electronics sets
     breathtaking pace of development", Can. Electron. Eng., $\underline{17}$ (2),
     28-31 (1973).
283. Bell, C. G. and Kaman, C., "Microprocessor - another member of
     the mini(mal) computer family", IEEE International Convention
     Technical Papers, 1974, Paper 24/4.
284. Birkner, J., "Minis versus micros: Getting the right machine
     for the job", Electron Prod., $\underline{17}$ (10), 45-7 (1975).
285. Bloch, A., "Hints for beginners in microprocessors", Electronics,
     $\underline{49}$ (20), 100-1 (1976).
286. Bowers, D. M., "Systems-on-a-chip. II. Microprocessor score-
     card", Mini-Micro Syst., $\underline{9}$ (7), 42-8 (1976), CCA12-8866.

287. Bruun, R., "Minis and micros: Birds of a feather or distant cousins", Infosystems, 21 (2), 32-4 (1974).
288. Byrd, J. S., "When your system's data rates differ, it's time for a microprocessor", EDN, 19 (22), 57-62 (1974).
289. Cassell, D. A., "Microprocessors and microcomputers: What to expect when you get them", Mod. Data, 7 (5), 34-5 (1974).
290. Cerulli, C. J., "Microprocessors - an unprecedented challenge to technical managers", IEEE Electron and Aerospace Systems Convention, EASCON '76, Paper 55, 5 pp.
291. Chung, D., "The monolithic microprocessor as a universal standard part", 1974 WESCON Technical Papers, Paper 19/3.
292. Coles, R. W., "Microprocessors explained", Prac. Electron., 13 (6), 426-30 (1977).
293. Connolly, R., "Microprocessors to get mil specs", Electronics, 50 (18), 76 (1977).
294. Coury, F. F., "Microprocessors - where they fit in", In: New Components and Subsystems for Digital Design: a Report, Technology Service, Santa Monica, CA, p. 7-9 (1975).
295. Cushman, R. H., "Microprocessors are changing your future. Are you prepared?", EDN, 18 (21), 26-32 (1973).
296. Cushman, R. H., "Understanding the microprocessor is no trivial task", EDN, 18 (22), 42-9 (1973).
297. Cushman, R. H., "Microprocessors are rapidly gaining on minicomputers", EDN, 19 (10), 16-19 (1974).
298. Cushman, R. H., "Understand the 8-bit microprocessor: You'll see a lot of it", EDN, 19 (2), 48-54 (1974).
299. Cushman, R. H., "What can you do with a microprocessor?", EDN, 19 (6), 42-8 (1974).
300. Cushman, R. H., "How to get acquainted with a microprocessor", EDN, 19 (18), 46-52 (1974).
301. Cushman, R. H., "First annual microprocessor directory", EDN, 19 (22), 31-3 (1974).
302. Cushman, R. H., "Newest microprocessors split into two divergent paths", EDN, 19 (24), 31-4 (1974).
303. Cushman, R. H., "Getting started in microprocessors on a shoestring", 1975 WESCON Technical Papers, Paper 10/4.
304. Cushman, R. H., "EDN's third annual microprocessor directory", EDN, 21 (21), 44-89 (1976), CCA12-3880.
305. Cushman, R. H., "To get a microcomputer project going, grab at corners of the problem", EDN, 22 (10), 84-90 (1977).
306. Davidow, W., "Microcomputers: Do they herald the beginning of the second industrial revolution?", Electron. Eng., 46 (559), 54-6 (1974).
307. Davie, H. and Scobie, D. C. H., "Undergraduate experiments in on-line computer control", Meas. Control, 8 (12), 480-3 (1975).
308. Davis, S., "Fresh view of mini- and microcomputers", Comput. Des., 13 (5), 67-79 (1974).
309. Dejka, W. J., "Workshop report: Microcomputer standardization concepts", Computer, 10 (2), 54-6 (1977).

310. Dethlefsen, H., "The microprocessor in focus", Radio Mentor Electron., 43 (1), 018-019 (1977), German, EEA80-24976.

311. Dorn, P. H., "Catching up with microprocessors", Datamation, 21 (5), 171 (1975).

312. Elliott, T. C., "Understanding microprocessors", Power, 121 (5), 25-32 (1977).

313. Faggin, F. and Hoff, M. E., Jr., "Standard parts and custom design merge in four-chip processor kit", Electronics, 45 (9), 112-16 (1972).

314. Falk, H., "Computer systems: Hardware/software", IEEE Spectrum, 12 (1), 38-43 (1975).

315. Feeney, H., "A new microcomputer family", 1974 WESCON Technical Papers, Paper 15/1.

316. Fertl, W., "Microcomputers - with or against minicomputers", Elektrotechnik, 56 (24), 18-19 (1974), German.

317. Fischer, W. A., "Synergistic combination of an oscilloscope and a microprocessor", Proceedings, National Computer Conference, 1975, p. 23-32.

318. Fisher, E. R., "Making microcomputer systems easy to use", 1975 WESCON Technical Papers, Paper 1.

319. Fisher, E. R., "Guidelines for selecting a microprocessor proto-typing kit", Report UCRL-76979, California Univ., Livermore Lawrence Lab., 1975, 4 pp.

320. Franson, P., "IC's and semiconductors - new devices pack every-thing you always wanted on a chip - and more", EDN, 21 (22), 86-8 (1976), CCA12-15560.

321. Garen, E. R., "Should a microprocessor be used at all", New Electron, 8 (3), 21 (1975).

322. Gilb, T., "Microcomputers - present properties and probable applications", Comput. & People, 24 (2), 32-5 (1975).

323. Hall, J., "Some microprocessor basics", Instrum. & Control Syst., 50 (2), 25-9 (1977).

324. Harrison, P., "The impact of the microprocessor", Data Process, 17 (3), 214-16 (1975).

325. Hatgood, D., "Considering a microprocessor", Elektronik, 23 (10), 379-82 (1974), German.

326. Hatvany, J., "New achievements in the development of integrated machine producing systems", Meres & Autom., 25 (2), 59-61 (1977), Hungarian, CCA12-15564.

327. Hickey, J., "Microprocessors fight back", Instrum. & Control Syst., 49 (10), 31-6 (1976), CCA12-8857.

328. Hofer, R., "Developments with microprocessors: Collective experience", Elektronik, 26 (1), 45-52 (1977), German.

329. Jones, D. F. and Martin, E. R., "Use μPs in minicomputer sys-tems. The mini can test and code the μP, and the μP can relieve the mini of time-consuming simple tasks", Electron. Des., 24 (22), 148-52 (1976).

330. Joseph, E. D., "Computer revolution can be predicted based on impact of microprocessors", Prof. Eng., 45 (9), 34-7 (1975).

331. Jurgen, R. K., "Microprocessor: in the driver's seat?", IEEE
Spectrum, <u>12</u> (6), 73-7 (1975).
332. Katsaros, J., "The universal microprocessor", Fairchild J.
Semicond. Progr., <u>3</u> (4), 3-9 (1975).
333. Kaye, D. N., "How to pick a microprocessor, a mini, or anything
in between", Electron. Des., <u>23</u> (16), 26-8 (1975).
334. Khanna, V. and Daly, T. C., "How to use your microcomputer for
all it's worth - during data handling", 1974 WESCON Technical
Papers, Paper 23/1.
335. Khanna, V. and Daley, T., "Making the most of your micro",
Digital Des., <u>5</u> (7), 36-8 (1975).
336. Kleiman, H., "The microcomputer: Technological innovation and
transfer", Report AD-A026970/4SL, Battelle Columbus Labs.,
Ohio, 1975, 41 pp.
337. Klein, S., "Microcomputers on a chip come on strong", Mini-
Micro Syst., <u>10</u> (2), 22-4, 26 (1977), CCA12-15567.
338. Knowles, A., "The microprocessor selection process", Proceedings, 1st National Microprocessor Conference, Microprocessors:
Economics, Technology, Applications, 1974, Sponsor - Arthur D.
Little, Inc.
339. Kornstein, H., "The microprocessor reaches maturity", Systems,
<u>6</u> (10), 10-11 (1976), CCA12-15580.
340. Kovalcik, E. J., "Understanding small computers", Instrum.
Control Syst., <u>49</u>(1), 57-62 (1976).
341. Krummel, L. and Schultz, G., "Advances in microcomputer development systems", Computer, <u>10</u> (2), 13-19 (1977).
342. Labudda, H. J., "The microprocessor - the other side of the
coin", Elektronik, <u>26</u> (1), 55-7 (1977), German, CCA12-15419.
343. Lally, J., "Intellec microcomputer development system", 1975
WESCON Technical Papers, Paper 15/3.
344. Lamond, F. E., "Structural consequences of the LSI revolution",
Proceedings, European Computer Congress, 1974, p. 395-409.
345. Langley, F. J., "Commercial micro computer chips for integrated
phased array control", International IEE/AP-S Symposium Digest,
1974, p. 406-9.
346. Lapidus, G., "Computer: Now it's a component!", IEEE Spectrum,
<u>11</u> (3), 80-2 (1974).
347. Larsen, D. G. and Rony, P. R., "Microcomputers: How to put them
to work", 11th IEEE Computer Society International Conference,
COMPCON 75, Digest of Papers, p. 159-62.
348. Lee, I., "LSI microprocessors and microprograms for user-
oriented machines", 7th Workshop on Microprogramming, Supplement
to the Conference Proceedings, Micro 7, 1974, p. S-1 to S-13.
349. Lee, I., "LSI microcomputer and digital control", ASME Paper
No. 75 - AUT-V, 1975, 8 pp.
350. Lewing, V., "Exorciser", 1975 WESCON Technical Paper, Paper 15/2.
351. Lewis, D. R. and Siena, W. R., "Clear the hurdles of microprocessors", Electron. Des., <u>21</u>(20), 76-80 (1973).

352. Liles, W. C. and Teener, M. D., "Development aid survey", In:
New Components and Subsystems for Digital Design: a Report,
Technology Service, Santa Monica, CA, p. 187-9 (1975).

353. Lunch, F. and Showen, C., "Choosing microprocessors for reduced
parts counts", Digital Des., 7 (11), 18, 20, 22, 24 (1977).

354. McDermott, J., "That lowly calculator is turning into a vest-
pocket computer", Electron. Des., 21 (13), 28-30 (1973).

355. Mandell, L. J., "Pitfalls to avoid in applying microprocessors",
EDN, 20 (2), 22-6 (1975).

356. Manning, P., "Need more 'intelligence'? Plug in a microcomputer",
Automation, 21 (10), 53-5 (1974).

357. Maples, M. D., "Microprocessors in computing?", 10th IEEE
Computer Society International Conference, COMPCON 75, Digest
of Papers, p. 69-71.

358. Maples, M. D. and Fisher, E. R., "Microprocessor prototyping
kits", Proc. IEEE, 64 (6), 932-6 (1976).

359. Metzger, J., "Forum: It's go for microprocessors", Electron.
Prod. 16 (6), 114-27 (1973).

360. Miskolczi, J., "Microprogrammable microprocessors", Meres &
Autom., 25 (1), 20-8 (1977), Hungarian, CCA12-15563.

361. Monrad-Krohn, L., "The micro vs. the minicomputer", Mini-Micro
Syst., 10 (2), 28, 30, 32-3 (1977), CCA12-15568.

362. Murphy, J. A., "Microprocessors and microcomputers: What's
available?", Mod. Data, 7 (5), 36-8 (1974).

363. Neye, A., "Microprocessors pushing forward", Elektronik J.,
9 (11), 72-4 (1974), German.

364. Nichols, J., "Source-destination matrix for instruction set
presentation", 10th IEEE Computer Society International Conference, COMPCON 75, Digest of Papers, p. 61-4.

365. Nicoud, J. D., "Introduction to microprocessors", Proceedings,
Symposium on Electronics, 1974, p. 6-18.

366. Ogdin, J., "Microprocessors scorecard", Proceedings, Symposium
on Electronics, 1974, p. 74-108.

367. Ogdin, J., "Getting started in microprocessors", New Logic
Notebook, 1 (5), 1-24 (1975).

368. Ogdin, J. L., "Microcomputers: Promises and practices", IEEE
International Convention Technical Papers, 1974, Paper 17/1.

369. Ogdin, J. L., "Survey of microprocessors reveals limitless
variety", EDN, 19 (8), 38-43 (1974).

370. Ogdin, J. L., "Survey of 8-bit microprocessors reveals wide
choice for users", EDN, 19 (12), 44-50 (1974).

371. Ogdin, J. L., "Getting started in microprocessors and microcomputers", Instrum. Technol., 22 (1), 35-43 (1975).

372. Ogdin, J. L., "Microprocessors have their problems too",
Instrum. & Control Syst., 49 (8), 35-9 (1976), CCA12-15561.

373. Olken, H., "Microcomputers - from the user's viewpoint", Res.
Develop., 25 (10), 16-20 (1974).

374. Phillips, A. S., "Inside the microprocessors", Mod. Data, 8
(1), 37-41 (1975).

375. Phillips, J. C., "Getting involved with microprocessors", RCA
Eng., <u>22</u> (5), 7-11 (1977).
376. Powner, E. T., Escuder, M. A. and Depledge, P. G., "Introduction
to microprocessor systems - 1.", Int. J. Electr. Eng. Educ.,
<u>14</u> (1), 73-80 (1977).
377. Prommmer, A., "Microprocessors, a new dimension for logic circuits", Mess Pruef, <u>12</u> (3), 90, 93 (1976), German.
378. Queyssac, D., "Understanding microprocessors", Mes. Regul.
Autom., <u>41</u> (9), 29-35 (1976), French, CCA12-4058.
379. Raphael, H. A., "Join micros into intelligent networks", Electron.
Des., <u>23</u> (5), 52-7 (1975).
380. Raphael, H., "Simplify low-cost μP selection", Electron. Des.,
<u>25</u> (5), 60-5 (1977).
381. Reyling, G., Jr., "Microprocessors: The next generation in
digital design", EE/Syst. Eng. Today, <u>32</u> (11), 86-90 (1973).
382. Reyling, G. F., "Considerations in choosing a microprogrammable
bit-sliced architecture", Computer, <u>7</u> (7), 26-9 (1974).
383. Reyling, G. F., "Extend LSI-processor capabilities", Electron.
Des., <u>22</u> (22), 90-5 (1974).
384. Riezeman, M. J., "Systems are getting smarter", Electronics,
<u>47</u> (14), 99-102 (1974).
385. Riley, W. B., "Peripherals now, mainframes later", Electronics,
<u>47</u> (14), 96-8 (1974).
386. Riviere, C. J. and Nichols, P. J., "Microcomputers unlock the
next generation", Data Commun., <u>3</u> (3), 21-8 (1974).
387. Robinson, A. L., "Computers: First the maxi, then the mini,
now it's the micro", Science, <u>186</u> (4169), 1102-4 (1974).
388. Rudnick, S., "Microprocessors unmasked", Digital Des., <u>4</u> (6),
32-6 (1974).
389. Saba, M. M. and Grimes, J. D., "Microprocessors: A component
for all seasons", 1974 WESCON Technical Papers, Paper 23/3.
390. Sarrazin, Y., "Microprocessor: Revolution or evolution?", Automatique & Informatique Industrielles, no. 33, 17-20 (1975),
French.
391. Sawin, D. H., "Selection of a microprocessor for system design",
Proceedings, Symposium on Electronics, 1974, p. 109-20.
392. Sawyer, G., "Microprocessor system made easy", 10th IEEE Computer Society International Conference, COMPCON 75, Digest of
Papers, p. 91-4.
393. Schade, P. A., "4-bit μP's - should you be using one?", EDN,
<u>20</u>(21), 144, 146 (1975).
394. Schmid, H., "Monolithic processors", Comput. Des., <u>13</u> (10),
87-95 (1974).
395. Schroeder, K., "Microcomputer vs. minicomputer: selection
criteria", Annual Conference Proceedings - Industrial Applications of Microprocessors, 1977, p. 190-4.
396. Schultz, G. W. and Holt, R. M., "MOS LSI minicomputer comes of
age", AFIPS Conference Proceedings, Vol. 41, FJCC, 1972,
p. 1069-80.

397. Schultz, G. W. and others, "Guide to using LSI microprocessors",
     Computer, 6 (8), 13-19 (1973).
398. Schwartz, M. and Kute, B., "Anatomy of a microcomputer", Mach.
     Des., 48 (6), 60-6 (1976).
399. Schwartz, M. and Winter, K., "Using a microprocessor at speeds
     beyond its apparent intrinsic limit", Comput. Des., 15 (6),
     106, 108, 110 (1976).
400. Scrupski, S. E., "Data-handling gains flexibility", Electronics,
     47 (14), 88-91 (1974).
401. Sideris, G., "Microcomputers muscle in", Electronics, 46 (5),
     63-4 (1973).
402. Smith, H., "Impact of LSI on microcomputer and calculator chips",
     NEREM Record, 1972, p. 143-6.
403. Smith, H., "Anatomy of a microcomputer", EE/Syst. Eng. Today,
     32 (11), 91-4 (1973).
404. Snigier, P., "Microprocessor development systems which one is
     'best'?", EDN, 22(5), 68-78 (1977).
405. Snyder, F. G., "The microprocessor shake-out", Digital Des., 4
     (9), 20-2 (1974).
406. Stakem, P. H., "Using a calculator chip to extend a micropro-
     cessor's capabilities", Comput. Des., 14 (9), 98-9 (1975).
407. Starkweather, J. A., "Making computers easier to use through:
     Distributed mini/macro processors: PILOT", 11th IEEE Computer
     Society International Conference, COMPCON 75, Digest of Papers,
     p. 185-8.
408. Steger, J. P., "Development aids for microprocessor users. I.",
     Elektroniker, 15 (11), EL1-8 (1976), German, CCA12-6597.
409. Tallman, J. L., "Bringing up the pace µP: A detailed application
     story", EDN, 21 (2), 5-18 (1976).
410. Teener, M. and Liles, W., "Microcomputers: Where the action
     really is", Mod. Data, 8 (2), 49-53 (1975).
411. Torrero, E. A., "Microprocessors finding growing role between
     calculator chips and minis", Electron. Des., 21 (11), 80-5 (1973).
412. Torrero, E. A., "Focus on microprocessors", Electron. Des., 22
     (18), 52-69 (1974).
413. Vacroux, A. G., "Microcomputer selection", Notes for NEC Micro-
     computer Institute, April 1974.
414. Vacroux, A. G., "Introduction to microcomputers", Notes for NEC
     Microcomputer Institute, April 1974.
415. Van Naarden, R., "Microprocessor selection process", 1975
     WESCON Technical Papers, Paper 6/6.
416. Vuille, J. P., "Standards for µP systems. V.", Euromicro Newsl.,
     3 (1), 49-53 (1977), CCA12-12836.
417. Walker, G. M., "Microprocessors to public", Electronics, 47
     (14), 92-5 (1974).
418. Walker, R., "Microprocessors in perspective", Electron Power,
     20 (13), 528-30 (1974).
419. Weinstein, C., "Microprocessor development equipment", Electron.
     & Power, 23 (4), 291-8 (1977), CCA12-12923.

420. Weissberger, A. J., "MOS/LSI microprocessor selection",
     Electron. Des., 22 (12), 100-4 (1974).
421. Weissberger, A. J., "Application ideas for microprocessors",
     Instrum. & Control Syst., 48 (10), 19-24 (1975).
422. Weissman, J., "Changing personnel skills and new vendor/user
     responsibilities and relationships for microprocessor users",
     Proceedings, 1st National Microprocessor Conference, Micro-
     processors: Economics, Technology, Applications, 1974.
423. Welty, J., "The pervasive microprocessor", Dataquest, Inc.,
     Semiconductor Industry Seminar, 1976, p. 7-1-7-21.
424. Wiener, H., "Computers that fit in your pocket", Comput.
     Decis., 5 (8), 8-13 (1973).
425. Wolf, H., "Microcomputers - A new revolution", EE/Syst. Eng.
     Today, 33 (1), 80-3 (1974).
426. Yasaki, E., "Emerging microcomputer", Datamation, 20 (12),
     81 (1974).
427. Young, L., "Impact on product definition, design, and develop-
     ment for microprocessor users", Proceedings, 1st National Micro-
     processor Conference, Microprocessors: Economics, Technology,
     Applications, 1974.
428. Zueblin, G., "Most 'micros' are wrongly utilized", Elektroniker,
     51(4), EL13-17 (1976), German.

2.   Education

429. Booth, T. L. and Carey, B. J., "Impact of micro/minicomputers
     on electrical engineering curriculums", 9th IEEE Computer
     Society International Conference, COMPCON 74, Digest of Papers,
     p. 135-7.
430. Cushman, R. H., "Microprocessor instruction sets: the vocabulary
     of programming", EDN, 20 (6), 35-41 (1975).
431. Cushman, R. H., "2 1/2-Generation machines make learning µP's
     a snap", EDN, 20 (21), 117-22 (1975).
432. Cushman, R. H., "Even bare-bones (microcomputer) development
     systems make good learning tools", EDN, 22 (6), 115-22 (1977),
     CCA12-15426.
433. DiGirolamo, G. B., "Learning to use microprocessors", RCA Eng.,
     22 (5), 12-14 (1977).
434. Eccles, W. J., "Training for using microprocessors", Anal.
     Instrum., 14, 95-6 (1976).
435. Gebler, P., "Introducing microprocessors", Electron. Engineer-
     ing, 49 (589), 61-3 (1977), CCA12-10635.
436. Jones, G. D., "MST-80 microprocessor trainer", Report UCID-
     17145, California Univ., Livermore, Lawrence Lab., 1976, 26 pp.
437. Larsen, D. G. and Rony, P. R., "Microcomputers: How to put them
     to work", 11th IEEE Computer Society International Conference,
     COMPCON 75, Digest of Papers, p. 159-62.

438. Larson, A. L., "Microcomputer education at the Air Force
     Academy", 10th IEEE Computer Society International Conference,
     COMPCON 75, Digest of Papers, p. 115-17.
439. Lee, I., "'Hands-on' approach to microcomputer education",
     Proc. IEEE, 64 (6), 1002-7 (1976).
440. Mitchell, S. G., "Microprocessor technology for managers",
     Report AD-A033919/2SL, Defense Systems Management School, Fort
     Belvoir, Va., 1976, 58 pp.
441. Nelson, G. A., "System design using multiple microprocessor: A
     tutorial", IEEE Electromagnetic Compatibility Symposium Record,
     1975, Sess. 5C, Paper IC.
442. Niederjohn, R. J., Larsen, G. W. and Ritt, D. M., "Microcomputer
     interfacing course", IEEE Trans. Educ., E-20 (2), 114-15 (1977).
443. Ogdin, C. A., "EDN microcomputer design course", EDN, 21 (21),
     127-36 (1976), CCA12-4056.
444. Pirri, F. and Becattini, G., "Child: A system for the develop-
     ment and didactic introduction to microprocessors", Report
     N77-12768/6SL, Florence Univ., Italy, Ist. D'Elettronica, 1976,
     15 pp., Italian.
445. Schade, C. M., "Computer as a laboratory instrument", 8th IEEE
     Computer Society International Conference, COMPCON 74, Digest
     of Papers, p. 209-10.
446. Sloan, M. E., "Potential trends in microcomputer education",
     9th IEEE Computer Society International Conference, COMPCON 74,
     Digest of Papers, p. 131-4.
447. Williman, A. O. and Jelinek, H. J., "Introduction to LSI micro-
     processor developments", Computer, 9 (6), 34-46 (1976).

3. Costs

448. Anon., "Low cost system of initiation to microprocessors",
     Autom. & Inf. Ind., no. 52, 36-7 (1976), French, CCA12-12916.
449. Anon., "The microprocessor/microcomputer industry", Industry
     Analysis Service, Creative Strategies Incorporated, 4340 Stevens
     Creek Blvd., San Jose, Calif., 95129, February 1977, 44 pp.
450. Barnich, R. G., "'Hidden' costs in designing with microcom-
     puters", Mach. Des., 48 (15), 80-3 (1976).
451. Bromley, A. G., "An economical serial micro-computer", Proceed-
     ings, 6th Australian Computer Conference, Vol. 2, 1974, p. 627-37.
452. Bryant, J. D. and Longley, R., "16-bit microcomputer is seeking
     a big bite of low-cost controller tasks", Electronics, 50 (13),
     118-24 (1977).
453. Conway, J. and Snigier, P., "Plummeting prices, numerous
     features and an ultrawide selection make this year's μC's a
     bumper harvest", EDN, 21 (21), 95-6, 121 (1976), CCA12-4055.
454. Cushman, R. H., "Don't overlook the 4-bit microprocessor:
     They're here and they're cheap", EDN, 19 (4), 44-50 (1974).
455. Evans, G., "Microprocessors - a low conversion rate", New
     Electron., 10 (1), 19, 21 (1977).

456. Frankenberger, D. and Agarwal, D. P., "Microprogrammed mini
     reduces complexity and cost", Electron. Prod., $\underline{16}$ (7), 167-70,
     173 (1973).
457. Gallacher, J., "Miniaturization makes microcomputers a practical
     proposition", Control Instrum., $\underline{7}$ (3), 16-17 (1975).
458. Johnson, T., "Microprocessors - an era assessment of LSI
     computer components", Comput. & Autom., no. 6, 45 (1973).
459. Kaufman, P. A., "How about the low cost ($1000) minicomputer?",
     1973 WESCON Technical Papers, Paper 11/3.
460. Kovacs, M. and Saufert, J., "Technical and economical questions
     of the large scale integration circuits microprocessors", Hira-
     dastechnika, $\underline{28}$ (1), 1-13 (1977), Hungarian, EEA80-28897.
461. Lapidus, G., "MOS/LSI launches the low-cost processors", IEEE
     Spectrum, $\underline{9}$ (11), 33-40 (1972).
462. McCrea, P. G., "Economical high speed display processor",
     Aust. Comput. J., $\underline{7}$ (1), 3-6 (1975).
463. Raphael, H., "Simplify low-cost μP selection", Electron. Des.,
     $\underline{25}$ (5), 60-5 (1977).
464. Schlieper, H. U., "Cost and efficiency of the MCS-8008 and MCS-
     8080 microprocessor system", Proceedings, Symposium on Elec-
     tronics, 1974, p. 149-60.
465. Sherman, E. N., "Microprocessors - money savers for remote
     computing", Comput. Decis., $\underline{5}$ (12), 10 (1973).
466. Swales, N. P. and Weisbecker, J. A., "COSMAC - a microprocessor
     for minimum cost systems", RCA Eng., $\underline{21}$ (3), 64-9 (1975).
467. Williams, W. D., "Costing computer control", Control Eng., $\underline{21}$
     (11), 48-50 (1974).
468. Zander, E. J., "Are microcomputers always a bargain?", Instrum.
     & Control Syst., $\underline{50}$ (1), 47-9 (1977), CCA12-15562.

4.   Markets

469. Anon., "Microcomputers aim at a huge new market", Business Week,
     no. 2279, 180-2 (1973).
470. Anon., "Microprocessor markets 1975-1980 - a user based study of
     microprocessor requirements", Report IDC-1569, International
     Data Corp., Waltham, Mass., 1975, 5 pp.
471. Anon., "The markets for minicomputers, microcomputers, and
     miniperipherals - Volume 1 - Applications", Report, Dataquest,
     Inc., Palo Alto, Calif., 1975, 3 pp.
472. Anon., "The markets for minicomputers, microcomputers, and
     miniperipherals - Volume 2 - Industry overview", Report, Data-
     quest, Inc., Palo Alto, Calif., 1975, 3 pp.
473. Anon., "The markets for minicomputers, microcomputers, and mini-
     peripherals - Volume 3 - Miniperipherals", Report, Dataquest,
     Inc., Palo Alto, Calif., 1975, 4 pp.
474. Anon., "The changing world of advanced LSI", Comput. Des.,
     $\underline{15}$ (11), 134-40 (1976), CCA12-4043.
475. Anon., "The microprocessor/microcomputer industry", Report,
     Creative Strategies, Inc., San Jose, CA, 1977, 44 pp.

476. MacAdam, D., "Microcomputers: Key to new markets for small companies", Can. Electron. Eng., 19 (2), 34-5 (1975).

477. Riley, J. F., "Microprocessor status", Report, Dataquest, Inc., Menlo Park, Calif., 1976, 6 pp.

478. Riley, J. F., Klesken, D. L. and Zieber, F. L., "Semiconductor Industry Services (SIS) - MOS microprocessor shipments", Report, Dataquest, Inc., Menlo Park, Calif., 1977, 12 pp.

479. Riley, J. F., Zieber, F. L. and Klesken, D. L., "Semiconductor Industry Services (SIS) Newsletter - MOS microprocessor shipments", Report, Dataquest, Inc., Menlo Park, Calif., 1976, 9 pp.

480. Schreier, P., "Modules and subassemblies - refining the building blocks", EDN, 21 (22), 122-4 (1976), CCA12-15422.

481. Wickham, R. F., "Microprocessor market - present and future", 1974 WESCON Technical Papers, Paper 11/1.

482. Wickham, R. F., "Microprocessors - markets and technology", SME Technical Papers, MS75-728, 1975, 7 pp.

5. Manufacturers
a. Intel

483. Abbott, D. L., "Indexed jump feature for Intel 8008 microprocessor", Comput. Des., 13 (16), 112 (1974).

484. Adams, G., Morgan, T. and Zarrella, J., "A microcomputer tailored for multiprocessor control applications", Control Eng., 23 (9), 58-60 (1976), CCA12-15549.

485. Anon., "A 8080 microprocessor microcomputer for non-specialists", Autom. & Inf. Ind., no. 51, 25 (1976), French, CCA12-15545.

486. Anon., "Three-chip microcomputer family cuts cost and improves reliability", Des. Eng., 25 (February 1977), CCA12-12918.

487. Anon., "Micro-practice. II.", Micomp, 2 (1), 36-9 (1977), German, CCA12-15572.

488. Ball, M., "Number systems: Cross-reference guide MCS-8 microcomputer", Report UCID-17127, California Univ., Livermore, Lawrence Livermore Lab., 1974, 8 pp.

489. Banerji, D. K. and Raymond, J., "A cross-assembler and interpreter for Intel 8008 microprocessor", Proceedings, Symposium on Electronics, 1974, p. 177-82.

490. Bourret, S., "LIL8/V2: A list interpretive language for the MCS-8 microcomputer", Report UCID-17130, California Univ., Livermore, Lawrence Livermore Lab., 1974, 21 pp.

491. Cope, S. N., "Floating-point arithmetic routines and macros for an Intel 8008 microprocessor", Report N76-11756/3SL, Oxford Univ., England, Dept. of Engineering Science, 1975, 30 pp.

492. Cushman, R. H., "Intel 8080: The first of the second-generation microprocessors", EDN, 19 (9), 30-6 (1974).

493. Deuel, D. R. and Gault, J. W., "Integrated hardware/software microprocessor based development system", 11th IEEE Computer Society International Conference, COMPCON 75, Digest of Papers, p. 163-5.

494. Faggin, F. and others, "MCS-4 - a LSI microcomputer system",
     IEEE Region 6 Conference Record, 1974, p. 1-6.
495. Foster, C. C., "Something new - the Intel MCS-4 microcomputer
     set", Computer Architecture News, 1 (2), 16-17 (1972).
496. Gallice P. and Mathis, M., "Autonomous controller (JCAM 10) for
     CAMAC crate", 2nd ISPRA Nuclear Electron Symposium, Proceedings,
     1975, p. 337-40.
497. Gladstone, B. and Page, P. D., "Programming hints ease use of
     familiar microprocessor", Comput. Des., 15 (8), 77-83 (1976).
498. Jensen, C. W., "Understanding RAM and ODT operation in the
     MCS-8", Report UCID-17133, California Univ., Livermore,
     Lawrence Livermore Lab., 1974, 11 pp.
499. Kildall, G. A., "An Intel 8080-based floppy disc system", 1975
     WESCON Technical Papers, Paper 10/1.
500. Larsen, D. G., Titus, J. A. and Rony, P. R., "Minicomputer
     interfacing: The 8080 logical instructions", Comput. Des., 16
     (5), 136-8 (1977).
501. Lindheimer, M. and Wold, I., "A shared memory interface for
     an Intel 8080 and Nova 1200", Proceedings, Symposium on Elec-
     tronics, 1974, p. 195-206.
502. Magnuson, W. G., Jr., "Simulator program for the Intel MCS-8
     8008 CPU", Report UCID-16351, 1973, 36 pp.
503. Magnuson, W. G., Jr. and Allison, T. G., "PL/M: A high level
     language for the Intel MCS-8 8008 CPU", Report UCID-16350,
     1973, 8 pp.
504. Maples, M. D., "Floating-point package for Intel 8008 and 8080
     microprocessors", Report UCRL-51940, California Univ., Liver-
     more, Lawrence Livermore Lab., 1975, 43 pp.
505. Maud, A., "Emulation of the processor for distributed processor/
     memory system on Intel 3000 microprocessor chip set", Report
     AD-A035288/4SL, Air Force Inst. of Tech., Wright-Patterson AFB,
     Ohio, School of Engineering, 1976, 197 pp.
506. Osvalds, G., "Two-phase clock generator and driver synchronize
     the 8080 microprocessor", Electron. Des., 23 (9), 88 (1975).
507. Overman, W. and Estrin, G., "Developing a SARA building block -
     the 8080", Proceedings of the Symposium on Design Automation
     and Microprocessors, 1977, p. 77-86.
508. Peterson, R. L., "Adapting a minicomputer programming system
     to the Intel MCS-4 and MCS-8 microcomputer systems", Report
     UCID-16281, 1972, 19 pp.
509. Pletscher, W., "Hit-Kit from the USA", Micomp, 2 (1), 41-3
     (1977), German, CCA12-15573.
510. Rosenfeld, P. and Hanna, S. J., "Developing modular software
     for the 8080A", Electronics, 49 (20), 83-7 (1976).
511. Seifert, W. M., "Software for the Intel 8080 microprocessor
     resident on machine M (0)", Report LA-6440-MS, Los Alamos
     Scientific Lab., N. Mex. Energy Research and Development Admini-
     stration, 1976, 27 pp.

512. Seim, T. A., "Automation of a brazing process with an Intel 8008 microprocessor", IEEE Trans. Ind. Electron. Control Instrum., IECI-$\underline{22}$ (3), 303-7 (1975).

513. Spann, J. M., "Display octal debugging technique (DODT-80) for the MCS-80 microcomputer", Report UCRL-51970, California Univ., Livermore, Lawrence Livermore Lab., 1976, 21 pp.

514. Szymanski, E. S., "Charts guide 8080 users through the software jungle", EDN, $\underline{21}$ (4), 100, 102 (1976).

515. Thomae, I. H., "Memory mapped I/O and vectored interrupts for Intel 4-bit microcomputer systems", New Electron., $\underline{9}$ (24), 32, 35 (1976), CCA12-12847.

516. Veldstra, B. and Dassen, J. M. H., "Small operating system for a 8080 micro-computer system", Euromicro Newsl., $\underline{2}$ (4), 54-64 (1976).

517. Wahlstrom, B. and Uotila, E., "Interactive program for the simulation of Intel 8080 on a minicomputer system", Proceedings of International Symposium and Course on Mini- and Microcomputers and Their Applications, MIMI 76, p. 94-6.

518. Wakerly, J. F., "Circuit steps program for 8080 debugging", Electronics, $\underline{49}$ (16), 110-11 (1976).

b.   Motorola

519. Fylstra, D. H., "PL/M6800: A compatible high-level language for the Motorola microprocessor", Symposium on Trends and Applications 1976: Micro and Mini Systems, p. 103-6.

520. Lewing, V. C., "Motorola's M6800 microcomputer system", 1974 WESCON Technical Papers, Paper 15/2.

521. Moore, A. W., "The M6800 compatible family for MPU system design", 9th IEEE Computer Society International Conference, COMPCON 74, Digest of Papers, p. 235-8.

522. Sawyer, G., "Microprocessor system made easy", 10th IEEE Computer Society International Conference, COMPCON 75, Digest of Papers, p. 91-4.

523. Schaltegger, P., "The alternative - the Motorola polyvalent development system", Micomp, $\underline{2}$ (1), 32-4 (1977), German, CCA12-15571.

524. Siebert, H. P., '8-bit microcomputer concept - M6800", Elektronik, $\underline{23}$ (10), 387-90 (1974), German.

c.   National Semiconductor

525. Anon., "INS8080A 8-bit N-channel microprocessor", Compute, $\underline{2}$ (9), 10 pp. (1976), CCA12-15585.

526. Anon., "The National Semiconductors MM 5799, 5734 and 5781/82 microprocessors", Autom. & Inf. Ind., no. 50, 38-41 (1976), French, CCA12-12914.

527. Anon., "Functional analysis of National Semiconductor SC/MP microprocessor system", Compute, $\underline{2}$ (10), 1-4 (1976), CCA12-15586.

528. Brown, D., "Getting to grips with microprocessors", Pract.
     Electron., 12 (12), 963-5 (1976), CCA12-15576.
529. Chamberlain, H., "An IMP-16 microcomputer system. II.", Compute,
     2 (7), 17-23 (1976), CCA12-15584.
530. Cushman, R. H., "Take a common sense approach to your µP appli-
     cation", EDN, 21 (19), 85-94 (1976), CCA12-12814.
531. Gerstohofer, R., "Simple logic single-steps SC/MP microprocessor",
     Electronics, 50 (1), 107, 109 (1977).
532. Hughes, P., "A simple 8-bit microprocessor: SC/MP", Elektro-
     technik, 58 (22), 16-18, 22, 24 (1976), German, CCA12-10705.
533. Schmitt, H., Hughes, P. and Zerling, H., "Programming micro-
     processors: software training on SC/MP. I.", Elektrotechnik,
     58 (22), 27-8, 31-3 (1976), German, CCA12-10706.
534. Schmitt, H., Hughes, P. and Zerling, H., "Programming micro-
     processors. Software training on the SC/MP. II.", Elektrotechnik,
     58 (24), 25-8, 30 (1976), German, CCA12-13050.

d.   RCA

535. Callaghan, J. D., Stobbe, W. and Stonaker, W. M., "Studio II -
     Using COSMAC in a home entertainment product", RCA Eng., 22 (5),
     68-70 (1977).
536. Longo, A., Russo, P. M. and Lippman, M. D., "Leased channel with
     microprocessor control", RCA Eng., 21 (3), 60-3 (1975).
537. Rajagopalan, V., Veillette, D. and Provencher, M., "Micropro-
     cessor in thyristor control application", IECI Annual Conference
     Proceedings - Industrial Applications of Microprocessors, 1977,
     p. 13-17.
538. Solomon, L. A., "COSMAC resident software development aids",
     RCA Eng., 22 (5), 37-42 (1977).
539. Swales, N. P. and Tweddle, H., "Register-based output function
     for the COSMAC microprocessor", New Electron., 10(3), 18
     (1977), EEA80-28394.
540. Swales, N. P. and Weisbecker, J. A., "COSMAC - a microprocessor
     for minimum cost systems", RCA Eng., 21 (3), 64-9 (1975).
541. Winder, R. O., "COSMAC - a COS/MOS microprocessor", 1974 Inter-
     national Solid-State Circuits Conference, Digest of Technical
     Papers, p. 64-5.
542. Young, A., "Getting to know the COSMAC microprocessor is simple",
     Electron. Des., 24 (22), 136-45 (1976), CCA12-8917.

e.   Rockwell

543. Anon., "Rockwell parallel processing system (PPS) microcomputers",
     Proceedings, Symposium on Electronics, 1974, p. 327-39.
544. Consolini, L., "The Rockwell PPS4/2 microcalculator system",
     Elettron. Oggi, no. 10, 1299-1303 (1976), Italian, CCA12-12941.
545. Sonn, E. D., "Four-bit chip set cuts cash register's cost and
     size", Electronics, 49 (8), 154, 157 (1976).

546. Wickes, W. E., "PPS – a MOS/LSI 4-bit parallel microprocessor",
     IEEE Region 6 Conference on Minicomputers and Their Applications
     Conference Record, 1973, p. 202-6.
547. Yutzi, G. N., "1200 BPS (FSK) MOS/LSI modem for microprocessor
     or stand-alone applications", IEEE 1975 Region 6 – West USA –
     Conference, Proceedings, 1975, p. 75-7.

f.   Signetics

548. Greenfield, P., Kirk, J. and Zemel, D., "Signetics 2650 micro-
     processor", 9th IEEE Computer Society International Conference,
     COMPCON 74, Digest of Papers, p. 229-34.
549. Hoffmann, H., "Microprocessor 2650, an 8-bit system that can be
     easily handled", Elektronik, $\underline{25}$ (4), 51-2 (1976), German.
550. Jackson, P. J., "Product information: system design aids for
     the 2650 microprocessor", Mullard Tech. Commun., $\underline{14}$ (134),
     168-72 (1977), EEA80-28908.
551. Kroeger, J. H., "Design decisions for the 2650 integrated pro-
     cessor", 1974 WESCON Technical Papers, Paper 15/6.
552. Moran, B. K., "My experience with the 2650", Byte, $\underline{2}$ (11),
     66-7 (1977).
553. Rowe, J., "The Signetics 2650", Electron. Aust., $\underline{38}$ (6), 64-7
     (1976), CCA12-12924.
554. Schmidt, W. P., "Development system for the microprocessor
     2650", Elektronik, $\underline{25}$ (11), 73-4 (1976), German, CCA12-4054.
555. Uimari, D., "Using the 2650 microprocessor", Electron. Des.,
     $\underline{24}$ (18), 70-8 (1976).

g.   Texas Instruments

556. Anon., "From microprocessor to minicomputer: the family of
     Texas Instruments", Autom. & Inf. Ind., no. 50, 27 (1976),
     French, CCA12-12912.
557. Baer, W. J., "Architectural details of a third generation
     microprocessor (for Texas Instruments TMS9900)", Proceedings
     of International Symposium and Course on Mini- and Microcom-
     puters and Their Applications, MIMI 76, p. 49-51.
558. Del Corso, D., "Note on TMS 5501 usage", Euromicro Newsl., $\underline{3}$
     (1), 54-8 (1977).
559. Lofthus, A. and Ogden, D., "Sixteen bit processor performs
     like minicomputer", Electronics, $\underline{49}$ (11), 99-105 (1976).

h.   Zilog

560. Anon., "Third-generation microprocessor from Zilog", Micomp,
     $\underline{1}$ (2), 47, 49, 65 (1976), German, CCA12-4059.
561. Anon., "Modifications to microsoft 8k BASIC 3.1 for Cromemco
     Z-80", Dr. Dobb's J., $\underline{2}$ (2), 20 (1977).

562. Blomeyer-Bartenstein, H. P., "A new microcomputer concept: component – modules – development system", Elektronik, <u>25</u> (11), 83-7 (1976), German, CCA12-6687.

563. Blomeyer-Bartenstein, H. P., "Use of standard memory modules with microcomputer systems", Elektronik, <u>26</u> (3), 75-82 (1977), German, CCA12-15420.

564. Shima, M., Faggin, F. and Ungerman, R., "The Z-80-the third generation microcomputer family", 13th IEEE Computer Society International Conference, COMPCON 76, Digest of Papers, p. 14-19, CCA12-1459.

565. Shima, M. and Ungerman, R., "Z-80 chip set heralds third microprocessor generation", Electronics, <u>49</u> (17), 89-93 (1976).

i.   Others

566. Almes, G. T., Drongowski, P. J. and Fuller, S. H., "Emulating the nova on the PDP 11/40: a case study", 11th IEEE Computer Society International Conference, COMPCON 75, Digest of Papers, p. 53-6.

567. Anon., "The Mostek approach to microprocessors", New Electron., <u>8</u> (3), 22-3 (1975), EEA78-15413.

568. Anon., "The fastest of the microprocessors: the M10800 in ECL and the 'juxtaposible slice' concept", Autom. & Inf. Ind., no. 50, 34-7 (1976), French, CCA12-12913.

569. Biewer, M., "Designers guide to programmed logic for PLS400 systems", Report, Pro-Log Corp., 852 Airport Road, Monterey, CA, 1973.

570. Biewer, M., "Designers guide to programmed logic for MPS800 systems", Report, Pro-Log Corp., 852 Airport Road, Monterey, CA, 1974.

571. Briner, R., "Too cheap? The new computer system HP-3000/Series. II.", Micomp, <u>2</u> (1), 24, 26, 29, 31 (1977), German, CCA12-15570.

572. Colin, A., "Intel 3000 and AM2900 microprocessors - a comparison", Microprocessors, <u>1</u> (5), 287-92 (1977).

573. Edwards, D., "The Mostek F8 (microprocessor)", Electron. Aust., <u>38</u> (9), 86-7 (1976), EEA80-28373.

574. Eklund, M. H. and Trubrisky, L. G., "NUBLU - a microprocessor", Honeywell Comput. J., <u>8</u> (1), 3-5 (1974).

575. Fertl, W., "MPS10 - a modular microcomputer", Elektronik, <u>23</u> (10), 383-6 (1974), German.

576. Gruppuso, F. M. and Sideline, A., "16-bit single chip microprocessor", 10th IEEE Computer Society International Conference, COMPCON 75, Digest of Papers, p. 83-6.

577. Keith, G. and others, "SKX-3000 LSI computer", 1972 WESCON Technical Papers, Paper 26/2.

578. Kuznia, C., Kober, R. and Kopp, H., "SMS 101 - a structured multimicroprocessor system with deadlock-free operation scheme", 3rd Annual Symposium on Computer Architecture, 1976, p. 122, CCA12-3906.

579. Lilley, R. W., "An assembler for the MOS technology 6502 micro-processor as implemented in JOLT (TM) and RIM-1 (TM)", Report N77-12030/1SL, Ohio Univ., Athens, Dept. of Electrical Engineering, 1976, 41 pp.

580. Mick, J. R., "AM2900 bipolar microprocessor family", 8th Workshop on Microprogramming, Proceedings, Micro 8, 1975.

581. Pfeiffer, E. A., "Control logic for µP enables single-cycle operation", Electron. Des., 24 (21), 78 (1976), CCA12-3897.

582. Rivers-Latham, P. W., "Microprocessor-based fast Fourier transform microsystem", Minicomputers and Small Business Systems, Forum, 1976, p. 403-29.

583. Rodler, O., "A comprehensive 16-bit microcomputer", Elektronik, 25 (11), 88-91 (1976), German, CCA12-6613.

584. Rothfuss, H., "Microcomputer problems stabilized KIM 1 - micro-computer module", Und-oder-Nor & Steuerungstech., no. 10, 53-4 (1976), German, CCA12-12935.

585. Rowe, J., "The Fairchild F8", Electron. Aust., 38 (8), 84-5, 87 (1976), CCA12-15558.

586. Springer, J., "The use of the AM2901 microprocessor", Autom. & Inf. Ind., no. 53, 13-17 (1977), French, CCA12-15547.

VII. PERSONAL/HOBBY COMPUTERS

587. Anon., "Microcomputer in kit form: a solution to put information processing at the disposal of everyone?", Autom. & Inf. Ind., no. 52, 31-3 (1976), French, CCA12-12915.

588. Curran, L., "Personal computers mean business", Electronics, 50 (7), 89-96 (1977), CCA12-12921.

589. Davis, S., "Simplified hardware design and realistic sound effects characterize µC-based games", EDN, 22 (16), 20-7 (1977).

590. Garland, H., "Design innovations in personal computers", Computer, 10 (3), 24-7 (1977).

591. Gilder, J. H., "Home computers: from a bag of parts to a system you simply plug in", Electron. Des., 25 (19), 36-8 (1977).

592. Kagan, C. A. B. and Schear, L. G., "The home reducer - a scenario on the home use of computers", Proceedings, National Computer Conference, 1973, p. 759-63.

593. Kane, G. R., "Personal minicomputers to aid undergraduate instruction in computer system design", IEEE Trans. Educ., E-20 (1), 10-13 (1977).

594. Kaplan, A. R., "Home computers versus hobby computers", Datamation, 23 (7), 72-5 (1977).

595. Kay, A. C., "Microelectronics and the personal computer", Sci. Am., 237 (3), 230-44 (1977).

596. Krausman, D. T., "The microcomputer kit: An excellent small system development tool", Behav. Res. Methods & Instrum., 8 (6), 501-7 (1976), CCA12-15548.

597. Masek, A., "Towards personalised computing systems", Neue Tech., 18 (11), 727-9 (1976), German, CCA12-15574.

598. Matlack, R. J. and Paul, D. K., "Minicomputer industry service (MCIS) - specialized industry areas - the hobby computer market", Report, Dataquest, Inc., Menlo Park, CA, 1977, 5 pp.

599. Negroponte, N., "Idiosyncratic systems toward personal computers and understanding context", Report, MIT, Cambridge, Mass., 1975, 23 pp.

600. Rotheneuecher, O. H., "The IBM 5100 portable computer and competitors' products", Report L751201, Arthur D. Little, Inc., Cambridge, Mass., 1975, 6 pp.

601. Steger, J. P., "Development aids for users of microprocessors. II.", Elektroniker, 15 (12), EL26-33 (1976), German, CCA12-10703.

602. Terrell, P., "Neighborhood computer stores - the answer to microcomputing marketing", Proceedings of the 1977 National Computer Conference, AFIPS, p. 999-1004.

603. Warren, J. C., Jr., Peppe, M. E. and Fry, J. P., "Personal computing - an overview for computer professionals", Proceedings of the 1977 National Computer Conference, AFIPS, p. 493-8.

604. Weisbecker, J., "A practical, low-cost, home/school microprocessor system", Computer, 7 (8), 20-31 (1974).

605. Yasaki, E. K., "Microcomputers: for fun and profit", Datamation, 23 (7), 66-71 (1977).

VIII. PROGRAMMING

606. Allen, B., "Standard process controller can be programmed in field", Electronics, 49 (8), 164, 169-70 (1976).

607. Allen, P. G., "Comparison of a microprocessor basic to other basic implementations", Microcomputer '77, Conference Record, 1977, p. 64-6.

608. Allison, D. R., "Software issues in LSI microprocessor design", 12th IEEE Computer Society International Conference, COMPCON 76, Digest of Papers, p. 15-18.

609. Anderson, L. H., "Development of a portable compiler for industrial microcomputer systems", Proceedings, National Computer Conference, 1975, p. 33-40.

610. Andrews, M., McCormick, S. F. and Taylor, G. D., "Evaluation of the square root function on microprocessors", Proceedings of the Annual Conference ACM, 1976, p. 185-91.

611. Anon., "Portable microcomputer with APL language offers large computing power", Comput. Des., 12 (11), 122-3 (1973).

612. Anon., "Software becomes the real challenge", Electronics, 49 (8), 104-8 (1976).

613. Anon., "System languages for microprocessors: considerations and trends", Comput. Des., 15 (7), 87-96 (1976).

614. Anon., "New control options - 1. Programmability closes the gap between sequence and process control", Mod. Plast., 53 (6), 42-4 (1976).

615. Baig, W. G., "Microprocessor program development support", Microprocessors, 1 (4),259-62 (1977).

616. Bass, C. and Brown, D., "Perspective on microcomputer software", Proc. IEEE, 64 (6), 905-9 (1976).

617. Bleasdale, E., "Development of high-level languages", Microprocessors, 1 (4), 241-5 (1977).

618. Bond, J., "Designer's guide to software for the hardware designer - 1, 2", EDN, 19 (11), 40-4 (1974); ibid, 19 (15), 51-6 (1977).

619. Bourret, S. C., "Write your own macroinstructions and simplify µC programming", EDN, 21 (9), 103-7 (1976).

620. Brookshire, J. R., "Multipurpose digital microprocessor emulator", Report AD-A031782/6SL, Army Missile Research Development and Engineering Lab., Redstone Arsenal, Ala., 1976, 119 pp.

621. Burger, P., "System integration and testing with microprocessors - 2. Software aspects", IEEE Trans. Ind. Electron Control Instrum., IECI-22 (3), 364-7 (1975).

622. Cannon, L. E. and Kreager, P. S., "Using a microprocessor: A real-life application. Part 2 - Software", Comput. Des., 14 (10), 81-9 (1975).

623. Carter, L., "Microprocessor programming trade-offs - which language is best?", EDN, 21 (18), 101-3 (1976).

624. Casilli, G. and Kirn, W., "Microcomputer software development", Digital Des., 5 (7), 50-2 (1975).

625. Cassell, D. A., "Microcomputer programming", Mod. Data, 8 (1), 49-51 (1975).

626. Cassell, D. A., "Only small, clever programs need apply", Digital Des., 5 (3), 24-6 (1975).

627. Chu, Y., "Design of a microprogrammed lexical microprocessor", Micro 8: 8th Annual Workshop on Microprogramming, Proceedings, 1975, p. 26-9.

628. Chu, Y. and Cannon, E. R., "Interactive high-level language direct-execution microprocessor system", IEEE Trans. Software Eng., SE-2 (2), 126-34 (1976).

629. Clark, J., "Microprocessor software - from a user's point of view", 13th IEEE Computer Society International Conference, COMPCON 76, Digest of Papers, p. 378-80.

630. Clarke, D. W., "High-level languages and simulators for microprocessors", Minicomputers and Small Business Systems, 1976, p. 207-19.

631. Cleaveland, J. C. and Satten, C. D., "The design and implementation of a simple programming language for microcomputers", Proceedings of the 1977 National Computer Conference, AFIPS, Vol. 46, p. 629-36.

632. Cohen, D. and Lio, M. T., "Emulation of computer networks by microprogrammable microcomputers", 7th Annual Workshop on Microprogramming, Conference Record, 1974, p. 159-67.

633. Conley, S. W., "Portable microcomputer cross-assemblers in BASIC", Computer, 8 (10), 32-40 (1975).

634. Crenshaw, J. W., "Matrix chart brings order to the 8080 set", EDN, 22 (11), 206 (1977).

635. Dawes, N. W., "A simple network interacting programs' executive
     (SNIPE)", Software-Pract. Exper., 7 (3), 341-5 (1977).
636. Dawson, D. E. and Parrish, E. A., "Programming system for a
     microcomputer using a time-shared computer", IEEE Trans.
     Instrum. Meas., IM-24 (3), 272-3 (1975).
637. de la Guardia, M. F. and Field, J. A., "A high level language
     oriented multiprocessor", Proceedings of the 1976 International
     Conference on Parallel Processing, p. 256-62, CCA12-12869.
638. Denker, J., "Direct microprocessor link loads timeshared
     programs", Electronics, 50 (6), 110-11 (1977).
639. DesRochers, G., "Microprogramming helps squeeze more from your
     equipment dollar", EDN, 21 (17), 102-3, 105 (1976), CCA12-6620.
640. Deuel, D. R. and Gault, J. W., "Integrated hardware/software
     microprocessor based development system", 7th IEEE Computer
     Society International Conference, COMPCON 75, Digest of Papers,
     p. 163-5.
641. Dolhoff, T., "Microprocessor software: how to optimize timing
     and memory usage", Digital Des., 7 (3), 44-6, 50-6 (1977).
642. Donaghey, L. F., Bobba, G. M. and Rubin, X. K., "Decision-making
     with flags in process control", Comput. Des., 15 (12), 77-83
     (1976).
643. Falk, H., "Microcomputer software makes its debut", IEEE
     Specttrum, 11 (10), 78-84 (1974).
644. Feger, O., "Basic procedure for microprocessor programming",
     Elektronik, 26 (3), 59-61 (1977), German, CCA12-13049.
645. Ferguson, M. E., "Pittman 6800 tiny BASIC mods for SWTPC PR-40
     printer", Dr. Dobb's J., 2 (2), 7 (1977).
646. Fisher, E. R., "High level languages in microcomputer automa-
     tion", Report UCRL-77868, California Univ., Livermore, Lawrence
     Livermore Lab., 1976, 6 pp.
647. Gallacher, J., "Software for hardware engineers", Microproc-
     cessors, 1 (3), 169-79 (1977).
648. Gibbons, J., "When to use higher-level languages in microcom-
     puter-based systems", Electronics, 48 (16), 107-11 (1975).
649. Gibbons, J. H., "Software components for microcomputer systems",
     13th IEEE Computer Society International Conference, COMPCON 76,
     Digest of Papers, p. 372-5.
650. Goksel, K. and Parrish, E. A., "Microcomputer programming
     system", IEEE Trans. Instrum. Meas., IM-23 (2), 177-8 (1974).
651. Gordon, A., "Process control language for use with industrial
     microcomputers", SME Technical Papers, Series MS76-204, 1976,
     8 pp.
652. Hashizume, B., "Floating point arithmetic", Byte, 2 (11), 76-8,
     180-2, 184-8 (1977).
653. Hastings, A. and Kozlay, D., "A software development instrument",
     11th IEEE Computer Society International Conference, COMPCON 75,
     Digest of Papers, p. 201-4.
654. Hawkins, C. J. B., "Host development systems - the software path
     to tomorrow", Microprocessors, 1 (4), 246-50 (1977).

655. Hennessy, J. L., Kieburtz, R. B. and Smith, D. R., "TOMAL: A
     task-oriented microprocessor applications language", IEEE
     Trans. Ind. Electron Control Instrum., IECI-22 (3), 283-9 (1975).
656. Holt, O. D., Pokoski, J. L. and Cordell, D. L., "Software
     development system for microcomputers", IEEE Trans. Ind. Elec-
     tron Control Instrum., IECI-22 (3), 279-82 (1975).
657. Holt, R. C., "ZED: A very modest implementation language",
     Infor. J., 14 (2), 169-82 (1976).
658. Jaworsky, A., "System for aiding microprocessor programming",
     Rev. Tech. Thomson CSF, 8 (2), 255-87 (1976), French.
659. Jayakumar, M. S. and McCalla, T. M., Jr., "Simulation of micro-
     processor emulation using GASP-PL/P", Computer, 10 (4), 20-6
     (1977).
660. Jones, A. K., Chansler, R. J., Jr., Durham, I., Feiler, P. and
     Schwans, K., "Software management of Cm* - A distributed multi-
     processor", Proceedings of the 1977 National Computer Conference,
     AFIPS, p. 657-64.
661. Jones, L. H., "Instruction sequencing in microprogrammed com-
     puters", AFIPS National Computer Conference, Proceedings, Vol.
     44, 1975, p. 91-8.
662. Jones, L. H., "Survey of current work in microprogramming",
     Computer, 8 (8), 33-7 (1975).
663. Kaegi, T., "Programming of microcomputers", Elektroniker,
     14 (2), EL19-27 (1975), German.
664. Kildall, G. A., "High-level languages simplifies microcomputer
     programming", Electronics, 47 (13), 103-9 (1974).
665. Kildall, G. A., "Systems languages: Management's key to con-
     trolled software evolution", 1974 WESCON Technical Papers,
     Paper 19/2.
666. Kildall, G. A., "Microcomputer software design - a checkpoint",
     Proceedings, National Computer Conference, 1975, p. 99-106.
667. Kirk, B. R., "Microprocessor software development method using
     a specification language", Microprocessors, 1 (6), 353-61 (1977).
668. Korn, G. A., "Proposed method for simplified microcomputer
     programming", Computer, 8 (10), 43-52 (1975).
669. Kornstein, H., "Software and programming", In: Microcomputers:
     Fundamentals and Applications, MINICONSULT, London, p. 73-88
     (1974), Publ. 1975.
670. Lebovici, D., "Microcomputer software development", Proceedings,
     1st National Microprocessor Conference. Microprocessors:
     Economics, Technology, Applications, 1974.
671. Lee, H. L., "String handling for Pittman's 6800 tiny BASIC",
     Dr. Dobb's J., 2 (2), 5 (1977).
672. Leonard, M. G., "Microcomputer software", Mach. Des., 48 (27),
     70-6 (1976).
673. Leventhal, L., "Put microprocessor software to work", Electron.
     Des., 24 (16), 58-64 (1976).

674. Levine, S. T., "With new, powerful software, designers can 'see'
     a system work before it's actually built", Electron. Des., $\underline{23}$
     (10), 80-7 (1975).
675. Levine, S. T., "Assembly language for µPs", Electron. Des., $\underline{23}$
     (26), 58-63 (1975).
676. Mallett, P. W. and Lewis, T. G., "Considerations for implement-
     ing a high level microprogramming language translation system",
     Computer, $\underline{8}$ (8), 40-51 (1975).
677. Martinez, R., "Look at trends in microprocessor/microcomputer
     software systems", Comput. Des., $\underline{14}$ (6), 51-7 (1975).
678. Matelan, M. N., "Generation of firmware for control micropro-
     cessors", Report UCRL-77580, California Univ., Livermore,
     Lawrence Livermore Lab., 1975, 9 pp.
679. Merrill, R. M., "Programming a microprocessor", Proceedings,
     Symposium on Electronics, 1974, p. 31-44.
680. Mueller, R. A. and Johnson, G. R., "Generator for microprocessor
     assemblers and simulators", Proc. IEEE, $\underline{64}$ (6), 910-20 (1976).
681. Nauful, E. S., "Software support for microprocessors poses new
     design choices", Comput. Des., $\underline{15}$ (10), 93-8 (1976), CCA12-3866.
682. Ogdin, J. L., "Survey of microprogrammable microprocessors
     reveals ultimate software flexibility", EDN, $\underline{19}$ (14), 69-74
     (1974).
683. Ogdin, J. L., "PL/M vs. assembly language - Part 1", New Logic
     Notebook, $\underline{1}$ (3), 1-25 (1974).
684. Ogdin, J. L., "PL/M vs. assembly language - Part 2", New Logic
     Notebook, $\underline{1}$ (4), 1-27 (1974).
685. Ogdin, J. L., "Programming techniques", New Logic Notebook,
     $\underline{1}$ (7), 1-33 (1975).
686. Patterson, G. W., "Microprocessor software - a manufacturer's
     perspective", 13th IEEE Computer Society International Con-
     ference, COMPCON 76, Digest of Papers, p. 368-71.
687. Pedroso, L. R. B., "ML80: A structured machine-oriented micro-
     computer programming language", Report AD-A020055/0SL, Naval
     Postgraduate School, Monterey, Calif., 1975, 216 pp.
688. Peterson, N. E., "Avoiding missteps in programming and memory",
     Electronics, $\underline{50}$ (3), 110-12 (1977), CCA12-10648.
689. Pokoski, J. L. and Holt, O., "Developing software for micro-
     computer applications", Comput. Des., $\underline{14}$ (3), 88-90 (1975).
690. Popper, C., "Advanced microcomputer software techniques",
     Notes for NEC Microcomputer Institute, 1974.
691. Popper, C., "SMAL - a structured macro-assembly language for
     a microprocessor", 9th IEEE Computer Society International
     Conference, COMPCON 74, Digest of Papers, p. 147-51.
692. Potton, A., "The software challenge for microprocessor engineers",
     Electron. Engineering, $\underline{48}$ (586), 49, 53-5 (1976), CCA12-3891.
693. Pracht, C. P., "Distributed microprocessor-based control system
     programmable by the process engineer", Advances in Instrumenta-
     tion, Vol. 31, 1976, Proceedings of 31st Annual ISA Conference
     and Exhibition, Pt. 1, Paper 515, 7 pp.

694. Pyeron, J. J., Jr. and Rootenberg, J., "Multi-byte reverse
     direction multiplication routine for the Fairchild F8 micro-
     processor", Symposium on Trends and Applications 1976: Micro
     and Mini Systems, p. 46-58.
695. Razouk, R., "The graph model of behavior simulator", Proceedings
     of the Symposium on Design Automation and Microprocessors, 1977,
     p. 67-76.
696. Reyling, G., Jr., "Extend LSI-processor capabilities", Electron.
     Des., 22 (22), 90-5 (1974).
697. Riley, W. B., "Blending hardware and software", Electronics,
     47 (14), 103-4 (1974).
698. Rose, C., "N.mPc: An adjustable software system to support the
     development of microprocessor-based systems - an overview",
     Proceedings of the Symposium on Design Automation and Micro-
     processors, 1977, p. 1.
699. Roybal, P., "Overview of available microcomputer software",
     Proceedings, 1st National Microprocessor Conference, Micro-
     processors: Economics, Technology, Applications, 1974.
700. Roybal, P., "μP system software: Who, what, when, where, why
     and . . . huh?", EDN, 20 (21), 90-4 (1975).
701. Seim, T. A., "Assembling microprocessor software with mini-
     computers", Symposium on Trends and Applications 1976: Micro
     and Mini Systems, p. 93-102.
702. Steger, J. P., "Introduction to microprocessor programming",
     Electron. Eng., 47 (572), 43-7 (1975).
703. Thornley, A., "Matching the program to a microcomputer-based
     control system", Control Instrum., 8 (3), 38-9, 41 (1976).
704. Titus, J. A., "Starting μP software", Electron. Des., 24 (25),
     74-7 (1976), CCA12-10650.
705. Townzen, D. A., "A task-scheduling executive program for micro-
     computer systems", Comput. Des., 16 (6), 194, 198-9, 202 (1977).
706. Trifari, J., "Running minicomputer programming through micros",
     Electron. Prod., 18 (11), 41-4 (1976).
707. Ulrickson, R. W., "Software for microprocessors - 1. Software
     is a vital part of any computer", Electron. Des., 25 (1), 90-6
     (1977).
708. Ulrickson, R. W., "Your microcomputer's I/O software may well
     determine its over-all performance. And in real-time applications,
     I/O timing often means everything", Electron. Des., 25 (7),
     76-83 (1977).
709. Weaver, A. C., Tindall, M. H. and Danielson, R. L., "A BASIC
     language interpreter for the Intel 8008 microprocessor", Report
     PB 235 874/5SL, 1974, 53 pp.
710. Weiss, C. D., "Microcomputer programming I: Basics", Notes
     for NEC Professional Growth in Engineering Institute, 1973.
711. Weiss, C. D., "Microcomputer programming II: Producing software",
     Notes for NEC Professional Growth in Engineering Institute, 1973.
712. Weiss, C. D., "Software for MOS/LSI microprocessors", Electron.
     Des., 22 (7), 50-7 (1974).

713. Weiss, C. D., "MOS/LSI microcomputer coding", Electron. Des.,
     22 (8), 66-71 (1974).
714. Weiss, C. D., "Basic microcomputer software", Electron. Des.,
     22 (9), 142-6 (1974).
715. Weissberger, A. J., "User microprogram developments for an LSI
     processor", EDN, 19 (24), 39-44 (1974).
716. Williams, T. L. and Conley, S., "MACL1 - a portable macro
     language for microprocessors", Microcomputer '77, Conference
     Record, p. 54-9.
717. Wong, K. P. and Humpage, W. D., "Simulating microprocessor
     elements on a host computing system", IFAC Symposium - Auto-
     mation Control and Protection of Electrical Power Systems, 1977,
     p. 109-13.

     IX.   SOFTWARE TOOLS

718. Anderson, J. A. and Lipovski, G. J., "Virtual memory for micro-
     processors", 2nd Annual Symposium on Computer Architecture,
     Conference Proceedings, 1975, p. 80-4.
719. Anon., "Programming station for computers", Radio Electron.,
     24 (17), 552 (1976), Dutch, CCA12-12939.
720. Anon., "8080 subroutine package", Dr. Dobb's J.,   (2), 47 (1977).
721. Avivi, N., "Hardware to assist the program debugging effort of
     microprocessor based system", 9th Convention of Electrical
     and Electronic Engineers in Israel, 1975, p. B-1-4/1-5, Hebrew,
     CCA11-31566.
722. Barnes, J. and Bergquist, R., "Unite µP hardware and software",
     Electron. Des., 24 (7), 74-6 (1976).
723. Barnes, J. and Gregory, V., "Basic microprocessor test tool:
     the portable debugger", EDN, 20 (15), 51-6 (1975).
724. Baxter, W. F., "Microprocessor controller/debug system",
     Report AD-A029010/6SL, Army Missile Research Development and
     Engineering Lab., Redstone Arsenal, Ala., 1976, 57 pp.
725. Bourque, J., "Case study: Using microprocessors in numerical
     control tools", Proceedings, 1st National Microprocessor Con-
     ference. Microprocessors: Economics, Technology Applications,
     1974.
726. Brand, H., "MCS-8 microcomputer system display octal debugging
     technique. DODT for MCS-8 microcomputer", Report UCID-17129,
     California Univ., Livermore, Lawrence Livermore Lab., 1974,
     17 pp.
727. Campos, I., "Specialization of SARA for software synthesis",
     Proceedings of the Symposium on Design Automation and Micro-
     processors, 1977, p. 87-94.
728. De Marchin, P., "Multi-level nesting of subroutines in a one-
     level microprocessor", Comput. Des., 15 (2), 118, 120, 122,
     124 (1976).
729. Dusan, R., "Debug that microcomputer system with a mini",
     Electron. Des., 23 (12), 74-7 (1975).

730. Falkoff, D., Kerllenevich, N. and Kreiker, P., "Exploit existing Nova software", Electron. Des., 25 (19), 54-64 (1977).

731. Farnbach, W. A., "Logic state analyzers: a new instrument for analyzing sequential digital processes", IEEE Trans. Instrum. Meas., IM-24 (4), 353-6 (1976).

732. Frenzel, L. E., "Dip switches and diodes form programmable ROM", Electronics, 47 (11), 126-7 (1975).

733. Gladstone, B., "Software development: It pays to have the right tool at the right time -1.", EDN, 21 (12), 91-9 (1976).

734. Gladstone, B., "Software development: Some tools do big jobs automatically - 2.", EDN, 21 (15), 45-54 (1976).

735. Goksel, K. and Parrish, E. A., "Microcomputer programming system", IEEE Trans. Instrum. Meas., IM-23 (2), 177-8 (1974).

736. Gordon, H. T., "Stringout mods for Espinosa's 6502 routine", Dr. Dobb's J., 2 (2), 8 (1977).

737. Harrison, J. M., "Small program ensures data validity; light-emitting memory aids microprocessor debugging", EDN, 22 (9), 106-7 (1977).

738. Hastings, A. and Kozlay, D., "Software development instrument", 11th IEEE Computer Society International Conference, COMPCON 75, Digest of Papers, p. 201-4.

739. Holt, O. D., Pokoski, J. L. and Cordell, D. L., "Software development system for microcomputers", IEEE Trans. Ind. Electron Control Instrum., IECI-22 (3), 279-82 (1975).

740. Lewis, J. W. and Davis, J. E., "Development of a microprocessor-based clinical analyzer controller", Advances in Instrumentation, Vol. 31, 1976, Proceedings of 31st Annual ISA Conference and Exhibition, Pt. 1, Paper 510, 4 pp.

741. Madea, G., "Advanced software tools - how should they be provided?", 1975 WESCON Technical Papers, Paper 6/1.

742. Morgan, D. R., "Microprocessor applications in laboratory testing and debugging aids", 7th IEEE International Symposium on Circuits and Systems, Proceedings, 1974, p. 293-5.

743. Neuman, C. P. and Dickey, T. E., "SIMIC - a microcomputer development and evaluation tool", Symposium on Trends and Applications 1976: Micro and Mini Systems, p. 57-71.

744. Raphael, H. A. and Hou, S., "Test equipment for microcomputers", Mach. Des., 48 (19), 78-81 (1976).

745. Schneider, P., "Software aids for the development of user software for microprocessors", Proceedings, Symposium on Electronics, 1974, p. 183-94.

746. Shaw, C. B., "Making microcomputer programming easier", Digital Des., 7 (11), 66, 68, 70, 72 (1977).

747. Stockton, J. R., "Some simple microprocessor kit enhancements", Microprocessors, 1 (7), 415-20 (1977).

748. Tallman, J. L., "How 'debug' software can make a μP breadboard intelligent", EDN, 21 (4), 53-62 (1976).

749. Ulrickson, R., "Software modules are the building blocks", Electron. Des., 25 (3), 62-6 (1977).

750. Ward, S. A. and Jessel, P. G., "Domain specific systems – an
     approach to microprocessor software development", Report,
     MIT, Cambridge, Mass., 1976, 8 pp.
751. Watson, I. M., "Comparison of commercially available software
     tools for microprocessor programming", Proc. IEEE, $\underline{64}$ (6),
     910-20 (1976).
752. Weissberger, A. J., "User microgram development for an LSI
     processor", EDN, $\underline{19}$ (24), 39-44 (1974).
753. Wozniak, S., "SWEET16: the 6502 dream machine", Byte, $\underline{2}$ (11),
     150-2, 154, 159 (1977).
754. Wyland, D. C., "Employ µP software tools properly", Electron.
     Des., $\underline{23}$ (26), 50-5 (1975).
755. Zinniker, R., "Microprocessor practice: When does it pay to
     use a subroutine?", Elektroniker, $\underline{15}$ (12), EL24 (1976),
     German, CCA12-10702.
756. Zurkow, J. L., "Hardware helps in tracing microprocessor pro-
     gram", Electronics, $\underline{49}$ (13), 105, 107 (1976).

X.   HARDWARE TOOLS

757. Anderson, L. H., "Development of a portable compiler for in-
     dustrial microcomputer systems", AFIPS National Computer
     Conference, Proceedings, Vol. 44, 1975, p. 33-40.
758. Baum, A. and Senzig, D., "Hardware considerations in a micro-
     computer multiprocessing system", 10th IEEE Computer Society
     International Conference, COMPCON 75, Digest of Papers, p. 27-30.
759. Bisset, S., "Structuring an efficient and economical interrupt
     system using LSI hardware", 10th IEEE Computer Society Inter-
     national Conference, COMPCON 75, Digest of Papers, p. 57-9.
760. Broderick, W. J., "MDS: an advanced development tool", 11th
     IEEE Computer Society International Conference, COMPCON 75,
     Digest of Papers, p. 209-10.
761. Cavlan, N., "Structure and applications of field programmable
     logic arrays", Microelectron. & Reliab., $\underline{15}$ (4), 285-95 (1976).
762. Cushman, R. H., "Are you taking advantage of available µP
     development tools?", EDN, $\underline{21}$ (6), 77-83 (1976).
763. Davie, H., "Microprocessor paper-tape editor", Microprocessors,
     $\underline{1}$ (7), 421-3 (1977).
764. Davies, A. C. and Fung, Y. T., "Interfacing a hardware multi-
     plier to a general-purpose microprocessor", Microprocessors,
     $\underline{1}$ (7), 425-32 (1977).
765. Drongowski, P. J., "Capability requirements in a multimicro
     processor, hardware/software simulation environment", Proceed-
     ings of the Symposium on Design Automation and Microprocessors,
     1977, p. 2-5.
766. Dukarski, R., "Microcomputer system development hardware",
     Proceedings, 1st National Microprocessor Conference. Micro-
     processors: Economics, Technology, Applications, 1974.

767. Grodowski, R. A. and Struger, O. J., "Microprocessor in pro-
     grammable logic/computing controllers for the industrial en-
     vironment", IEEE Trans. Ind. Electron Control Instrum., IECI-22
     (3), 318-26 (1975).
768. Johnson, D. W., "Go from flow chart to hardware", Electron.
     Des., 24 (18), 90-5 (1976).
769. Judmann, K. P., "BH 76 - a good value-for-money µP small design
     equipment. An aid in hard- and software development", Radio
     Elektron. Sehau, 52 (11), 28-31 (1976), German, CCA12-12850.
770. Kaestner, O., "Implementing branch instructions with polynomial
     counters", Comput. Des., 14 (1), 69-75 (1975).
771. Kline, B., Maerz, M. and Rosenfeld, P., "In-circuit approach
     to the development of microcomputer-based products", Proc. IEEE,
     64 (6), 937-42 (1976).
772. Lee, E. D., "When programming microprocessors, use your hard-
     ware background", Electronics, 49 (14), 93-100 (1976).
773. Levan, D. and Majithia, J. C., "Iterative algorithms for function
     computations", Proc. IEEE, 62 (6), 861-2 (1974).
774. Logan, J. D. and Kreager, P. S., "Using a microprocessor: a
     real-life application. Part I - Hardware", Comput. Des., 14 (9),
     69-77 (1975).
775. Moon, J. B. and Gray, L. C., "Hardware aids software overtones",
     11th IEEE Computer Society Conference, COMPCON 75, Digest of
     Papers, p. 205-8.
776. Morton, J., "µP timing is a snap with this new hardware building
     block", EDN, 20 (17), 55-9 (1975).
777. Nichols, J. L., "Hardware versus software for microprocessor
     I/O", Comput. Des., 15 (6), 102-4, 107 (1976).
778. Parasuraman, B., "Hardware multiplication techniques for micro-
     processor systems", Comput. Des., 16 (4), 75-82 (1977).
779. Risch, D. M., "Counter keeps track of microprocessor interrupts",
     Electronics, 47 (26), 102 (1974).
780. Rosner, R., "Do-it-yourself PROM programming simplifies µP
     development systems", EDN, 21 (1), 33-6 (1976).
781. Schmid, H., "Speed microcomputer multiplication", Electron. Des.,
     23 (9), 44-51 (1975).
782. Schwartz, M. and Winter, K., "Using a microprocessor at speeds
     beyond its apparent intrinsic limit", Comput. Des., 15 (6), 106,
     108, 110 (1976).
783. Smith, M. F., "Using interrupts for real time clocks", Byte, 2
     (11), 50, 52-3 (1977).
784. Sneed, J. R., "Adding an interrupt driven real time clock", Byte,
     2 (11), 72-4 (1977).
785. Sohrabji, N., "Macro processor simplifies microcomputer pro-
     gramming", Comput. Des., 15 (8), 108-10, 112 (1976).
786. Tenny, R., "Hexadecimal encoder debounces keyboard", Electronics,
     49 (5), 101, 103 (1976).
787. Tenny, R., "Need a µP-TTY program development tool? Build a
     'RAM loader'", EDN, 21 (9), 81-5 (1976).

788. Thomae, I. H., "Add vectored interrupts and memory-mapped I/O to 4-bit µP's", EDN, <u>21</u> (18), 55-8 (1976).

789. Torku, K. and Zaky, S. G., "Implementation of a standard interface for control of experiments using a microprocessor", Proceedings of International Symposium on Mini and Macro Computers: MIMI '76, p. 26-30, Publ. 1977.

790. Travis, T., "Patching a program into a ROM", Electron. Des., <u>24</u> (18), 98-101 (1976).

791. Wagner, T. J., "Verification of hardware designs thru symbolic manipulation", Proceedings of the Symposium on Design Automation and Microprocessors, 1977, p. 50-3.

792. Wyland, D. C., "Increase microcomputer efficiency", Electron. Des., <u>23</u> (23), 70-5 (1975).

793. Zorkow, J. L., "Hardware helps in tracing microprocessor program", Electronics, <u>49</u> (13), 105, 107 (1976).

XI.   INTERFACING

794. Abbott, D. L., "Microprocessors and CAMAC - modular peripheral interfaces with distributed intelligence", 8th IEEE Computer Society International Conference, COMPCON 74, Digest of Papers, p. 307-10.

795. Ahlgren, D. R., "Restored processor function saves logic and improves performance in microcomputer system", Comput. Des., <u>14</u> (8), 88-9 (1975).

796. Anon., "Microcomputer with keyboard and display unit", Elektronik, <u>25</u> (11), 91-2 (1976), German, CCA12-6688.

797. Aspinall, D., "Interfacing of micro-processors to I/O services", Proceedings, Symposium on Electronics, 1974, p. 19-29.

798. Bainter, J., "Distribute the I/O management task with intelligent programmable interface devices", In: New Components and Subsystems for Digital Design: a Report, Technology Service, Santa Monica, CA, p. 81-93 (1975).

799. Baradello, C., Krutz, R. and Sauer, W., "Interfacing a microcomputer to a minicomputer for control software development", Advances in Instrumentation, Vol. 31, 1976, Proceedings of 31st Annual ISA Conference and Exhibition, Pt. 1, Paper 551, 10 pp.

800. Campbell, C. K. and Sarna, S. K., "Minicomputer interfacing for electrical engineering undergraduates", IEEE Trans. Educ., E-<u>20</u> (1), 13-17 (1977).

801. Carlo-Stella, G. R. and Keiles, Y., "Process, operator and computer interface for a distributed microprocessor-based system", Proceedings of the Computer Interface Instrumentation Symposium, 1976, p. 91-5.

802. Chroust, G., "Microprogramming - an interface property", Euromicro Newsl., <u>2</u> (4), 48-53 (1976), CCA12-6618.

803. Davies, P., "How to interface the printer to your mini or microcomputer", 1975 WESCON Technical Papers, Paper 21/2.

804. De Shon, W. E., "Use TTY or CRT interchangeably on μP system", Electron. Des., <u>24</u> (24), 174-5 (1976), CCA12-12833.

805. Dufty, P. O. and Vranesic, Z. G., "Microcomputer interfacing module", Electron. Lett., <u>7</u> (19), 589-90 (1971).

806. Ebright, A., "Programmable peripheral interface IC's boost your μC's flexibility", EDN, <u>22</u> (7), 79-84 (1977).

807. Evans, B. B., "Advantages and problems of interfacing micro and mini computers", Proceedings, Symposium on Electronics, 1974, p. 359-68.

808. Franson, P., "The world of interfacing - microcomputers don't live by microprocessors alone", EDN, <u>21</u> (14), 38-47 (1976), EEA80-3006.

809. Fronheiser, K., "Simplify add-on peripheral controllers with LSI data-communications circuits", Electron. Des., <u>23</u> (10), 122-7 (1975).

810. Fullagar, D., Bradshaw, P., Evans, L. and O'Neill, B., "Interfacing data converters and microprocessors", Electronics, <u>49</u> (25), 81-9 (1976).

811. Gusman, D., "Marry your μP to monolithic A/Ds", Electron. Des., <u>25</u> (2), 82-6 (1977).

812. Haas, D., "Programmable interface components make microprocessor systems more powerful", Elektronik, <u>25</u> (11), 100-2 (1976), German, EEA80-9008.

813. Hammond, D., "Chip scans keyboard without hardware interface", Electronics, <u>50</u> (1), 110-12 (1977).

814. Hardy, M. E., Jr., "Microprocessor technology to extend the utility of computer peripherals", 8th IEEE Computer Society International Conference, COMPCON 74, Digest of Papers, p. 299-302.

815. Haynes, G., "New analogue component concepts for digital processing systems", Microelectron. & Reliab., <u>16</u> (1), 57-67 (1977).

816. Heid, J. P., "A hybrid computer interface for microprocessors", Proceedings of the 1977 National Computer Conference, AFIPS, Vol. 46, p. 207-16.

817. Hill, S. A., "Multiprocess control interface makes remote μP command possible", EDN, <u>21</u> (3), 87-9 (1976).

818. Hordos, B., "Calculator converts to a μC I/O keypad/display", EDN, <u>22</u> (3), 97-101 (1977).

819. Isaacson, R., "Utilising microprocessors in analogue environments", Electron. Engineering, <u>49</u> (590), 57-9, 61 (1977), CCA12-12954.

820. Johnson, B. A., "Interfacing microprocessor weighing systems", Proceedings of the Computer Interface Instrumentation Symposium, 1976, p. 97-103.

821. Juliussen, J. E., "Why is peripheral interfacing so expensive?", 13th IEEE Computer Society International Conference, COMPCON 76, Digest of Papers, p. 274-6.

822. Kim, S., "Inexpensive audio cassette recorder interface the μPs", EDN, <u>21</u> (5), 83-6 (1976).

823. Kimble, G. E., "Manufacturer/user interface for microprocessor based systems", Advances in Instrumentation, Vol. 31, 1976, Proceedings of 31st Annual ISA Conference and Exhibition, Pt. 1, Paper 512, 4 pp.

824. Klein, R. D., "Do-it-yourself data display - 1, 2", Elektronik, 26 (1, 2), 59-62, 74-6 (1977), German.

825. Lane, S., "Selection of a printer for mini and microcomputer applications", 1975 WESCON Technical Papers, Paper 21/1.

826. Larsen, D. G. and Rony, P. R., "Designing with microcomputers - a new approach to interfacing and breadboarding", 1975 WESCON Technical Papers, Paper 1/2.

827. Larsen, D. G., Rony, P. R., Titus, C. and Titus, J. A., "Microcomputer interfacing: Interfacing analog-to-digital converters", Comput. Des., 16 (8), 124-6 (1977).

828. Larsen, D. G., Titus, J. A. and Rony, P. R., "Computer interfacing: Anatomy of a microcomputer", Comput. Des., 15 (2), 128, 130 (1976).

829. Lewis, D. R. and Siena, W. R., "Clear the hurdles of microprocessors with orderly design procedures that provide the right interface and develop the software efficiently", Electron. Des., 21 (20), 76-80 (1973).

830. McPhillips, S., "Microprocessor input/output", New Logic Notebook, 1 (16), 1-31 (1975).

831. Martin, D. P. and Berland, K. S., "IC's interface keyboard to microprocessor", Electronics, 48 (5), 83, 85 (1975).

832. Matic, B., "Mate microprocessors with CRT displays", Electron. Des., 25 (19), 68-74 (1977).

833. Mattera, L., "Data converters latch into microprocessors", Electronics, 50 (18), 81-90 (1977).

834. Millet, D. F., "Microcomputer interfaces, peripheral chips, ROM, RAM", Proceedings, 1st National Microprocessor Conference, Microprocessors: Economics, Technology, Applications, 1974.

835. Mitzen, P., "The microprocessor and its interfaces", 1975 WESCON Technical Papers, Paper 24/1.

836. Mount, R. D., "Octal display for microprocessors", Wireless World, 83 (1495), 41 (1977), CCA12-10660.

837 Murray, J. M., "Universal peripheral interface device", Annual Conference Proceedings - Industrial Application of Microprocessors, 1977, p. 52-7.

838. Nichols, A. J. and McKenzie, K., "Build a compact microcomputer", Electron. Des., 24 (10), 84-92 (1976).

839. Nichols, J. L., "Hardware versus software for microprocessor I/O", Comput. Des., 15 (8), 102-4, 107 (1976).

840. Nicoud, J. D., "Peripheral interface standards for microprocessors", Proc. IEEE, 64 (6), 896-904 (1976).

841. Olsen, S. C., "μP I/O expansion scheme uses standard parts", EDN, 22 (1), 36-7 (1977).

842. Prazak, P. and Mrozowski, A., "Analog I/O hybrids simplify microprocessor interfaces", Electronics, 50 (11), 106-10 (1977).

843. Queyssac, D., "Understanding microprocessors. III.", Mes. Regul.
     Autom., 41 (11), 49-53 (1976), French, CCA12-12846.
844. Roberts, S. K., "Interfacing a teletypewriter with an IC micro-
     processor", Electronics, 47 (15), 96 (1974).
845. Rony, P. R., Larsen, D. G. and Titus, J. A., "Microcomputer
     interfacing: The substitution of software for hardware", Comput.
     Des., 15 (7), 120-1 (1976).
846. Russo, P. M., "An interface for multimicrocomputer systems",
     13th IEEE Computer Society International Conference, COMPCON 76,
     Digest of Papers, p. 277-82.
847. Sahakian, A., "Computer/cassette interface takes tone from
     clock", Electronics, 50 (1), 115 (1977), CCA12-3984.
848. Sawicki, F. J., "Shift registers act as control interface for
     microprocessor", Electronics, 50 (3), 106-7 (1977), CCA12-10606.
849. Schmid, H. and Mrozowski, G., "Mating microprocessors with con-
     verters calls for treating the A/Ds and D/As as memory loca-
     tions", Electron. Des., 23 (18), 76-81 (1975).
850. Schueberl, E. and Marschik, W., "Microcomputer, CAMAC, crate,
     and controller", Report SGAE-2414, Oesterreichische Studienge-
     sellschaft Fuer Atomenergie G.M.B.H. Seibersdorf. Forschungs-
     zentrum, 1975, 21 pp., German.
851. Scott, D. T., "Microprocessor reads BCD on only three lines",
     Electronics, 50 (3), 118-19 (1977), CCA12-10649.
852. Sebern, M. J., "Minicomputer-compatible microcomputer system:
     the DEC LSI-11", Proc. IEEE, 64 (6), 881-8 (1976).
853. Sidline, A. B., "Operating system software interfaces for a
     microprocessor", 10th IEEE Computer Society International Con-
     ference, COMPCON 75, Digest of Papers, p. 166-9.
854. Starzynski, A., "Modular microcomputer in the CAMAC system",
     CAMAC Bull., no. 9, 17-19 (1974).
855. Stewart, C. R., "Operator interface in distributed micropro-
     cessor control system", Advances in Instrumentation, Vol. 31,
     1976, Proceedings of 31st Annual ISA Conference and Exhibition,
     Pt. 1, Paper 544, 8 pp.
856. Thompson, R., "Displays don't trouble 8-bit μPs", Electron.
     Des., 24 (7), 68-71 (1976).
857. Titus, J. A., Rony, P. R. and Larsen, D. G., "Microcomputer
     interfacing: register pair instructions", Comput. Des., 16 (3),
     122, 124, 126 (1977), CCA12-12949.
858. Titus, C., Rony, P. R., Larsen, D. G. and Titus, J. A., "Micro-
     computer interfacing: using digital-to-analog converters",
     Comput. Des., 16 (7), 116-17 (1977).
859. Titus, J. A., Titus, C., Tony, P. R. and Larsen, D. G., "Micro-
     computer interfacing: interfacing a 10-bit DAC", Comput. Des.,
     16 (6), 203-5 (1977).
860. Toth, A., "The typical peripherals and peripheral interfacing
     methods of microcomputer systems", Inf. Elektron., 12 (2),
     70-5 (1977), Hungarian, CCA12-12960.
861. Trottier, L., "Video RAM's add a new dimension to microcomputer
     interfacing", EDN, 22 (5), 81-5 (1977).

862. Vacroux, A. G., "Microcomputer interface techniques", Notes for
     NEC Microcomputer Institute, 1974.
863. Vacroux, A. G., "Explore microcomputer I/O capabilities",
     Electron. Des., 23 (10), 114-19 (1975).
864. Weissberger, A. J., "LSI around the MPU", Microcomputer '77
     Conference Record, p. 5-9, EEA80-28902.
865. Wickes, W. E., "8-bit microprocessor aims at control applica-
     tions", Electronics, 49 (12), 101-5 (1976).
866. Yutzi, G. N., "1200 BPS (FSK) MOS/LSI modem for microprocessor
     or stand-alone applications", IEEE 1975 Region 6 - West USA -
     Conference Proceedings, 1975, p. 75-7.
867. Zissos, D. and Duncan, F. G., "Microprocessor interfaces",
     Electron. Lett., 12 (23), 624-5 (1976), CCA12-6614.

XII. PERIPHERAL AND SUPPORT CIRCUITS

868. Anon., "Microcomputer package provides high speed processing
     capabilities", Comput. Des., 15 (8), 122 (1976), CCA11-28493.
869. Bainter, J. R., "Dual-555-timer circuit restarts microprocessor",
     Electronics, 49 (6), 106-7 (1976).
870. Bass, J. E., "Peripheral-oriented microcomputer system", Proc.
     IEEE, 64 (6), 860-73 (1976).
871. Bennett, R. N., "2.4-V battery backup protects microprocessor
     memory", Electronics, 50 (3), 109 (1977), CCA12-10687.
872. Berstis, K. A., "Roving display checks microprocessor I/O",
     Electronics, 50 (7), 118-19 (1977), CCA12-12811.
873. Bliss, C., "Stopping the 8080 (microprocessor)", Electron.
     Ind., 2 (10), 19-21 (1976), CCA12-3898.
874. Briner, R. G., "µP and Co. -4-bit and 8-bit microprocessor
     systems with extensive peripherals", Micomp., 2 (1), 10-12,
     15 (1977), German, CCA12-15569.
875. Burgett, K. and O'Neil, E. F., "An integral real-time executive
     for microcomputers", Comput. Des., 16 (7), 77-82 (1977).
876. Cassell, D. A., "The 'Do-it-yourself' minicomputer- a highly
     modular microcomputer packaging scheme", 9th IEEE Computer
     Society International Conference, COMPCON 74, Digest of Papers,
     p. 59-60.
877. Ciarcia, S., "Memory mapped I/O", Byte, 2 (11), 10, 12, 14,
     16 (1977).
878. Davis, S., "Update on magnetic tape memories", Comput. Des.,
     13 (8), 127-40 (1974).
879. Deus, S., "Build a glitchless microprocessor clock with only a
     two-chip divider", Electron. Des., 24 (6), 102 (1976).
880. Drongowski, P. J. and Rose, C. W., "N.mPc: An adapatable system
     to support the development of microprocessor-based systems",
     13th IEEE Computer Society International Conference, COMPCON 76,
     Digest of Papers, p. 180-4.

881. Ethridge, C. D., "High-speed multiple-channel analog to digital
     data acquisition module for microprocessor systems", Report,
     LA-UR-76-994, Los Alamos Scientific Lab., N. Mex., 1976, 18 pp.
882. Felber, C. K. and Gauger, D. R., "This manual DMA will get your
     μP up and running quickly", EDN, 21 (12), 117-9 (1976),
     CCA12-1435.
883. Fischer, W. A., "Synergistic combination of an oscilloscope and
     a microprocessor", AFIPS National Computer Conference Proceed-
     ings, Vol. 44, 1975, p. 23-32.
884. Fronek, D. K. and Lane, R. A., "Microprocessor printed circuit
     boards for digital prototype systems", Report AD-A0312303/3SL,
     Army Missile Research Development and Engineering Lab., Red-
     stone Arsenal, Ala., 1976, 54 pp.
885. Gersthofer, R., "Simple logic single-steps SC/MP microprocessor",
     Electronics, 50 (1), 107-9 (1977), CCA12-3876.
886. Hilford, M. H., "Using intelligent peripherals for boosting
     microprocessor speed", Mach. Des., 49 (1), 70-4 (1977).
887. Jones, W. T., Budde, S. R. and Anderson, S. W., "2-pass cross-
     assembler for the 8080 microprocessor", Microprocessors, 1 (5),
     293-7 (1977).
888. Juliussen, J. E., "Magnetic bubble systems approach practical
     use", Comput. Des., 15 (10), 81-91 (1976).
889. Juliussen, J. E., Lee, D. M. and Cox, G. M., "Bubbles appearing
     first as microprocessor mass storage", Electronics, 50 (16),
     81-6 (1977).
890. Katsaros, J. J., "Microcomputers and process control: Where's
     the missing link", Advances in Instrumentation Vol. 31, 1976,
     Proceedings of 31st Annual ISA Conference and Exhibition, Pt.
     1, Paper 511, 6 pp.
891. Kehl, T. H. and Dunkel, L., "Simplified floppy-disc controller
     for microcomputers", Comput. Des., 15 (6), 91-7 (1976).
892. Kuzdrall, J. A., "Memory, peripherals share microprocessor ad-
     dress range", Electronics, 48 (24), 105, 107 (1975).
893. Kyogoku, T., Kawakami, K., Watanabe, T. and Konno, T., "CMOS
     high-speed full-adder for microprocessors", Natl. Tech. Rep.,
     22 (5), 601-6 (1976), Japanese, CCA12-10597.
894. Longacre, A., Jr., "Voltage doublers power microprocessor
     PROMs", Electronics, 49 (19), 105 (1976).
895. Louie, G., Wipfli, J. and Ebright, A., "A dual processor serial
     data controller chip", 1977 International Solid-State Circuits
     Conference, Digest of the Technical Papers, p. 144-5.
896. Mattera, L., "Data converters latch into microprocessors",
     Electronics, 50 (18), 81-90 (1977).
897. Michel, C., "Hardwired logic and microprocessors", Euromicro
     Newsl., 3 (1), 73-80 (1977), CCA12-12839.
898. Morrison, R. L. and Wiatrowski, C. A., "Adapter simplifies de-
     velopment of microprocessor systems", Comput. Des., 15 (5),
     175-7 (1976).

899. Morton, J., "μP timing is a snap with this new hardware building block", EDN, 20 (17), 55-9 (1975).

900. Murray, J. M., "Universal peripheral interface device", IECI Annual Conference Proceedings - Industrial Applications of Microprocessors, 1977, p. 52-7.

901. Olsen, S. C., "μP I/O expansion scheme uses standard parts", EDN, 22 (1), 36-7 (1977).

902. Osvalds, G., "Two-phase clock generator and driver synchronize the 8080 microprocessor", Electron. Des., 23 (9), 88 (1975).

903. Parasuraman, B., "Hardware multiplication techniques for microprocessor systems", Comput. Des., 16 (4), 75-82 (1977).

904. Pfeiffer, E. A., "Control logic for μP enables single-cycle operation", Electron. Des., 24 (21), 78 (1976), EEA80-6258.

905. Pohm, A. V., Covault, M. L. and Doctor, S. R., "Bubble memories for microcomputers and minicomputers", IEEE Trans. Magn., MAG-12 (6), 636-8 (1976), CCA12-6652.

906. Ramos, I. and Aramberri, J., "Cross-macroassembler for micro MCS-80 in a HP-2116-B microcomputer", Euromicro Newsl., 2 (4), 65-71 (1976).

907. Raphael, H., "How to expand a microcomputer's memory", Electronics, 49 (26), 67-9 (1976), CCA12-3892.

908. Redding, F. I. and Snyder, F. G., "Hybrid/LSI package yields a single-component microcomputer", Comput. Des., 15 (4), 114-16, 118 (1976).

909. Reichel, D. R., "Flexible discs a look at the latest mini-peripheral", EDN, 20 (2), 37-40 (1975).

910. Reyling, G. F., Jr., "Single-chip microprocessor employs minicomputer word length", Electronics, 47 (26), 87-93 (1974).

911. Rowe, J., "Simple power supply for microprocessor systems", Electron. Aust., 38 (6), 68-9 (1976), CCA12-15557.

912. Sahni, R. N., "Keyboard circuit saves time, needs no microprocessor scanning software", Electron. Des., 23 (26), 70 (1975).

913. Sallee, G. F., "Microprocessor terminal includes a 1200 BPS modem", IEEE 1975 Region 6 - West USA - Conference, Proceedings, 1975, p. 78-80.

914. Schmid, H., "Speed microcomputer multiplication with a CPU complementary circuit or peripheral multiplier", Electron. Des., 23 (9), 44-51 (1975).

915. Sebern, M. J., "Minicomputer-compatible microcomputer system: The DEC LSI-11", Proc. IEEE, 64 (6), 831-6 (1976).

916. Shriver, B. D. and Kornerup, P., "UNRAU: a unified numeric representation arithmetic unit", 3rd Symposium on Computer Arithmetic, 1975, p. 179-84.

917. Swales, N. P. and Tweddle, H., "Register-based output function for the COSMAC microprocessor", New Electron., 10 (1), 18 (1977).

918. Tenny, R., "Build a programmable sequencer with a broad operating range", EDN, 21 (10), 90-2 (1976).

919. Tireford, H., "Addressing microprocessors", Elektronik, 25 (12), 49-53 (1976), German, CCA12-3895.

920. Vacroux, A. G., "Explore microcomputer I/O capabilities",
     Electron. Des., 23 (10), 114-19 (1975).
921. Yates, V., "Microprocessors. III.", Rev. Esp. Electron., 23
     (263), 48-50 (1976), Spanish, CCA12-3900.

XIII. MULTIPLE MICROPROCESSORS

922. Anon., "Designing a multi-µP-based computer family", Digital
     Des., 6 (12), 14, 16, 18-20 (1976), CCA12-15440.
923. Arnold, R. G. and Page, E. W., "A hierarchical, restructurable
     multi-microprocessor architecture", 3rd Annual Symposium on
     Computer Architecture, 1976, p. 40-5.
924. Aspelund, J. and Hartimo, I., "Multimicrosystem - a low-cost
     way to fault-tolerant data processing", Proceedings of Inter-
     national Symposium and Course on Mini- and Microcomputers and
     Their Applications, MIMI 76, p. 76-8.
925. Baker, K., "Improve complex software by using multiple micro-
     processors", Microprocessors, 1 (3), 165-8 (1977).
926. Britton, R. L., "Use of multiple microprocessors", 1975 WESCON
     Technical Papers, Paper 1/4.
927. Burton, E., "Multi-µP systems combine the efficiency", Electron.
     Des., 25 (16), 68-71 (1977).
928. Drimak, E. G., "Multiprocessor input/output interrupt mechanism",
     IBM Tech. Disclosure Bull., 19 (1), 48-50 (1976), CCA11-22621.
929. Dromard, F. and Noguez, G., "Asynchronous network of specific
     microprocessors", 6th Workshop on Microprogramming Conference
     Record, 1973, p. 44-9.
930. Drysdale, K. W. P., "A systems architecture for multiprocessor
     systems using microprocessors", Proceedings, European Computing
     Congress, 1974, p. 1179-90.
931. Earle, J., "Centralized microcomputers", 13th IEEE Computer
     Society International Conference, COMPCON 76, Digest of Papers,
     p. 95-6.
932. Gebler, P., "Linking microprocessors to increase system through-
     put", Electron. Engineering, 49 (587), 51-5 (1977), CCA12-6686.
933. Lauhlin, R. S., "The galaxy/5: A large computer composed of
     multiple microcomputers", 13th IEEE Computer Society Inter-
     national Conference, COMPCON 76, Digest of Papers, p. 90-4.
934. Ohmori, K., Koike, N., Nezu, K. and Suzuki, S., "MICS - a
     multi-microprocessor system", NEC Res. Develop., no. 36, 32-6
     (1975).
935. Pathak, J., "Software setup eases traffic flow for multipro-
     cessors", Electronics, 50 (7), 108-12 (1977), CCA12-15556.
936. Pease, M. C., "The indirect binary n-cube microprocessor array",
     IEEE Trans. Comput., C-26 (5), 458-73 (1977), CCA12-12840.
937. Reyling, G., Jr., "Performance and control of multiple micro-
     processor systems", Comput. Des., 13 (3), 81-6 (1974).
938. Sippl, C. J., "Stacked microprocessors: A better way to go?",
     Digital Des., 7 (8), 68-72 (1977).

939. Spetz, W. L., "Microprocessor networks", Computer, <u>10</u> (7), 64-70 (1977).

940. Swan, R. J., Bechtolsheim, A., Lai, K. W. and Ousterhout, J. K., "The implementation of the Cm* multi-processor", Proceedings of the 1977 National Computer Conference, AFIPS, Vol. 46, p. 645-55.

941. Swan, R. J., Fuller, S. H. and Siewiorek, D. P., "Cm* - A modular multi-microprocessor", Proceedings of the 1977 National Computer Conference, AFIPS, Vol. 46, p. 637-44.

942. Widdoes, L. C., Jr., "The minerva multi-microprocessor", 3rd Annual Symposium on Computer Architecture, 1976, p. 34-9, CCA11-31569.

XIV.   PERFORMANCE EVALUATION

943. Alford, C. O. and Sledge, R. B., "Microprocessor architecture for discrete manufacturing control, part IV: Computer system performance evaluation", IEEE Trans. Manuf. Technol., MFT-<u>6</u> (1), 9-15 (1977).

944. Barrett, R. C., Bork, R. H., Fox, G., Hattendorf, D. W. and Vahey, M. D., "High speed microprocessor study", Report AD-A018702/1SL, Hughes Aircraft Co., Culver City, Calif., 1975, 73 pp.

945. Conte, G., Del Corso, D. and Giordana, M., "Software monitors versus hardware console: Price/performance analysis", Proceedings of International Symposium and Course on Mini- and Microcomputer and Their Applications, 1976, p. 113-16.

946. Cushman, R. H., "Exposing the black art of microprocessor benchmarking", EDN, <u>20</u> (8), 41-6 (1975).

947. Cushman, R. H., "Microprocessor benchmarks: How well does the microprocessor move data?", EDN, <u>20</u> (10), 43-8 (1975).

948. Cushman, R. H., "Beware of the errors that can creep into microprocessor benchmark programs", EDN, <u>20</u> (12), 105-10 (1975).

949. Derman, S., "Benchmark testing of μP's is still popular despite its limitations", Electron. Des., <u>24</u> (12), 92-5 (1976).

950. Ford, M. A., "Case study: Development of the 1SI-12/16 micro-computer; a study of using microprocessors to achieve mini-computer performance", Proceedings, 1st National Microprocessor Conference, Microprocessors: Economics, Technology, Applications, 1974.

951. Leung, C. K. C., Misunas, D. P., Neczwid, A. and Dennis, J. B., "Computer simulation facility for packet communication architecture", 3rd Annual Symposium on Computer Architecture, Conference Proceedings, 1976, p. 58-63.

952. Monico, J. A., "Microprocessors - a case history of an integrated approach", Presented at IEEE International Convention, March 1974.

953. Ravindran, V. K. and Thomas, T., "Characterization of multiple
     microprocessor networks", IEEE Computer Society International
     Conference, Digest of Papers, 1973, p. 133-7.
954. Reed, M. and Mergler, H. W., "Microprocessor-based control
     system", IECI Annual Conference Proceedings - Industral Appli-
     cations of Microprocessors, 1977, p. 211-17.
955. Tarui, T., Namimoto, K. and Takahashi, Y., "Twleve-bit micro-
     processor nears minicomputer's performance level", Electronics,
     47 (6), 111-16 (1974), EEA77-14585.

XV. DESIGN

956. Almes, G. T., Dronogowski, P. J. and Fuller, S. H., "Emulating
     the Nova on the PDP 11/40: a case study", 11th IEEE Computer
     Society International Conference, COMPCON 75, Digest of Papers,
     p. 53-6.
957. Amendt, T., "Microprocessors - the revolution in control system
     design", Control Instrum., 8 (2), 28-9, 31 (1976).
958. Anon., "Designers gain new freedom as options multiply", Elec-
     tronics, 49 (8), 78-100 (1976).
959. Anon., "Automotive electronics: Challenge to designers", Auto-
     mot. Eng., 84 (8), 52-7 (1976).
960. Anon., "Marry your µP to monolithic A/Ds", Electron. Des., 25
     (2), 82-6 (1977).
961. Anon., "Microprocessors and CAD - a catalyst to aid expansion?",
     Comput. Aided Des., 9 (2), 78 (1977), CCA12-13254.
962. Arato, P., Grantner, J. and Lantos, B., "On-line computer-aided
     design method and developing system for microprogrammed systems
     specified by flow chart", Euromicro Newsl., 3 (1), 34-8 (1977).
963. Barnich, R. G., "Designing microcomputers into your products",
     Advances in Instrumentation, Vol. 31, 1976, Proceedings of 31st
     Annual ISA Conference and Exhibition, Pt. 1, Paper 508, 3 pp.
964. Barrett, R. C., Bork, R. H., Fox, G., Hattendorf, D. W. and
     Vahey, M. D., "High speed microprocessor study", Report AD-
     A018702/1SL, Hughes Aircraft Co., Culver City, Calif., 1974,
     73 pp.
965. Bass, J. E., "System-oriented microcomputers", 10th IEEE Com-
     puter Society International Conference, COMPCON 75, Digest of
     Papers, p, 87-90.
966. Bond, J., "Designer's guide to: Software for the hardware
     designer - Part II", EDN, 19 (15), 51-6 (1974).
967. Bowra, J. W. and Torng, H. C., "Modeling and design of multiple
     function-unit processors", IEEE Trans. Comput., C-25 (3),
     210-21 (1976).
968. Brannan, S. G., "Take a minute to consider how you'll time your
     next µP design", EDN, 20 (17), 53-5 (1975).
969. Bray, D. W. and Mayer, R. A., "Towards a universal micropro-
     cessor", Microprocessors, 1 (3), 159-64 (1977).

970. Bur, P. W., "Design considerations for achieving reliable con-
     trol within shared microprocessor-based digital controllers",
     Advances in Instrumentation, Vol. 31, 1976, Proceedings of 31st
     Annual ISA Conference and Exhibition, Pt. 1, Paper 541, 9 pp.
971. Callan, J. D. and Baskin, R., "Microprocessor system design",
     Digital Des., 5 (2), 40-2 (1975).
972. Carey, B. J., "Formal methods expedite microprocessor-based
     system design", Comput. Des., 16 (2), 97-105 (1977), CCA12-10643.
973. Carey, B., "Automated design based upon microprogrammable bit-
     slice microprocessors", Proceedings of the Symposium on Design
     Automation and Microprocessors, 1977, p. 20-4.
974. Casilli, G. S., "When time is of the essence (microcomputer
     development)", Mini-Macro Syst., 9 (12), 42-5 (1976),
     CCA12-12819.
975. Catterton, R. D. and Casilli, G. S., "'Universal' development
     system is aim of master-slave processors", Electronics, 49
     (19), 91-6 (1976).
976. Chu, Y., "Design of a microprogrammed lexical microprocessor",
     Micro 8: 8th Annual Workshop on Microprogramming, Proceedings,
     1975, p. 28-9.
977. Chung, D. H., "Four design principles get the most out of micro-
     processor systems", Electronics, 50 (2), 102-10 (1977),
     EEA80-9006.
978. Coker, D., "16-k RAM eases memory design for maintenance and
     microcomputers", Electronics, 50 (9), 115-19 (1977), CCA12-15490.
979. Collins, D. C., "Microprocessor design and application",
     Computer, 8 (10), 20-1 (1975).
980. Cook, R. W. and Flynn, M. J., "System design of a dynamic micro-
     processor", IEEE Trans. Comput., C-19 (3), 213-22 (1970).
981. Cosley, J. and Vasa, S., "Block transfer with DMA augments
     microprocessor efficiency", Comput. Des., 16 (1), 81-5 (1977).
982. Crane, D. P., "Microprocessor: The general solution to system
     design", IEEE 1975 Region 6 - West USA - Conference, Proceed-
     ings, 1975, p. 213-14.
983. Cushman, R. H., "Single-chip microprocessors move into the 16-bit
     arena", EDN, 20 (4), 24-30 (1975).
984. Cushman, R. H., "How development systems can speed up the μP
     design process", EDN, 21 (8), 63-72 (1976).
985. Cushman, R. H., "Development system solves a microprocessor
     riddle", EDN, 21 (10), 69-74 (1976).
986. Cushman, R. H., "Sharpen your μC design skills quickly on
     'μSystem' projects", EDN, 22 (4), 89-92 (1977), CCA12-15423.
987. Cushman, R. H., "Imaginative planning leads to sound μC-based
     product design", EDN, 22 (12), 79-84 (1977).
988. Davidow, W., "8-bit processor: more power on a single chip",
     EE/Syst. Eng. Today, 33 (5), 48 (1974).
989. Davidow, W., "Coming merger of hardware and software design",
     Electronics, 48 (11), 91-4 (1975).

990. Ebertin, M. A., "8-bit parallel processing system", 1974
     WESCON Technical Papers, Paper 15/5.
991. Edwards, M. and Dagless, E., "LSI microprogrammable micro-
     processors", Microprocessors, $\underline{1}$ (7), 407-14 (1977).
992. Emmerichs, J., "Designing the 'tiny assembler' defining the
     problem", Byte, $\underline{2}$ (4), 60-7 (1977).
993. Estrin, G., "Modeling by synthesis - the gap between intent
     and behavior", Proceedings of the Symposium on Design Auto-
     mation and Microprocessors, 1977, p. 54-9.
994. Fields, S. W., "Chip set makes a powerful processor", Elec-
     tronics, $\underline{45}$ (8), 121, 123 (1972).
995. Fisher, E., "Speed microprocessor responses", Electron. Des.,
     $\underline{23}$ (23), 78-82 (1975).
996. Fox, W. A. and Reyling, G. F., Jr., "Single chip 16-bit micro-
     processor", New Electron., $\underline{8}$ (3), 24-6 (1975).
997. Fuller, S. H., Siewiorek, D. P. and Swan, R. J., "The design
     of a multi-micro-computer system", 3rd Annual Symposium on
     Computer Architecture, 1976, p. 123, CCA12-3882.
998. Gaddis, R. and Moore, C. F., "Building a microprocessor net-
     work", Instrum. & Control Syst., $\underline{48}$ (5), 37-40 (1975).
999. Gallacher, J. "Design of a modular systems applications micro-
     computer", Microprocessors, $\underline{1}$ (6), 345-52 (1977).
1000. Gardner, R. I., "State of the implementation of SARA", Proceed-
     ings of the Symposium on Design Automation and Microprocessors,
     1977, p. 60-2.
1001. Gardner, R. I., "Multi-level modeling in SARA", Proceedings of
     the Symposium on Design Automation and Microprocessors, 1977,
     p. 63-6.
1002. Gebler, P., "New generation of single-chip microcomputers",
     Electron. Eng., $\underline{49}$ (588), 59, 61 (1977).
1003. Gilmore, J. and Huntington, R., "Designing with the 6820
     peripheral interface adapter", Electronics, $\underline{49}$ (26), 85-6
     (1976), CCA12-3875.
1004. Gladstone, B., "Designing with microprocessors instead of
     wired logic asks more of designers", Electronics, $\underline{46}$ (21),
     91-104 (1973).
1005. Goss, L. and Raphael, H., "Simplify your next microcomputer",
     Electron. Des., $\underline{25}$ (15), 64-73 (1977).
1006. Grudowski, R. A. and Struger, O. J., "Microprocessor in pro-
     grammable logic/computing controllers for the industrial
     environment", IEEE Trans. Ind. Electron Control Instrum.,
     IECI-$\underline{22}$ (3), 318-26 (1975).
1007. Gruppuso, F. M. and Sidline, A., "16-bit single chip micro-
     processor", 10th IEEE Computer Society International Con-
     ference, Digest of Papers, 1975, p. 83-6.
1008. Hackmeister, R., "Intelligent memory holds the key to a limit-
     less array of computers", Electron. Des., $\underline{25}$ (16), 30, 32 (1977).

1009. Hoff, M. E., Jr., "The one chip CPU - computer or component?",
      Proceedings, Computer Systems Design Conference, 1972, p. 16-23.
1010. Hoff, M. E., Jr., "Designing central processors with bipolar
      microcomputer components", AFIPS National Computer Conference,
      Proceedings, Vol. 44, 1975, p. 55-62.
1011. Hoff, M. E., Jr. and Mazor, S., "Processor design for LSI
      implementation", Proceedings, 4th International Microelectronics
      Conference, 1970, p. 24.
1012. Hollon, H. J., "Microprocessors", Commun. Broadcast, $\underline{2}$ (3),
      9-13 (1976).
1013. Hunt, J., "Engineering innovations in minicomputer design",
      Electronic Engineering, $\underline{45}$ (549), 69-71 (1973).
1014. Jackson, R. E., "Designing a microprocessor-based system",
      EE/Syst. Eng. Today, $\underline{32}$ (12), 62-7 (1973).
1015. Jaeger, R., "Designing microprocessors with standard-logic
      devices, Part 1 and 2", Electronics, $\underline{48}$ (2), 90-5 (1975); ibid:
      $\underline{48}$ (3), 102-7 (1975).
1016. Johnson, E. L., "The impact of microprocessors on continuous
      simulation", Proceedings, Winter Simulation Conference, Vol.
      2, 1974, p. 751-2.
1017. Jones, T. and Thomas, P., "Challenges in microprocessor system
      design", Comput. Des., $\underline{15}$ (11), 109-18 (1976), CCA12-4042.
1018. Jordan, B. W., Jr., "Microprocessors in a digital system design
      curriculum", Proc. IEEE, $\underline{64}$ (6), 1000-2 (1976).
1019. Kildall, G. A., "Microcomputer software design - a checkpoint",
      AFIPS National Computer Conference, Proceedings, Vol. 44, 1975,
      p. 99-106.
1020. Kodres, U. R. and McCracken, W. L., "Design study of an
      Avionics navigation microcomputer", 2nd Annual Symposium on
      Computer Architecture, Conference Proceedings, 1975, p. 99-105.
1021. Kuhn, K., Pose, R. and Troger, B., "Microprocessors - a micro-
      computer on a single chip", Radio Fernsehen Elektron., $\underline{24}$ (2),
      57-8 (1975), German.
1022. Lane, A., "Microprocessor - system design", Digital Des., $\underline{5}$
      (8), 62-6 (1975).
1023. Lau, S. Y., "Design high-performance processors", Electron.
      Des., $\underline{25}$ (7), 86-95 (1977).
1024. Lavoie, F. J., "Computer on a chip", Mach, Des., $\underline{45}$ (31),
      42-7 (1973).
1025. Lea, R. M., "Micro-APP: a building block for low-cost high-speed
      associative parallel processing", Radio & Electron. Eng., $\underline{47}$
      (3), 91-100 (1977), CCA12-15517.
1026. Lederrey, J., "Microcomputer system design and implementation",
      In: Microcomputers: Fundamentals and Applications, MINICONSULT,
      London, 1974, Publ. 1975.
1027. Lee, E., "Designing with logic processors for control applica-
      tions", 1975 WESCON Technical Papers, Paper 1/3.
1028. Lee, E., "An alternative to development systems (for microcom-
      puters)", Mini-Micro Syst., $\underline{9}$ (12), 45-7 (1976), CCA12-13060.

1029. Liccardo, M. A., "A microcomputer designed for control", In: New Components and Subsystems for Digital Design: a Report, Technology Service, Santa Monica, CA., p. 51-60 (1975).

1030. Lindgren, T. P., "Age of the computer on a chip", Proceedings, National Aerosapce Electronics Conference, 1973, p. 199-202.

1031. Linn, E. Y., Schoeffler, J. D. and Rose, C. W., "Distributed microcomputer data acquisition", Instrum. Technol., 22 (1), 55-61 (1975).

1032. Little, J. and Thomas, A. T., "Operator's console considerations in microprocessor system design", Comput. Des., 14 (11), 87-91 (1975).

1033. Logan, J. D. and Kreager, P. S., "Using a microprocessor: a real-life application - 1.", Comput. Des., 14 (9), 69-77 (1975).

1034. Louie, G., "Design a disc controller with bipolar computer chips", Systems, 3 (4), 22-6 (1975), CCA11-9612.

1035. McKenzie, K., "A structured approach to microcomputer system design", Behav. Res. Methods & Instrum., 8 (2), 123-8 (1976), CCA11-31556.

1036. McWilliams, T. M., Fuller, S. H. and Sherwood, W. H., "Using LSI processor bit-slices to build a PDP-H - A case study in microcomputer design", Proceedings of the 1977 National Computer Conference, AFIPS, Vol. 46, p. 243-53.

1037. Mandell, L. J., "Pitfalls to avoid in applying microprocessors", International Microelectronic Conference, Proceedings of the Technical Program, 1975, p. 200-3.

1038. Manning, P., "Making microprocessors by the module", Instrum. & Control Syst., 47 (11), 57-9 (1974).

1039. Mazor, S., "New single chip CPU", 8th IEEE Computer Society International Conference, Digest of Papers, 1974, p. 177-80.

1040. Mazor, S. and Goss, L., "New single chip microcomputer for control application", IECI Annual Conference Proceedings - Industrial Applications of Microprocessors, 1977, p. 109-12.

1041. Mazur, T., "Put together a complete microcomputer", Electron. Des., 24 (15), 66-77 (1976).

1042. Meisel, W. S., "New components for digital design: Fad or revolution?", In: New Components and Subsystems for Digital Design: a Report, Technology Services, Santa Monica, CA., p. 3-6 (1975).

1043. Miller, M. S., "Design a low-cost CRT terminal around a single-chip μP", EDN, 22 (9), 88-94 (1977).

1044. Morling, R. C. S., "Elements of logic design", In: Microcomputers: Fundamentals and Applications, MINICONSULT, London, 1974, Publ. 1975.

1045. Muething, G. F., "Designing the maximum performance into bit-slice minicomputers", Electronics, 49 (20), 91-6 (1976).

1046. Nichols, A. J. and McKenzie, K., "Build a compact microcomputer", Electron. Des., 24 (10), 84-92 (1976).

1047. Nichols, J. L., "Economic alternatives in microcomputer design", Comput. Des., 15 (4), 110, 112 (1976).

1048. Ogdin, J. L., "Other microcomputer chips", Mod. Data, $\underline{8}$ (2), 43-7 (1975).
1049. Oswald, R. S., Laurance, N. L. and Devlin, S. S., "Design considerations for an on-board computer system", SAE Special Publication No. 393, 1975, p. 79-84.
1050. Pease, M. C., "The indirect binary n-cube microprocessor array", IEEE Computer Society Repository No. 75-100, 40 pp.
1051. Peatman, J. B., "Designing in self-test capability", In: Microcomputers: Fundamentals and Applications, MINICONSULT, London, 1974, Publ. 1975.
1052. Raphael, H. A., "Distributed intelligence microcomputer design", 10th IEEE Computer Society International Conference, COMPCON 75, Digest of Papers, p. 21-6.
1053. Raphael, H. A., "Putting a microcomputer on a single chip", Comput. Des., $\underline{15}$ (12), 59-65 (1976), CCA12-10647.
1054. Reynolds, B., "Microprocessors and innovation", Control Instrum., $\underline{8}$ (4), 46-7, 49 (1976).
1055. Rigas, H. B., "Design considerations for microprocessor-based systems", Proceedings of the 1976 IEEE Conference on Decision and Control including the 15th Symposium on Adaptive Processes, p. 1146-8, CCA12-6611.
1056. Rivers-Latham, P. W., "Microprocessor-based fast Fourier transform microsystem", Minicomputer and Small Business Systems, Forum, 1976, p. 403-29.
1057. Rudenberg, H. G., "Approaching the minicomputer on a silicon chip - progress and expectations for LSI circuits", AFIPS Conference Proceedings, Vol. 40, SJCC, 1972, p. 775-81.
1058. Santoni, A., "Peripheral chips add functions - and problems - to µP systems", Electron. Des., $\underline{25}$ (10), 32, 34, 36, 38 (1977).
1059. Schultz, G. W., "Designing optimized microprogrammed control sections for microprocessors", Comput. Des., $\underline{13}$ (4), 119-24 (1974).
1060. Sherwood, W., "Some applications of the Stanford University drawing system for LSI microprocessor", Proceedings of the Symposium on Design Automation and Microprocessors, 1977, p. 15-19.
1061. Sherwood, W., "Simulation hierarchy for microprocessor design", Proceedings of the Symposium on Design Automation and Microprocessors, 1977, p. 44-9.
1062. Short, K. L., "Digital systems design with programmed logic", IEEE Trans. Educ., E-$\underline{20}$ (1), 21-6 (1977).
1063. Smith, H., "Impact of microcomputers on the designer", 1973 WESCON Technical Papers, Paper 11/1.
1064. Smith, R. J., II and Matelan, M. N., "Practical considerations in implementing a real-time controller design automation system", Proceedings of the Symposium on Design Automation and Microprocessors, 1977, p. 25-7.
1065. Snigier, P., "Microprocessor development systems - which one is best?", EDN, $\underline{22}$ (5), 68-78 (1977).

1066. Sohn, D. W. and Volk, A., "Third-generation microcomputer set packs it all into 3 chips", Electronics, 50 (10), 109-13 (1977).
1067. Stamm, D. and Budde, D., "A single chip, highly integrated, user-programmable microcomputer", 1977 International Solid State Circuits Conference, Digest of the Technical Papers, p. 142-3.
1068. Sullivan, L., "Microcomputer needn't take many IC's", Electron. Des., 24 (12), 126-9, 131 (1976).
1069. Thomas, A. T., "Design techniques for microprocessor memory systems", Comput. Des., 14 (8), 73-8 (1975).
1070. Titus, J., "How to design a microprocessor-based controller system", EDN, 19 (16), 49-56 (1974).
1071. Toong, H. M. D., "Automatic design of multiprocessor microprocessor systems", Proceedings of the Symposium on Design Automation and Microprocessors, 1977, p. 6-14.
1072. Torng, H. C. and Wilhelm, N. C., "The optimal interconnection of circuit modules in microprocessor and digital system design", IEEE Trans. Comput., C-26 (5), 450-5 (1977), CCA12-12815.
1073. Torrero, E. A., "Microprocessor emphasis swings from chip specs to aids for designers", Electron. Des., 23 (22), 74-6, 78 (1975).
1074. Vasquez, R. M., "Design worksheet can generate least-part system, best addressing", Electronics, 49 (11), 111-15 (1976).
1075. Wagner, W. E., "Unraveling problems in the design of microprocessor-based systems", Hewlett-Packard J., 26 (12), 12-16 (1975).
1076. Weissberger, A. J., "Keeping pace with a single-chip 16-bit microprocessor", AFIPS National Computer Conference, Proceedings, Vol. 44, 1975, p. 9-14.
1077. Werner, W. A. and Minnick, W., "User requirements for digital design verification simulators", Proceedings of the Symposium on Design Automation and Microprocessors, 1977, p. 36-43.
1078. Wickes, W. E., "A microprocessor chip designed with the user in mind", Computer, 10 (1), 18-22 (1977), CCA12-15487.
1079. Wilson, R., "Input multiplexing for a four-bit microprocessor", Electron. Eng., 48 (578), 19, 21 (1976).
1080. Wray, B., "Microprocessor control systems and design techniques", SME Technical Paper Series, MS76-201, 1976, 13 pp.
1081. Wyland, D. C., "Design your own microcomputer", Electron. Des., 23 (20), 72-8 (1975).

XVI.   TESTING

1082. Anderson, R. E., "Test methods change to meet complex demands", Electronics, 49 (8), 125-8 (1976).
1083. Anon., "Microprocessor tests itself in tester built from the user's end product", EDN, 21 (15), 27-8 (1976), CCA12-6601.

1084. Anon., "μP fails test, so talking calculator's designers opt
for customized circuitry", Digital Des., 6 (7), 18, 20 (1976),
CCA12-3872.
1085. Anon., "A universal logic analyser for microprocessor systems",
Electron. & Microelectron. Ind., no. 226, 24-5 (1976),
French, EEA80-6259.
1086. Anon., "Functional data modules let IC tester examine all
microprocessors and related LSI", EDN, 22 (6), 69-71 (1977).
1087. Anon., "Field servicing microprocessor-based systems", Digital
Des., 7 (9), 78-82 (1977).
1088. Barnes, J. and Bergquist, B., "Unite μP hardware and software",
Electron. Des., 24 (7), 74-6 (1976).
1089. Barraclough, W., Chiang, A. C. L. and Sohl, W., "Techniques
for testing the microcomputer family", Proc. IEEE, 64 (6),
943-50 (1976).
1090. Berstis, K. A., "Roving display checks microprocessor I/O",
Electronics, 50 (7), 118-9 (1977).
1091. Blomeyer, P., "Logic testers for microprocessor systems",
Elektronik, 25 (9), 66-8 (1976), German, CCA11-31525.
1092. Bruni, L., "Design a μP analyzer", Electron. Des., 24 (18),
82-7 (1976).
1093. Bruni, L., "Keep your microcomputer alive", Electron. Des.,
25 (16), 80-1 (1977).
1094. Cahoon, S. A., "Schmoo/stress technique to determine design
sensitivities of a MOSFET microprocessor unit chip", 1975
Semicondutor Test Symposium, Digest of Papers, p. 69-74.
1095. Chiang, A., Ferraro, E. and Liu, B., "Microprocessor unit
chip testing using low cost tester", 5th Semiconductor Test
Symposium, Digest of Papers, 1974, p. 239-61.
1096. Chiang, A. C. L., "Test schemes for microprocessor chips",
Comput. Des., 14 (4), 87-92 (1975).
1097. Chiang, A. C. L. and McCaskill, R., "Two new appraoches
simplify testing of microprocessors", Electronics, 49 (2),
100-5 (1976).
1098. Chrones, C., "Testing microprocessor chip: A large-scale
challenge", Electron. Packag. & Prod., 15 (4), 35-42 (1975).
1099. El Hamalaway, M. A. and Vanwormhoudt, M. C., "A standard
microcomputer system with diagnostic facilities", Euromicro
Newsl., 3 (1), 59-60 (1977), CCA12-12838.
1100. Farnbach, W. A., "Bring up your μP 'bit by bit'", Electron.
Des. 24 (15), 80-5 (1976).
1101. Foley, E. B., Jr. and Firman, A. H., "Testing microcomputer
boards automatically", Comput. Des., 15 (12), 92-4 (1976),
CCA12-10632.
1102. Gordon, G. and Nadig, H., "Hexadecimal signatures identify
troublespots in microprocessor systems", Electronics, 50 (5),
89-96 (1977).

1103. Grason, J., "Design aids and hardware testing of microprocessor system circuit packs", Proceedings of the Symposium on Design Automation and Microprocessors, 1977, p. 95-9.

1104. Hackmeister, R., "A characterization scheme for microprocessors", 1975 Semiconductor Test Symposium, Digest of Papers, p. 78-83.

1105. Hackmeister, D. and Chiang, A. C. L., "Microprocessor test technique reveals instruction pattern sensitivity", Comput. Des., 14 (12), 81-5 (1975).

1106. Hall, H. P., "Analog tests: the microprocessor scores", IEEE Spectrum, 14 (4), 36-40 (1977).

1107. Hatch, R. I., "Microprocessor multiplies a digital multimeter's functions", Electronics, 49 (19), 97-101 (1976).

1108, Hnatek, E. R., "Checking microprocessors?", Electron. Des., 23 (22), 102-5 (1975).

1109. Huston, R., "Microprocessor function test generation on the Sentry 600", 5th Semiconductor Test Symposium, Digest of Papers, 1974, p. 216-38.

1110. Huston, R., "Microprocessor testing: a testing turnaround smart DUT runs the tester", Proceedings of the Technical Program, International Microelectronics Conference, 1975, p. 16-24.

1111. Huston, R. E., "Microprocessor test equipment needs", 1975 Semiconductor Test Symposium, Digest of Papers, p. 59-68.

1112. Izumi, D., "The challenge of microprocessor chip testing", 1975 WESCON Technical Papers, Paper 27/1.

1113. Jackson, P., "Microelectronic methods", Insul./Circuits, 22 (8), 19-21 (1976).

1114. King, E. E. and Nelson, G. P., "Radiation testing of an 8-bit CMOS microprocessor", IEEE Trans. Nucl. Sci., NS-22 (5), 2120-1 (1975).

1115. King, G., "Testing microprocessor chips", Circuits Manuf., 15 (4), 52-7 (1975).

1116. Leatherman, J. and Burger, P., "System integration and testing with microprocessors - 1. Hardware aspects", IEEE Trans. Ind. Electron Control Instrum., IECI-22 (3), 360-3 (1975).

1117. Lee, E. C., "A simple concept in microprocessor testing", 1976 Semiconductor Test Symposium, Digest of Papers, p. 13-15, CCA12-4068.

1118. Lilley, R. W., "Test program for 4-k memory card, jolt microprocessor", Report N76-30845/1SL, Ohio Univ., Athens, 1976, 5 pp.

1119. Luciw, W., "Can a user test LSI microprocessors effectively?", IEEE Trans. Manuf. Technol., MFT-5 (1), 21-3 (1976).

1120. McCaskill, R., "Test approaches for four bit microprocessor slices", 1976 Semiconductor Test Symposium, Digest of Papers, p. 22-6, CCA12-3833.

1121. Mandel, W. J., "Techniques of microprocessor test development", 1974 WESCON Technical Papers, Paper 22/5.

1122. Morse, S. P., "A tutorial paper on software approaches to testing of microprocessor systems", 1975 Semiconductor Test Symposium, Digest of Papers, p. 53-8.

1123. Prasad, B. A., "Modeling techniques for dynamic logic and test pattern generation of a microprocessor", Proceedings of the Symposium on Design Automation and Microprocessors, 1977, p. 100-7.

1124. Raphael, H. A. and Hou, S., "Test equipment for microcomputers", Mach, Des., 48 (19), 78-81 (1976).

1125. Runyon, S., "Focus on logic and µP analyzer", Electron Des., 25 (3), 40-50 (1977).

1126. Santoni, A., "Testers are getting better at finding microprocessor flaws", Electronics, 49 (26), 57-66 (1976), CCA12-3874.

1127. Schusheim, B., "Flexible approach to microprocessor testing", Comput. Des., 15 (3), 67-72 (1976).

1128. Smith, D. H., "Microprocessor testing - method or madness", 1976 Semiconductor Test Symposium, Digest of Papers, p. 27-9, CCA12-3834.

1129. Smith, D. H., "Exercising the functional structure gives microprocessors a real workout", Electronics, 50 (4), 109-12 (1977), CCA12-15554.

1130. Sohl., W. E., "Microprocessor testing: overview of techniques", Circuits Manuf., 17 (6), 50-2 (1977).

1131. Vodovoz, E., "Testing microprocessors is a gamble", Electron. Prod., 18 (6), 26-9, 146-7 (1975).

XVII. RELIABILITY

1132. Barnes, J. and Gregory, V., "Use microprocessors to enhance performance with noisy data", EDN, 21 (15), 71-2 (1976), CCA12-6602.

1133. Foster, R. R., "Why consider screening, burn-in, and 100-percent testing for commercial devices?", IEEE Trans. Manuf. Technol., MFT-5 (3), 52-8 (1976).

1134. Hilow, R. C., "Microprocessors - meeting the reliability challenge", 1976 National Telecommunications Conference, Pt. I, Paper 5/1, CCA12-15595.

1135. Hnatek, E. R., "Microprocessor device reliability", Microprocessors, 1 (5), 299-303 (1977).

1136. Maki, G. K., "Design of microprocessors for self fault detection", Microcomputer '77, Conference Records, p. 236-40.

1137. Measel, P. R., Sivo, L. L., Quilitz, W. E. and Davidson, T. K., "Development of a hard microcontroller", IEEE Trans. Nucl. Sci., NS-23 (6), 1738-42 (1976), EEA80-9020.

1138. Srini, V. P., "Fault diagnosis of microprocessor systems", Computer, 10 (1), 60-5 (1977).

1139. Wakerly, J. F., "Reliability of microcomputer systems using
      triple modular redundancy", 12th IEEE Computer Society Inter-
      national Conference, COMPCON 76, Digest of Papers, p. 23-6.
1140. Wakerly, J. F., "Microcomputer reliability improvement using
      triple-modular redundancy", Proc. IEEE, 64 (6), 889-95 (1976).

XVIII.   FUTURE TRENDS

1141. Allan, R., "Components: microprocessors galore", IEEE
      Spectrum, 13 (1), 50-6 (1976), CCA11-12494.
1142. Anon., "Systems architecture - trends in mini and microcom-
      puters", Report T23-IIIB, Diebold Group, Inc., New York,
      N. Y., 1977, 45 pp.
1143. Blume, H. M., Jr., "Trends in intelligent peripheral chip
      design", 13th IEEE Computer Society International Conference,
      COMPCON 76, Digest of Papers, p. 20-2, CCA12-1460.
1144. Butler, M. K., "Prospective capabilities in hardware", Pro-
      ceedings of the Conference on Computer Support of Environmental
      Science and Analysis, 1975, p. 449-73.
1145. Cotter, G. E., Fitzgerald, F. A. and Driscoll, L. A., "Avionics
      multiplexing with smart terminals", IEEE Proceedings National
      Aerospace Electronic Conference, 1975, p. 672-8.
1146. Gilb, T., "Present and future application areas for microcom-
      puters", Manage Datamatics, 4 (5), 159-66 (1975).
1147. Gott, W., "Microprocessors make intelligent data terminal
      systems", EE/Syst. Eng. Today, 33 (1), 78-80 (1974).
1148. Hirsch, P., "Singer's cheap, smart terminal", Datamation, 20
      (12), 117 (1974).
1149. Krummel, L. and Schultz, G., "Advances in microcomputer
      development systems", Computer, 10 (2), 13-19 (1977).
1150. Laliotis, T. A., "Microprocessors: Present and future",
      Computer, 7 (7), 20-4 (1974).
1151. Madnick, S. E., "Important areas in the future of micropro-
      cessors", Proceedings, 1st National Microprocessor Conference,
      Microprocessors: Economics, Technology, Applications, 1974.
1152. Martino, J. P., "Forecast of electronics - overview", IEEE
      Proceedings of National Aerospace Electronic Conference, 1975,
      p. 727-30.
1153. Moore, G. E., "Microprocessors and integrated electronic
      technology", Proc. IEEE, 64 (6), 837-41 (1976).
1154. Nelson, J. C., "Economic implications of microprocessors on
      future computer technology and systems", Proceedings of the
      National Computer Conference, Vol. 44, AFIPS, 1975, p. 629-32.
1155. Neth, J. and Forsberg, R., "Microprocessors and microcomputers:
      What will the future bring?", EDN, 19 (22), 24-9 (1974).
1156. Nicholas, A. J., "Overview of microprocessor applications",
      Proc. IEEE, 64 (6), 951-3 (1976).

1157. Noyce, R. N. and Davidow, W. H., "State of the art of solid
      state memories and microprocessors", Proceedings of the Inter-
      national Colloquium on Automatic Electron Technology, 1974,
      p. 15-18, Publ. 1975.
1158. Petritz, R. L., "The pervasive microprocessors: trends and
      prospects", IEEE Spectrum, 14 (7), 18-24 (1977).
1159. Scrupski, S. E., "Processors moving from computer room",
      Electronics, 48 (21), 76-80 (1975).
1160. Svensson, C. M. and Jeppson, K., "Microprocessor - the brain
      of control systems of the 80's", Tek Tidskr, 106 (18), 17-19
      (1976), Swedish.
1161. Weissberger, A. J., "LSI microprocessors in telecommunications",
      IEEE 1975 Region 6 - West USA - Conference, Proceedings,
      p. 186-91.
1162. Wirsching, J. E., "Computer of the 1980's - is it a network
      of microcomputers?", Proceedings of the IEEE Computer Society
      Conference, COMPCON 75, Digest of Papers, p. 23-6.
1163. Wolf, H. F. and Socolovsky, A., "Microcomputers - an applica-
      tions forecast", Proceedings of the International Microelec-
      tronics Conference, Technical Program, 1975, p. 195-9.
1164. Zinschlag, H. P., Lord, W. E. and Conover, J. A., "LSI and
      process control - is there a microprocessor in your future?",
      Instrum. Technol., 22 (11), 47-52 (1975).

XIX. APPLICATIONS
      1. General

1165. Anderson, H. R., "Case study: The use of a microprocessor in
      commercial products", Proceedings, 1st National Microprocessor
      Conference, Microprocessors: Economics, Technology, Applica-
      tions, 1974.
1166. Anon., "The mini computer's quiet revolution", EDP Anal., no.
      12, 1-13 (1972).
1167. Anon., "Will they revolutionize computing", Can. Data Syst.,
      no. 7, 30-2 (1974).
1168. Anon., "Microcomputer applications and developments", Meas.
      Control, 8 (1), 10 (1975).
1169. Barr, R. G., Becker, J. A., Lidinsky, W. P. and Tantillo, V.
      V., "Research-oriented dynamic microprocessor", IEEE Trans.
      Comput., C-22 (11), 976-85 (1973).
1170. Bemer, R. W., "The frictional interface between computers and
      society", Comput. & People, 24 (1), 14-19 (1975).
1171. Blomeyer-Bartenstein, H. P., "Microprocessor and microcomputer",
      Data Rep., 11 (5), 36-40 (1976), German, CCA12-6595.
1172. Bollinger, J. and Mills, J., "Role of microprocessors in future
      CNC systems", Ann. Cirp., 25 (1), 323-8 (1976).

1173. Conte, G., del Corso, D., Giordana, M. and Pozzolo, V., "A modular project with auxiliary equipment to develop the general use of microprocessor systems", Alta Freq., $\underline{45}$ (11), 640-8 (1976), Italian, CCA12-8910.

1174. Cushman, R. H., "Successful µsystems combine user needs and µC technology", EDN, $\underline{22}$ (8), 104-11 (1977).

1175. David, J. R., "Microcomputers - new technology brings new capabilities", Proceedings, 34th ASIS Annual Meeting, 1974, p. 41-2.

1176. Davidow, W., "How microprocessors boost profits", Electronics, $\underline{47}$ (14), 105-8 (1974).

1177. Dittmar, E., "Microcomputers in automation - fundamentals of hardware and software, system planning. I.", Electrontechnik, $\underline{58}$ (23), 10-13 (1976), German, CCA12-8919.

1178. Falk, H., "Linking microprocessors to the real world", IEEE Spectrum, $\underline{11}$ (9), 59-67 (1974).

1179. Fox, W. A. and Reyling, G. F., Jr., "A single chip 16-bit microprocessor for general applications", Microelectron. & Reliab., $\underline{14}$ (4), 387-97 (1975).

1180. Francis, B., "Microprocessors: the mini computer/random logic alternative?", Electron. Eng., $\underline{47}$ (565), 45-8 (1975).

1181. Garen, E. R., "Applying microprocessors and microcomputers", Mod. Data, $\underline{8}$ (2), 54-7 (1975).

1182. Hallgren, B., "Swiss conference on micro-computers: Many fine applications - but the manufacturers have difficulties", Elteknik med aktuell Elektronik, $\underline{18}$ (1), 40-2 (1975), Swedish.

1183. Hoertt, D. J., "Microcircuits: the brain and the brawn", IEEE Trans. Ind. Electron Control Instrum., IECI-$\underline{22}$ (2), 159-63 (1975).

1184. Hovey, R. L. and Hansen, A. L., "Application of minis and micros in AFOS", Symposium on Trends and Applications 1976: Micro and Mini Systems, p. 39-42.

1185. Inguaggiato, B., "Processing units for scientific applications employing Intel 8088-I", 2nd ISPRA Nuclear Electronic Symposium, Proceedings, 1975, p. 315-18.

1186. Kirk, B. R., "Microprocessors for everyday applications", Microelectron. & Reliab., $\underline{14}$ (4), 377-8 (1975).

1187. Knoop, D. E. and Loessel, M. C., "Microcomputers: A technique for electronic control", IEEE Trans. Ind. Appl., IA-$\underline{12}$ (4), 354-8 (1976).

1188. Laver, M., "Microprocessor, side effect and society", Microprocessors, $\underline{1}$ (5), 305-8 (1977).

1189. Lesser, V. R., "Introduction to the direct emulation of control structures by a parallel microcomputer", IEEE Trans. Comput., C-$\underline{20}$ (7), 751-64 (1971).

1190. Lilen, H., "Micros, minis, midis, and maxis: Computers with an industrial vocation", Automatique & Informatique Industrielles, no. 33, 22-32 (1975), French.

1191. McKay, C. W., "Technique for developing multipurpose, firm-
      ware-driven microcontrollers from eight bit microprocessors",
      Conference on Modeling and Simulation, Proceedings, 1976,
      p. 85-8.
1192. Maerz, M. and Rosenfeld, P., "A new approach to the develop-
      ment of microcomputer-based products", 1975 WESCON Technical
      Papers, Paper 6/4.
1193. Miller, G., "Using microprocessor development systems", Digital
      Des., 7 (9), 101-4 (1977).
1194. Namimoto, K., "Microcomputer and its applications", Toshiba
      Rev., no. 97, 19-23 (1975), EEA78-41698.
1195. Parasuraman, B., "Applications of LSI processors", 1973 WESCON
      Technical Papers, Paper 11/2.
1196. Reynolds, B., "Microprocessors and innovation", Control
      Instrum., 8 (4), 46-7, 49 (1976).
1197. Rosenbalatt, A. I., "Automatic control proliferates", Elec-
      tronics, 47 (14), 83-7 (1974).
1198. Seim, T. A., "Microprocessors and experimentation in scientific
      laboratory", Comput. Des., 15 (9), 83-9 (1976).
1199. Steel, M. W. R., "Microprocessor applications in the School of
      Engineering at the University of Bath UK", Microprocessors, 1
      (5), 309-12 (1977).
1200. Villwock, R. D., "A macro application for a microprocessor",
      In: New Components and Subsystems for Digital Design: a Report,
      Technology Service, Santa Monica, CA, p. 31-5 (1975).
1201. Walker, H. P., "Automatic selective level measuring set:
      design and application", Onde Electr., 55 (7), 391-7 (1975).
1202. Weissberger, A. J., "Microprocessors as intelligent remote
      controllers", 1974 WESCON Technical Papers, Paper 23/4.

2.   Data Processing
a.   General

1203. Abbott, D. L., "Modular ports for microcomputers in distributed
      systems", Proceedings, Symposium on Electronics, 1974,
      p. 277-86.
1204. Cohen, D. and Liu, M. T., "Emulation of computer networks by
      microprogrammable microcomputers", 7th Annual Workshop on
      Microprogram, Conference Record, 1974, p. 159-67.
1205. Haynes, G., "New analogue component concepts for digital pro-
      cessing systems", Microelectron. & Reliab., 16 (1), 57-67
      (1976), CCA12-15409.
1206. Hirt, K. A., "Prototype ring-structured computer network using
      microcomputers", Report AD 772877/7, 1973, 49 pp.
1207. Hoff, M. E., Jr., "Considerations for the use of microcomputers
      in small systems design", 1972 WESCON Technical Papers,
      Paper 26/3.
1208. Jackson, R. E., "Designing a microprocessor-based system",
      EE/Syst. Eng. Today, 32 (12), 62-7 (1973).

1209. Johnson, L. M., "A cellular computer utilizing microprocessors",
      Thesis, University of Minnesota, 1974, 162 pp., Order No.
      75-171.
1210. Jones, D. F. and Martin, E. R., "Use μPs in minicomputer sys-
      tems", Electron. Des., 24 (22), 148-52 (1976), CCA12-8918.
1211. Madnick, S. E., "New concepts in data-base management: Appli-
      cation of the new technology", Proceedings, ASIS Annual
      Meeting, 1975.
1212. Malinovsky, B. N., Bojun, V. P. and Kozlov, L. G., "Real-time
      information processing and special-purpose microprocessors",
      Euromicro Newsl., 3 (1), 67-72 (1977).
1213. Modell, H. S., Ward, R. G. and Sparr, T. M., "Coordination of
      parallel processes in PL/1", Proceedings of the International
      Conference on Parallel Processing, 1976, p. 247-53.

                    b.   Data Acquisition

1214. Coppo, N., De Grandi, G., Montagni, M. and Stanchi, L., "Micro-
      processor-based CAMAC data acqusition and processing system",
      2nd ISPRA Nuclear Electronic Symposium, Proceedings, 1975,
      p. 341-6.
1215. Finch, M. D., "Microprocessor controlled data acquisition
      systems", IEEE Electromagnetic Compatibility Symposium Record,
      1975, Session 5C, Paper IB, 3 pp.
1216. Linn, E. Y., Schoeffler, J. D. and Rose, C. W., "Distributed
      microcomputer data acquisition", Instrum. Technol., 22 (1),
      55-61 (1975).
1217. Rose, C. W. and Schoeffler, J. D., "Microcomputers for data
      acquisition", Instrum. Technol., 21 (9), 65-9 (1974).
1218. Sturesson, S., "ASEA's digital electronic system DS-8", ASEA
      J., 49(5), 103-6 (1976).
1219. Van Gerwen, P. J., Verhoeckx, N. A. M., Van Essen, H. A. and
      Snijders, F. A. M., "Microprocessor implementation of high-
      speed data modems", IEEE Trans. Commun., COM-25 (2), 238-50
      (1977).
1220. Weissberger, A. J., "Microprocessors expand industry applica-
      tions of data acquisition", Electronics, 47 (18), 107-10 (1974).
1221. Wilson, G. E. and Kerakovich, C. T., "Microprocessor-based
      system for laboratory data acquisition", Report PB 242 055/2WC,
      1975, 147 pp.
1222. Wilson, G. H., III and Williams, R. S., "Real-time data acquis-
      tion and reduction of critical dynamic fatigue parameters",
      Proc. IEEE, 63 (10), 1445-50 (1975).

                    c.   Signal Processing

1223. Bartram, J. F., Ramseyer, R. R. and Heines, J. M., "Fifth
      generation digital sonar signal processing", IEEE Electronic
      and Aerospace Systems Convention, EASCON 76, Record, Paper 91,
      7 pp.

1224. Britton, R. L., "Multi-microcomputer homogeneous pipeline processor architecture for signal processing applications", Symposium on Advances in Signal Processing Technol., 1975, p. 273-81.
1225. Cole, E. L., Jr. and Beaver, E. W., "Matrix controlled signal processing", IEEE Electronic and Aerospace Systems Convention, EASCON 74, Record, p. 593-9.
1226. Gault, J. W. and Stroh, R. W., "The application of microprocessors and other LSI devices to digital signal processing", Report AD-A033770/9SL, North Carolina State Univ., Raleigh, 1976, 67 pp.
1227. Hornbuckle, G. D. and Ancona, E. I., "LX-1 microprocessor and its application to real-time signal processing", IEEE Trans. Comput., C-$\underline{19}$ (8), 710-20 (1970).
1228. Parrish, E. A., Jr. and Lee, Y. C., "Microcomputer preprocessor/postprocessor for analog signals", IEEE Trans. Ind. Electron. & Control Instrum., IECI-$\underline{20}$ (1), 38-41 (1974).
1229. Pease, M. C., "Signal processing with a network of microprocessors", Proceedings, National Aerospace Electronics Conference, 1974, p. 292-8.
1230. Steber, G. R. and Northouse, R. A., "Image processing with mini and micro computers", 13th IEEE Conference on Decision and Control, Proceedings, 1974, p. 453-6.

d.   Terminals

1231. Andrews, D. I., "Line processor - a device for amplification of display terminal capabilities for text manipulation", Proceedings, National Computer Conference, 1974, p. 257-65.
1232. Anon., "Microprocessors in display system", Telecommun. J., $\underline{43}$ (12), 708-9 (1976).
1233. Anon., "Bit-slice microprocessor and IC display drivers implement full-graphics capability in plasma-panel terminal", EDN, $\underline{22}$ (11), 43-4 (1977).
1234. Bernardy, A., "Microprogrammed multiprocessor graphic controller", 6th Annual Workshop on Microprogramming, Conference Record, 1973, p. 131-3.
1235. Carson, J. H., Summers, J. K. and Welch, J. S., Jr., "A microprocessor selective encryption terminal for privacy protection", Proceedings of the 1977 National Computer Conference, AFIPS, Vol. 46, p. 35-8.
1236. Corbett, C., "Interactive graphics system using an overlay tablet", Digital Processes, $\underline{2}$ (4), 291-306 (1976).
1237. Cropper, L. C. and Whiting, J. W., "Microprocessors in CRT terminals", Computer, $\underline{7}$ (8), 48-53 (1974).
1238. Cropper, L. C. and Whiting, J., "Evaluation of the use of microprocessors in CRT display terminals", 8th IEEE Computer Society International Conference, COMPCON 74, Digest of Papers, p. 185-7.

1239. Dack, D., "Microprocessor display systems", In: Microcomputers:
      Fundamentals and Applications, MINICONSULT, London, p. 145-61
      (1974), Publ. 1975.
1240. Dromard, F., "Design of a microprogrammed alphanumeric terminal",
      7th Annual Workshop on Microprogramming, Conference Record,
      1974, p. 128-34.
1241. Emery, W., "Evolution of a remote data entry terminal", EE/Syst.
      Eng. Today, 32 (12), 68-9 (1973).
1242. Fishman, A., "Guidelines for point plotting on CRT's with mini
      and microcomputers", EDN, 20 (5), 65-9 (1975).
1243. Frick, D. F., "Application of microprocessors to the point of
      sale terminal", IEEE International Convention Technical
      Papers, 1974, Paper 24/2.
1244. Gray, M. T., "Microprocessors in CRT terminal applications:
      hardware/software tradeoffs", Computer, 8 (10), 53-9 (1975).
1245. Herr, J. R., "Microprocessor design for intelligent point-of-
      sale terminals", Computer, 7 (7), 30-4 (1974).
1246. Holderby, W. S., "Designing a microprocessor-based terminal for
      factory data collection", Comput. Des., 16 (3), 81-8 (1977).
1247. Irwin, G. and Kosman, K., "Compact computing power for terminals",
      Telesis, 3 (8), 239-44 (1974).
1248. Jessel, G. P. and Smoliar, S., "A microprocessor-based terminal
      system for undergraduate education", Presented at the ACM
      Computer Science Conference, Washington, D. C., February 1975.
1249. Kellner, R. and Maas, L. D., "Developmental system for micro-
      computer based intelligent graphics terminals", Report LA-UR-
      76-125, Los Alamos Scientific Lab., N. Mex., 1976, 13 pp.
1250. Kerr, H., "Microprocessors in graphic display systems", EE/Syst.
      Eng. Today, 32 (12), 70-1 (1973).
1251. Kirby, P. J. and Gardner, E. M., "Microcomputer controlled,
      interactive testing terminal development", Report AD-A035731/
      9SL, Air Force Human Resources Lab. Brooks, AFB, Tex., 1976,
      24 pp.
1252. Kohli, J. P., "Designing an application oriented terminal",
      AFIPS National Computer Conference, Proceedings, Vol. 44, 1975,
      p. 47-54.
1253. Miller, M. S., "Design a low-cost CRT terminal around a single-
      chip microprocessor", EDN, 22 (9), 88-94 (1977).
1254. Murray, D., "Minicomputers and microprocessors as intelligent
      remote terminals", Can. Control Instrum., 13 (12), 34-6 (1974).
1255. Nicoud, J. D., "Alphanumeric TV display interface", Micro-
      processors, 1 (4), 217-21 (1977).
1256. Reed, J. A., "Application of LSI microprocessors to display
      terminals", IEEE International Convention Technical Papers,
      1973, Paper 21/2.
1257. Sallee, G. F., "Microprocessor terminal includes a 1200 BPS
      modem", IEEE 1975 Region 6 - West USA - Conference, Proceed-
      ings, p. 78-80.

1258. Spademan, C. F., "Microprocessors and data terminals", IEEE
      Trans. Ind. Electron. Control Instrum., IECI-$\underline{22}$ (3), 326-8
      (1975).
1259. Spann, J. M., "Display terminal with alphanumeric and graphic
      display capability, for microcomputers", Report UCRL-75617,
      California Univ., Livermore, Lawrence Livermore Lab., 1974,
      8 pp.
1260. Stevens, F., "Microcomputer terminal controller for a data
      communciation network", Presented at IEEE International Con-
      vention, March 1974.
1261. Tani, T., "On-line display terminal fits many different systems",
      Electronics, $\underline{49}$ (8), 145, 148 (1976).
1262. Tibbals, H. F. and Curran, P., "Optimizing function distribu-
      tion in a terminal network", Microprocessors, $\underline{1}$ (6), 362-8
      (1977).
1263. Weaver, A., "Intelligent remote terminals utilizing micropro-
      cessors", Presented at the 2nd Computer Science Conference, 1974.
1264. Whiting, J. and Newman, S., "Microprocessors in CRT terminals",
      Proceedings, National Computer Conference, 1975, p. 41-5.

e.   Printers

1265. Barrus, G. B. and Mayne, D. W., "The Printonix 300 line printer",
      1975 WESCON Technical Papers, Paper 21/4.
1266. Bickoff, C., "Minicomputers play key role in low-cost line
      printer development", EDN, $\underline{21}$ (3), 98-101 (1976).
1267. Blacksher, R., "Control an intelligent teletypewriter", Electron.
      Des., $\underline{25}$ (2), 74-9 (1977).
1268. Hammond, D., "(Microprocessor) chip scans keyboard without
      hardware interface", Electronics, $\underline{50}$ (1), 110-12 (1977),
      CCA12-3893.
1269. Jackson, R. E., "Microprocessor control for a high-speed serial
      printer", Proc. IEEE, $\underline{64}$ (6), 960-5 (1976).
1270. Matsumoto, M., Honma, Y. and Tomita, H., "New serial printers",
      NEC Res. Dev., no. 43, 14-20 (1976).
1271. Moore, A. and Eidson, M., "Printer control - a minor task for a
      fast microprocessor", Electron. Des., $\underline{22}$ (25), 74-83 (1974).
1272. Pasco, R. C., "Converter lets processor drive teletypewriter",
      Electronics, $\underline{48}$ (22), 97, 99 (1975).
1273. Ratcliffe, D., "The first microprocessor controlled serial
      printer", 1975 WESCON Technical Papers, Paper 21/3.

f.   Disks/Tapes

1274. Anon., "Flexible-disc memory utilizes high performance cost-
      effective μC design", EDN, $\underline{22}$ (5), 43-4 (1977), CCA12-15498.
1275. Barlett, K. G., "Consider MSI for tape controllers", Electron.
      Des., $\underline{23}$ (16), 50-4 (1975).

1276. McCracken, J. A., "Tape time (position) locator using micro-
      processor technology", Audio Engineering Society Preprints,
      1976, for 54th Convention, Preprint 1107(E-3), 7 pp.
1277. Murine, G. E., "Extended on-board, real time, preprocessing
      of multispectral scanner data", Electro-Opt System Design
      Conference/International Laser Expo, Proceedings of the
      Technical Program, 1975, p. 158-65.

                    g.   Peripherals

1278. Anon., "Microcomputer in portable monitor conditions data and
      oversees I/O", EDN, 22 (9), 29-31 (1977).
1279. Avery, W. and Moro, S., "Making remote process input/output
      look like it's local", Control Eng., 22 (9), 62-5 (1975).
1280. Barron, E. T. and Glorioso, R. M., "Micro controlled peripheral
      processor", 6th Annual Workshop on Microprogramming, Conference
      Record, 1973, p. 122-30.
1281. Boehme, C. R., "Use of a microprocessor as a peripheral system
      controller", 8th IEEE Computer Society International Conference,
      COMPCON 74, Digest of Papers, p. 181-4.
1282. Burke, P. M., "Use of a micro-processor as an intelligent
      peripheral/instrument", Trends in On-Line Computer Control
      Systems, 1974, p. 162-9, Publ. 1975.
1283. Byrd, J. S., "When your system's data rates differ, it's time
      for a microprocessor", EDN, 19 (22), 57-62 (1974).
1284. Hardy, M. E., Jr., "Microprocessor technology to extend the
      utility of computer peripherals", 8th IEEE Computer Society
      International Conference, Digest of Papers, p. 299-302.

                    h.   Others

1285. Barnes, J. and Gregory, V., "Use microprocessors to enhance
      performance with noisy data", EDN, 21 (15), 71-2 (1976).
1286. Brown, G. E., Eckhouse, R. H., Jr. and Goldberg, R. P.,
      "Operating system enhancement through microprogramming",
      13th IEEE Computer Society International Conference, COMPCON
      76, Digest of Papers, p. 86-9.
1287. Chen, R. C., Jessel, P. G. and Patterson, R. A., "Microprocessor
      controlled data switch", 11th IEEE Computer Society Inter-
      national Conference, COMPCON 75, Digest of Papers, p. 221-4.
1288. Forney, G. D. and Vander May, J. E., "8-bit microprocessors
      can control data networks", Electronics, 49 (13), 110-12 (1976).
1289. Halsall, F. and Woollons, D. J., "Some aspects of a micro-
      processor-controlled inter-computer data link", Proceedings,
      Conference on Computer Systems and Technology, 1974, p. 117-22.
1290. Lindgard, A., Graae Sorensen, P. and Oxenboll, J., "Sharing a
      microcomputer between different experiment and plotting tasks,
      under control of a multiprogrammed computer", J. Phys. E., 10
      (3), 264-70 (1977).

1291. McGill, R. and Steinhoff, J., "A multimicroprocessor approach
      to numerical analysis: an application to gaming problems", 3rd
      Annual Symposium on Computer Architecture, 1976, p. 46-51,
      CCA12-3903.
1292. Madnick, S. E., "INFOPLEX: Hierarchical decomposition of a
      large information management system using a microprocessor
      complex", AFIPS National Computer Conference, Proceedings,
      Vol. 44, 1975, p. 581-6.
1293. Mann, K. L., "Precision analog computer operates in real or
      machine time", EDN, 21 (3), 81-4 (1976).
1294. Maples, M. D., "Microprocessors in computing?", 10th IEEE
      Computer Society International Conference, COMPCON 75,
      Digest of Papers, p. 69-71.
1295. Milich, C. P. and Stephenson, D. L., "A microprocessor transient
      data recording system", Proceedings, Symposium on Electronics,
      1974, p. 349-58.
1296. Morgan, D. R., "Microprocessor applications in laboratory
      testing and debugging aids", Proceedings, IEEE International
      Symposium on Circuits and Systems, 1974, p. 293-5.
1297. Nicoud, J. D. and Sommer, R., "Modular logic elements, micro-
      processors and peripherals improve efficiency of teaching
      and development", 10th IEEE Computer Society International
      Conference, COMPCON 75, Digest of Papers, p. 127-30.
1298. Roberson, D. A., "Microprocessor-based portable computer: the
      IBM 5100", Proc. IEEE, 64 (6), 994-9 (1976).
1299. Scott, R. C., III and Erich, H. P., "Microprocessors based
      correlator reference system", 13th IEEE Computer Society
      International Conference, COMPCON 75, Digest of Papers,
      p. 269-73.
1300. Searle, B. C. and Freberg, D. E., "Microprocessor applications
      in multiple processor systems", Computer, 8 (10), 22-30 (1975).

3.   Industrial Control
a.   General

1301. Alden, R. M., "Microcomputers for remote intelligent monitoring
      and control", Advances in Instrumentation, Vol. 31, 1976,
      Proceedings of 31st Annual ISA Conference and Exhibition, Pt.
      1, Paper 507, 4 pp.
1302. Anderson, L. H., "Development of a portable compiler for
      industrial microcomputer systems", AFIPS National Computer
      Conference Proceedings, Vol. 44, 1975, p. 33-40.
1303. Anderson, L. H., "Distributed intelligence microcomputer
      systems (DIMS)", ISA Trans., 15 (2), 127-30 (1976).
1304. Bailey, S. J., "Microprocessor: Candidate for distributed
      computing control", Control Eng., 21 (3), 40-4 (1974).
1305. Bailey, S. J., "LSI: the new look in controls", Control Eng.,
      21 (6), 50-3 (1974).

1306. Benedek, Z. M., "Microcomputer teams up with fast dedicated processor for real time control", Proceedings, Symposium on Electronics, 1974, p. 393-406.
1307. Biewer, M., "Microprocessors for dedicated control", 1974 WESCON Technical Papers, Paper 11/2.
1308. Burger, P. and Ronchinsky, S., "Microprocessor driven digital servo system", IECI Annual Conference Proceedings - Industrial Applications of Microprocessors, 1977, p. 159-63.
1309. Chung, D., "Multiprocessor control systems replace a single large processor that has a single memory. When several μPs are thus linked together, the combination is extremely power-ful", Electron. Des., $\underline{24}$ (12), 132-6 (1976).
1310. Cooke, P. A., "Microprocessor-based controller for the process control industry", Symposium on Minicomputer and Small Business Systems, 1976, p. 441-58.
1311. Davenport, D. A., "Using the micro-computer in the industrial environment", 28th IEEE Annual Conference on Electrical Engineering Problems in Rubber and Plastic Industry, 1976, p. 18-22.
1312. Deramo, A. D., "Microcomputers: A potential value for process control", Advances in Instrumentation, Vol. 31, 1976, Proceedings of 31st Annual ISA Conference and Exhibition, Pt. I, Paper 509, 4 pp.
1313. Dull, E., "Microprocessor in control engineering", Regelung stech Prax, $\underline{16}$ (11), 279-85 (1974), German.
1314. Ethridge, C. D., "Microprocessor unit for experimental process data", Report LA-UR-76-380, Los Alamos Scientific Lab., N. Mex., 1976, 17 pp.
1315. Glumsson, S. and Ocklind, J. P., "Micro-computers can give process techniques better control and more rapid development than hierarchical systems", Teknisk Tidskrift, $\underline{104}$ (4), 34-5 (1974), Swedish.
1316. Gordon, A., "Process control language for use with industrial microcomputers", SME Technical Paper Series, MS76-204, 1976, 8 pp.
1317. Grudowski, R. A. and Struger, O. J., "Microprocessor in pro-grammable logic/computing controllers for the industrial environment", IEEE Trans. Ind. Electron. Control Instrum., IECI-$\underline{22}$ (3), 318-26 (1975).
1318. Halling, H., "Distributed control system using the CAMAC serial highway and microprocessor modules", Proceedings of the IFAC-IFIP Workshop on Real-Time Programming, 1976, p. 13-27.
1319. Harris, H. E., "Second generation digital control system for total control of processing plants", Plast. Des. Process., $\underline{15}$ (11), 11-14 (1975).
1320. Hearn, D. W. and Saenz, J. A., "Microcomputer control of industrial processes", Report AD 786598/3WC, 1974, 136 pp.

1321. Hu, S. C., "Microprocessors - characteristics and role in process control", Advances in Instrumentation, Vol. 31, 1976, Proceedings of 31st Annual ISA Conference and Exhibition, Pt. I, Paper 552, 10 pp.
1322. Jacobs, M. R., "Track data with "smart" front ends", Instrum. Control Syst., $\underline{49}$ (11), 71-3 (1976).
1323. Kanaby, K. D. and Atkins, D. E., "A shared-memory micro-mini computer system for process control", 9th IEEE Computer Society International Conference, COMPCON 74, Digest of Papers, p. 5-10.
1324. Katsaros, J. J., "Microcomputers and process control: Where's the missing link", Advances in Instrumentation, Vol. 31, 1976, Proceeedings of 31st Annual ISA Conference and Exhibition, Pt. I, Paper 511, 6 pp.
1325. Kaufmann, J., "Let a µP keep track of your process", Electron. Des., $\underline{23}$ (26), 66-9 (1975).
1326. Kochhar, A. K., "Use of computers in manufacturing systems - 5.", Mach. Prod. Eng., $\underline{129}$ (3341), 666-73 (1976).
1327. Langill, A. W., Jr. and Borese, D. A., "Close look at two microprocessor control applications", Instrum. Control Syst., $\underline{50}$ (4), 43-6 (1977).
1328. Marcantonio, A. R. and Russo, P. M., "Automation-oriented microprocessor-based developmental system", IECI Annual Conference Proceedings - Industrial Applications of Microprocessors, 1977, p. 58-63.
1329. Maslanka, R. E., Mollis, R. C. and Wilson, D. R., "Microprocessor based control systems", GTE Autom. Electr. Tech. J., $\underline{15}$ (4), 178-90 (1976).
1330. Pooch, U. W., "Mini- and microcomputer controlled process applications", Symposium on Trends and Applications 1976: Micro and Mini Systems, p. 10-22.
1331. Riera, E., "Using microprocessors as substitutes for complex logic systems", Mundo Electron., no. 57, 65-9 (1976), EEA80-6263.
1332. Romeu, F. J., "Digital communication system for process control instrumentation", Advances in Instrumentation, Vol. 31, 1976, Proceedings of 31st Annual ISA Conference and Exhibition, Pt. 1, Paper 543, 11 pp.
1333. Schmid, H., "Speed microcomputer multiplication", Electron. Des., $\underline{23}$ (9), 44-51 (1975).
1334. Schmidt, G. and Birck, H., "Significance of microprocessors in process control", Regelungstech Prax, $\underline{16}$ (12), 308-15 (1974), German.
1335. Schoeberl, E. and Marschik, W., "Microcomputer crate controller with Intel SIM 8-01", Oesterreichische Studiengesellschaft Fuer Atomenergie G.M.B.H., Seibersdorf. Elektronikinstitut, 1974, 17 pp., German.
1336. Scrimgeour, J., "CAD/CAM expected to yield significant improvement in productivity", Eng. J., $\underline{59}$ (6), 12-13 (1976).

1337. Seim, T. A., "Microprocessor sampled data process controller",
      Report BNWL-1795, 1973, 117 pp.
1338. Seipp, W. H., "Controller has high speed, bit-manipulation
      capability", Electronics, 49 (13), 112-14 (1976).
1339. Smith, R., "Put your control where the action is", Instrum.
      Control Syst., 49 (4), 35-8 (1976).
1340. Stewart, C. R., "Operator interface in distributed micropro-
      cessor control system", Advances in Instrumentation, Vol. 31,
      1976, Proceedings of 31st Annual ISA Conference and Exhibition,
      Pt. 1, Paper 544, 8 pp.
1341. Struger, O. and Dummermuth, E., "Multiprocessor based program-
      mable controller", Advances in Instrumentations, Vol. 31, 1976,
      Proceedings of 31st Annual ISA Conference and Exhibition, Pt.
      1, Paper 556, 6 pp.
1342. Syrbe, M., "Higher reliability of process-control systems and
      lower peripheral costs by use of distributed microprocessors",
      Regelungstechnik, 22 (9), 264-8 (1974), German.
1343. Titus, J., "How to design a μP-based controller system", EDN,
      19 (16), 49-56 (1974).
1344. Walter, R. G., "Concept of factory automation", RCA Eng., 21
      (3), 36-8 (1975).
1345. Weissberger, A. J., "Microprocessors simplify industrial
      control", Electron. Des., 23 (22), 96-9 (1975).
1346. Weissberger, A. J., "Microprocessors in the processing plant",
      IEEE Trans. Ind. Electron. Control Instrum., IECI-22 (3),
      354-8 (1975).
1347. White, D. G., "A basic distributed microcomputer control system",
      Solid State Technol., 20 (7), 38-41 (1977).
1348. Wiatrowski, C. A. and Conant, B. K., "Parallel interface con-
      siderations for process control", Advances in Instrumentation,
      Vol. 31, 1976, Proceedings of 31st Annual ISA Conference and
      Exhibition, Pt. 4, Paper 842, 9 pp.
1349. Wiatrowski, C. A. and Teeple, C. R., "Add flexibility to your
      control system with distributed data processing", Instrum.
      Control Syst., 49 (3), 37-41 (1976).
1350. Wickes, W. E., "8-bit microprocessor aims at control applica-
      tions", Electronics, 49 (12), 101-5 (1976).
1351. Williams, R. J., "Distributed intelligence system", SME
      Technical Paper Series, MS76-197, 1976, 10 pp.
1352. Wilson, R. A. and Greene, A. M., "New markets for control via
      microprocessors", Iron Age, 214 (2), 53-6 (1974).
1353. Wise, K. D., "LSI, microprocessors and electronic automotive
      control", Proceedings of International Conference on Automotive
      Electronics and Electric Vehicles, 1976, p. 1/1-5, EEA80-9507.
1354. Wo, Y. K., "Versatile microcomputer-based controller with an
      IEEE instrumentation bus interface", IECI Annual Conference
      Proceedings - Industrial Applications to Microprocessors, 1977,
      p. 180-4.

1355. Wood, D., "De-centralisation of intelligence: the key to low
      cost control?", Control Instrum., $\underline{7}$ (9), 30-1, 33 (1975).

                    b.   Applications

1356. Anon., "New control options - 3. Dedicated 'micro' systems
      optimize performance of individual extruders", Mod. Plast.,
      $\underline{53}$ (8), 38-40 (1976).
1357. Anon., "Computer-controlled steelmaking: a one-time revolution
      that's maturing into the routine", 33 Mag. Met. Prod., $\underline{14}$
      (12), 25-33 (1976).
1358. Anon., "Move towards totally automated warehousing", Electr.
      Veh., $\underline{63}$ (1), 8, 10 (1977).
1359. Anon., "Distributed microprocessing cuts waste in extruded
      plastic control systems", EDN, $\underline{22}$ (10), 40-5 (1977).
1360. Banks, R. B., "Here's a look at minicomputer for automation
      projects", Oil Gas J., $\underline{70}$ (22), 46-50 (1972).
1361. Binck, H. J. and Zouck, J. H., "Microprocessor applied to
      supervisory control", Instrum. Technol., $\underline{22}$ (1), 45-52 (1975).
1362. Brinkman, A., Nijmeyer, J. and Van Poeteren, H., "Application
      of the all-purpose traffic processor", Philips Telecommun.
      Rev., $\underline{33}$ (1), 16-33 (1975).
1363. Cohen, T. J., "Microprocessor technology: an electronic
      revolution", Sea Technol., $\underline{17}$ (3), 20-1 (1976).
1364. Deramo, A. D., "Microcomputers: a potential value for process
      control", Advances in Instrumentation, Vol. 31, 1976, Pro-
      ceedings of 31st Annual ISA Conference and Exhibition, Pt. 1,
      Paper 509, 4 pp.
1365. Donaghey, L. F., "Microcomputer systems for chemical process
      control", Proc. IEEE, $\underline{64}$ (6), 975-87 (1976).
1366. Drazan, P. J. and Jeffery, M. F., "Microprocessor control and
      pneumatic drive of a manipulating arm", Proceedings of Con-
      ference on Industrial Robot Technology, 3rd, 1976, Paper D2,
      12 pp.
1367. Eikelberg, F., "Controls for molding: Which one's for me?",
      Mod. Plast., $\underline{54}$ (3), 46-9 (1977).
1368. Fisher, E. and others, "Stores stock microcomputer system: the
      MCS-8 8-bit processor controller", Report UCID-16341, 1973,
      15 pp.
1369. Fredriksen, T. R., "Microprocessor controls fully automatic
      wafer prober", Solid State Technol., $\underline{18}$ (9), 49-55 (1975).
1370. Godsey, E. E., "Microprocessor controls remote pipeline sites",
      Oil Gas J., $\underline{75}$ (5), 170, 175, 178 (1977).
1371. Goksel, K., Knowles, K. A., Jr., Parrish, E. A., Jr. and Moore,
      J. W., "Intelligent industrial arm using a microprocessor",
      IEEE Trans. Ind. Electron. Control Instrum., IECI-$\underline{22}$ (3),
      309-14 (1975).

1372. Goodwin, R. W., Kocanda, R. F. and Shea, M. F., "Microprocessor-based preaccelerator control system", Proceedings of Proton Linear Accelerator Conference, AECL, 1976, p. 264-8.

1373. Gordon, L. A., "Fundamentals of automated pollution control: Low-cost microprocessors provide solutions", Pollut. Eng., 9 (2), 20-6 (1977).

1374. Grano, N., "Microprocessor based control systems for paper-machines", 62nd Annual Meeting of the Technical Section, CPPA, Preprint of Papers, 1976, p. 135-9.

1375. Hasler, H. and Wuergler, H. J., "Microcomputer controls a gear hobbing machine", Brown Boveri Rev., 63 (9), 606-9 (1976).

1376. Herberg, D., Ratnakumar, N. and Lobban, P. E., "A microprocessor controller for an epitaxial reactor", Comput. Des., 16 (7), 108-10 (1977).

1377. Herrman, R. F., "Microprocessor controlled railway signalling interlock", IECI Annual Conference Proceedings - Industrial Applications of Microprocessors, 1977, p. 2-7.

1378. Hershberger, D. M. and Underwood, J. G., "Microprocessor-based gauge controller for metals rolling applications", IEEE Trans. Ind. Electron. Control Instrum., IECI-22 (3), 333-7 (1975).

1379. Hershberger, D. M. and Underwood, J. G., "Dedicated micropro-cessors for real-time analysis and control of rolled sheet processes", Analytical Instruments, Vol. 14, 1976, Proceedings of the ISA Analytic Instrumentation Symposium, 22nd, p. 97-101.

1380. Hill, M. L., "Digital fuel control of industrial gas turbines", ASME Paper No. 75-GT-106, 1975, 11 pp.

1381. Hills, J. B. and Fuller, H. J., "Digital actuation and inter-face of rotary control valves", Advances in Instrumentation, Vol. 31, 1976, Proceedings of 31st Annual ISA Conference and Exhibition, Pt. 4, Paper 838, 6 pp.

1382. Huff, E. C. and McArthur, K. G., "Microprocessor-based digital process controller", IEEE Trans. Ind. Electron. Control Instrum., IECI-22 (3), 345-53 (1975).

1383. Hundley, R. G., "Keyboard entry and digital display 3 mode process controller", Advances in Instrumentation, Vol. 29, 1974, Pt. 4, Paper 802, 7 pp.

1384. Hutt, M., "Microprocessor automated sputtering", Solid State Technol., 19 (12), 74-6 (1976).

1385. Jaillet, P. E., Borgatta, A. and Verrey, C., "The INTELFILE BG 7000: A microprocessor based controller for multiple 3M magnetic data cartridge", Proceedings, Symposium on Electronics, 1974, p. 265-76.

1386. Kaufman, B. A., "Memory-based microprocessor system for discrete machine control", IEEE Trans. Ind. Electron. Control Instrum., IECI-22 (3), 315-17 (1975).

1387. Keyes, M. A. and Gillespie, D. M., "Distributed digital control: System architecture for power, energy and utility applications", National Conference Publ. Inst. Eng. Aust. No. 77/1, 1977: IFAC Symp. - Autom. Control and Prot. of Electr. Power Syst., p. 162-8.

1388. Koehler, F. A. and Anderson, B. E., "Microprocessors in process control applications", Report MLM-23381(OP), Mound Lab., Miamisburg, Ohio, 1976, 7 pp.

1389. Krayenbrink, C. J. and Vlaanderen, A., "Philips all-purpose traffic processor system", Philips Telecommun. Rev., 33 (1), 8-15 (1975).

1390. Landis, D. G. and Maliakal, J. C., "Automatic diffusion furnace system using microprocessor control", Solid State Technol., 20 (7), 34-7 (1977).

1391. Maas, M. A., "Controls save energy, refine processes", Electronics, 48 (21), 96-100 (1975).

1392. McDevitt, D. B., "Microcomputer widens automatic control use", Oil Gas J., 73 (63), 55-8 (1975).

1393. McKay, C. W. and Gross, C., "Microprocessor based flow monitoring system", Advances in Instrumentation, Vol. 31, 1976, Proceedings of 31st Annual ISA Conference and Exhibition, Pt. 4, Paper 821, 4 pp.

1394. Mathias, R. A., "Adaptive control for machining centers", SME Technical Paper Series, MS76-196, 1976, 10 pp.

1395. Miller, J. L., "Programmable controllers, minicomputers, and microcomputers in manufacturing", RCA Eng., 21 (3), 7-10 (1975).

1396. Miroux, J., Tourres, L. and Tesseron, J. M., "Automation microprocessor-based system to be used in an EHV substation", Natl. Conf. Publ. Inst. Eng. Aust. No. 77/1, 1977: IFAC Symp. - Autom. Control and Prot. of Electr. Power Syst., p. 43-8.

1397. Molenaar, G., "Microcomputers in environmental simulation", Ind. Res., 19 (6), 74-8 (1977).

1398. Muller, M. C. and Fasang, P. P., "Microprocessor oriented data acquisition and control system for power system control", 3rd Annual Symp. on Computer Architecture, Conference Proceedings, 1976, p. 74-8.

1399. North, R. J., Jeffery, R. W., Dolman, J. A. and Tuck, A. N., "Digital computer aspects of the instrumentation and control of the new RAE 5 metre low speed tunnel", AGARD Conference Proceedings, No. 210, 1976: Numerical Methods and Windtunnel Test, for Fluid Dynamic Panel, Special Meeting, Paper 1, 10 pp.

1400. Pellerin, S., "Microprocessor control of a bridge crane", Digital Des., 7 (11), 78-80, 82, 84 (1977).

1401. Powell, R. D. and Burrage, R. G., "Microprocessor based controls for gas turbines", ASME Paper No. 76-GT-91, 1976, 12 pp.

1402. Raphael, H. A., "Motor control by PLL", Electron. Des., 23 (9), 54-7 (1975).

1403. Ridgeway, R. A., "Microprocessor utilization in hydraulic open-die forge press control", IEEE Trans. Ind. Electron. Control Instrum., IECI-22 (3), 307-9 (1975).

1404. Russell, B. D., "Application of microcomputers to the protection and control of power system substations", Proceedings of the IEEE Conference on Decision Control, 1975, p. 590-1.

1405. Sauer, W. J. and Paul, F. W., "Position control of machine tools
      using a microcomputer", Advances in Instrumentation, Vol. 31,
      1976, Proceedings of 31st Annual ISA Conference and Exhibition,
      Pt. 1, Paper 544, 10 pp.
1406. Scott, H. D. and Smoak, R. A., "Microcomputer controller for a
      nuclear pool reactor", IEEE Trans. Ind. Electron. Control
      Instrum., IECI-22 (1), 15-18 (1975).
1407. Seim, T. A., "Automation of a brazing process with an Intel
      8008 microprocessor", IEEE Trans. Ind. Electron. Control
      Instrum., IECI-22 (3), 303-7 (1975).
1408. Seipp, W. H., "Microprocessor control applied to injection
      molding operations", Mod. Plast., 53 (11), 126-8 (1976).
1409. Shunta, J. P., Tripathi, S. S. and Shah, M. K., "Distillation
      column control using a microprocessor controller", Proceedings
      of the Computer Interface Instrumentation Symposium, 1976,
      p. 105-10.
1410. Skrokov, M. R., "Microprocessor control benefits in Olefins
      plant design", ASME Paper No. 76-PET-63, 1976, 8 pp.
1411. Smith, K. D., Bachass, A., Huggins, R., Kirk, R. and Pease, L.,
      "Microprocessor controlled stacker crane", IEEE 1975 Region 6 -
      West USA - Conference, Proceedings, p. 181-3.
1412. Smyth, R. K., "Microcomputer applications to a telephone cable
      plant process controller", In: New Components and Subsystems
      for Digital Design: a Report, Technology Service, Santa Monica,
      CA, p. 41-50 (1975).
1413. Stratford, F. L. and Oghanna, W., "Microprocessor control of
      substations", IEEE Trans. Ind. Electron. Control Instrum.,
      IECI-24 (1), 30-4 (1977).
1414. Tuuli, R., Sarparanta, J. and Toimela, T., "System organiza-
      tion of data processing and remote control for electric
      utilities", 25th International Conference on Large High-Voltage
      Electrical Systems, Bull. No. 1, 1974, Paper 34-03, 11 pp.
1415. Uyetani, A., "Multiloop process controller based on micropro-
      cessor", Advances in Instrumentation, Vol. 31, 1976, Proceed-
      ings of 31st Annual ISA Conference and Exhibition, Pt. 4,
      Paper 823, 11 pp.
1416. Wager, J. and English, F. L., "Town border stations and inter-
      ruptible services monitoring", Am. Gas Assoc. Oper. Sect.
      Proc., 1976, Distrib. Conf., Paper 76-D-14, 2 pp.
1417. Wareham, R. L., "Trail use of microprocessor for process
      control a success", Oil Gas J., 74 (49), 95-6, 98 (1976).
1418. Wrobel, V., "A microcomputer-controlled seeder", Comput. Des.,
      16 (6), 184-6 (1977).

4. Instrumentation

1419. Aldridge, D., "A/D conversion systems: Let your µP do the
      working", EDN, 21 (9), 75-80 (1976).

1420. Amendt, T., "Microprocessors - the revolution in control system design", Control Instrum., $\underline{8}$ (2), 28-9, 31 (1976).

1421. Anderson, R. C., "A microprocessor controlled pressure scanning system", Report N76-21916/1SL, National Aeronautics and Space Administration, Lewis Research Center, Cleveland, Ohio, 1976, 13 pp.

1422. Andrews, M., "Real time adaptive filters: A least mean square approach", 13th IEEE Conference on Decision and Control, Proceedings, 1974, p. 631.

1423. Angel, W. T., "Microprocessor based controller for ultrasonic interferometer mercury manometer", Symposium on Trends and Applications 1976: Micro and Mini Systems, p. 31-3.

1424. Anon., "Automatic linear tester ICL: Now controlled by desktop calculator", New Rohde Schwarz, $\underline{16}$ (72), 11-13 (1976).

1425. Anon., "PABX uses microcomputer to provide digital switching function flexibility", EDN, $\underline{22}$ (5), 45-6 (1977).

1426. Armfield, S. C., "Time-code editing/synchronizing system for audio tape machines using microprocessor technology", 55th Convention of the Audio Engineering Society, Preprints, 1976, Preprint 1173 (P-4), 5 pp.

1427. Arnold, J. T. and Robbiano, P., "Portable vapor surveillance system", Report AD 782844/5WC, 1974, 93 pp.

1428. Bay, T., Kortright, H. J. and Muly, E. C., "Continuous particle size analysis and grinding control", 4th Annual ISA Min. and Metall. Group Symposium and Exhibition, Proc: Instrum. in the Min. and Metall. Ind., Vol. 3, 1975, Paper 668, 12 pp.

1429. Berg, A., "Convert all your synchro channels", Electron. Des., $\underline{24}$ (25), 78-82 (1976).

1430. Blomeyer-Bartenstein, H. P., "Microprocessors - key of the equipment generation of tomorrow", Siemens Components Report, Vol. XI, 1976, p. 33-4.

1431. Boose, E. F., "Lessons learned through a MIL-STD-1553 time division multiplex bus", IEEE Proceedings National Aerospace Electronic Conference, 1975, p. 634-41.

1432. Bretschi, J., "Microprocessor controlled visual sensor for industrial robots", Ind. Robot, $\underline{3}$ (4), 167-72 (1976).

1433. Burhans, R. W., "Low-cost high-performance VLF receiver front-end", IEEE Trans. Instrum. Meas., IM-$\underline{26}$ (1), 70-6 (1977).

1434. Burkitt, A., "Automatic testing flexibility and economy", Electron. Eng., $\underline{47}$ (570), 25-7 (1975).

1435. Byrd, J. S., "Testing teletypes with a microcomputer", Report DP-1406, Du Pont de Nemours (E.I.) and Co., Aiken, S. C., Savannah River Lab. Energy Research and Development Administration, 1976, 29 pp.

1436. Clark, R. J., "A microprocessor controlled electronic distance meter", Computer, $\underline{7}$ (8), 41-7 (1974).

1437. Clarke, D. W., Cope, S. W. and Gawthrop, P. J., "Feasibility
       study of the application of microprocessors to self-tuning
       controllers", Report N76-30851/9SL, Oxford Univ., England,
       Dept. of Engineering Science, 1975, 100 pp.
1438. Comella, T. M., "Instruments that think for themselves", Mach.
       Des., 47 (16), 50-4 (1975).
1439. Connors, J., Bell, H., Nordmann, B. and Wainland, D., "Get
       simultaneous analog outputs", Electron. Des., 24 (15), 88-91
       (1976).
1440. Cutler, S. E., "Micro computer networks in automobile traffic
       control", 10th IEEE Computer Society International Conference,
       COMPCON 75, Digest of Papers, p. 263-6.
1441. Dagless, E. L., "Micro-computer system structures for instru-
       mentation", Proceedings, Conference on Computer Systems and
       Technology, 1974, p. 64-72.
1442. Dahlquist, G., Ingemarsson, I. and Riesel, H., "Randomly
       generated program for automatic checking", Bit, 15 (4), 381-4
       (1975).
1443. Dessy, R. E., Starling, M., Van-Vuuren, P. and Titus, C., "In-
       board microprocessors and analytical equipment", Analytical
       Instruments, Vol. 13, 1975, 21st Annual ISA Analytic Instru-
       mentation Symposium, p. 101-2.
1444. Earle, W. E. and Fletcher, K. S., III, "Microprocessor con-
       trolled coulometric titrator", Chem. Instrum., 7 (2), 101-21
       (1976).
1445. Fisher, E. R., "High level languages in microcomputer automa-
       tion", Report UCRL-77868, California Univ., Livermore,
       Lawrence Livermore Lab., 1976, 6 pp.
1446. Franson, P., "Instruments - the trend is digital, and we're
       better off for it", EDN, 20 (13), 78-80 (1975).
1447. Fredriksen, T. R., "New wafer alignment technique", Micro-
       electron. & Reliab., 15 (2), 147-51 (1976).
1448. Fronek, D. K. and Green, D. G., "A microprocessor subsystem
       for automatic testing", Report AD-A031061/5SL, Alabama Univ.
       in Huntsville, Dept. of Electrical Engineering, 1976, 67 pp.
1449. Gabriele, T., "Multiprocessing can marry a radar", Electron.
       Des., 25 (16), 74-7 (1977).
1450. Geiger, D. F., "Binary to BCD conversion with μP's", EDN,
       21 (18), 110-11 (1976).
1451. Gere, D., "Reproducibility with a microprocessor-based LC",
       Rd. Res. Dev., 27 (10), 22-4 (1976).
1452. Gibbs, S. G. and Nolen, K. B., "Wellsite diagnosis of pumping
       problems using minicomputers", J. Pet. Technol., 25, 1319-23
       (1973).
1453. Gilmore, J. F. and Seifert, J. A., "System for remote testing",
       Bell Lab. Rec., 54 (6), 155-8 (1976).
1454. Harlan, K. H., "Microprocessor traffic control system includes
       1200 BPS modem", IEEE 1975 Region 6 - West USA - Conference,
       Proceedings, p. 184-5.

1455. Harmans, C. J. P. M. and Lassche, L., "High-precision method
      to study the Fermi surface and its pressure dependence using
      the De Haas-Van Alphen effect", J. Phys. E., $\underline{10}$ (2), 155-7
      (1977).
1456. Hnatek, E. R., "Microprocessor digital converter applications",
      Microprocessors, $\underline{1}$ (4), 222-30 (1977).
1457. Hoff, M. E., Jr., "Applications for microcomputers in instru-
      mentation", IEEE International Convention Technical Papers,
      1973, Paper 21/1.
1458. Imamura, M. S., Donovan, R. L., Oberg, J. L., Skelly, L. A.
      and Julseth, D. H., "Microprocessor-controlled battery protec-
      tion system", 10th Intersociety Energy Conversion Engineering
      Conference, Record, 1975, p. 1307-17.
1459. Iscoff, R., "Trends in wafer-probing systems", Electron. Packag.
      & Prod., $\underline{16}$ (4), 29, 38 (1976).
1460. Iwata, A., Queen, N. and Tanaka, F., "Automated gaging and
      sorting machine", Solid State Technol., $\underline{19}$ (8), 29-31 (1976).
1461. Jackson, P. C., "Examining the important trends in ATE",
      Electron. Packag. & Prod., $\underline{16}$ (1), 93-4, 96 (1976).
1462. Jaffa, K. C., "Computation of the mean and variance of light
      scattering with a microcomputer", IEEE 1975 Region 6 - West
      USA - Conference, Proceedings, p. 212.
1463. Kaegi, T. M., "The application of a microprocessor to test
      digital circuits", Proceedings: Symposium on Electronics, 1974,
      p. 341-8.
1464. Karasek, F. W., "MAD: a data system for instrumental analysis",
      Res/Develop., $\underline{27}$ (3), 40-2, 44 (1976).
1465. Kitai, R., Renyi, I. and Vajda, F., "Microprocessor application
      in a Walsh-Fourier spectral analyzer", IEEE Computer Society
      Repository No. R75-155, 23 pp.
1466. Krapka, M. J., "Microcomputer wake-up system", Proceedings,
      Symposium on Electronics, 1974, p. 253-63.
1467. Kulwiec, R., "Annunciator systems for monitoring plant equip-
      ment", Plant Eng., $\underline{30}$ (14), 62-7 (1976).
1468. Laliotis, T. A., "Application of microcomputers to low cost
      digital IC testers", Proceedings of the Technical Program,
      International Microelectronics Conference, 1975, p. 204-12.
1469. Laliotis, T. A., and Brumett, T. D., "Microprocessor-controlled
      DIC test system", Computer, $\underline{8}$ (10), 60-7 (1975).
1470. Langley, F. J., "Commercial micro computer chips for integrated
      phase array control", IEEE-S-MTT International Microwave
      Symposium, Digest of Technical Papers, 1974, p. 50-3.
1471. Lavry, D., "Microprocessor systems in electronic instrumenta-
      tion", Assc. Conf. Rec., Autom. Support Syst. Sump. for Adv.
      Maintainability, 1975, p. 9-10.
1472. Lilen, H., "Wired logic, microprocessors or custom circuits?",
      Mundo Electron., no. 60, 37-42 (1977), Spanish, EEA80-28387.

1473. Lin, D., Ulrich, F., Kinney, L. L. and Kumar, K. S. P., "Hier-
      archical techniques in traffic control", Proc. of the INFAC/
      IFIP/IFORS Int. Symp.: Control in Transp. Syst., 1976,
      p. 163-71.
1474. Lyons, N. P., "A microcomputer approach to automatic stimulus
      generation", 1975 WESCON Technical Papers, Paper 13/3.
1475. McDonnell, D., "Designing in microprocessors into linear con-
      trol loops", Electron. Eng., $\underline{48}$ (576), 47-51 (1976).
1476. McKay, H. D., "LSI microprocessor as a control element in a
      frequency agile-widrange receiver system", IEEE Electromagnetic
      Compatibility Symposium Record, 1975, Paper IA, 4 pp.
1477. Maples, M. D. and Barton, V. C., "Microprocessor-controlled
      Auger spectrometer", Microprocessors, $\underline{1}$ (4), 231-6 (1977).
1478. Marple, M. J., Land, W. H., Jr. and Lavin, R. D., "Extended
      area range instrumentation system utilizing DME links", IEEE
      Trans. Aerosp. Electron. Syst., AES-$\underline{12}$ (5), 590-9 (1976).
1479. Matsinger, J. H. and Sandy, F., "Microprocessor-controlled
      ion-machining apparatus", IEEE Trans. Ind. Electron. Control
      Instrum., IECI-$\underline{22}$ (3), 295-300 (1975).
1480. Matsumoto, Y., "Evaluation of the digital prediction filter
      applied to control a class of servomotor by microcomputers",
      IEEE Trans. Ind. Electron. Control Instrum., IECI-$\underline{23}$ (4),
      359-63 (1976).
1481. Melen, R. D., Shott, J. D., Walker, J. T. and Meindl, J. D.,
      "CCD dynamically focussed lenses for ultrasonic imaging
      system", International Conference on the Applications of
      Charge-Coupled Devices, Proceedings, 1975, p. 165-71.
1482. Meyer, C. R. and Sutherland, H. C., Jr., "Technique for
      totally automated audiometry", IEEE Trans. Bio-Med. Eng.,
      BME-$\underline{23}$ (2), 166-8 (1976).
1483. Middleton, F. H., LeBlanc, L. R. and Czarnecki, M. F., "Spectral
      tuning and calibration of a wave follower buoy", 8th Annual
      Offshore Technology Conference, Proceedings, 1976, p. 753-62.
1484. Mihelich, P. J., Jr., "Automatic fault isolation using a
      microprocessor", ASSC Conf. Rec. Autom. Support Syst. Symp.
      for Adv. Maintainability, 1975, p. 30-6.
1485. Morris, J. H., Patel, H. and Schwartz, M., "Scamp micropro-
      cessor aims to replace mechanical logic", Electronics, $\underline{48}$
      (19), 81-7 (1975).
1486. Murine, G. E., "Extended on-board, real time, preprocessing
      of multispectral scanner data", Electro-Opt. Syst. Des. Conf./
      Int. Laser Expo., Proc. of the Tech. Program, 1975, p. 158-65.
1487. Neubert, H. K. P., "Advances in strain gauging usher in a new
      transducer era", Control Instrum., $\underline{8}$ (7), 37, 39 (1976).
1488. Oliver, B. M., "The role of microelectronics in instrumenta-
      tion and control", Sci. Am., $\underline{237}$(3), 180-90 (1977).
1489. Orthner, F. H. and McKeown, D. M., Jr., "Packet switching net-
      work for minicomputers", 11th IEEE Computer Society Conference,
      COMPCON 75, Digest of Papers, p. 217-20.

1490. Ouchakov, V. B. and Petrov, G. M., "The use of microprocessors
      in conversion equipment of hybrid simulation systems", Pro-
      ceedings of the 8th AICA Congress on Simulation of Systems,
      1976, p. 483-5, Publ. by North-Holland, 1977.
1491. Pogge, R. D., "Simulate analog circuits with digital-filtering
      circuits", EDN, 21 (18), 93-9 (1976).
1492. Pratt, B., "Test A/D converters digitally", Electron. Des.,
      23 (25), 86-8 (1975).
1493. Pratt, G. W., Jr., "Opto-electronic torquemeter for engine
      control", SAE Prepr. No. 760070, 1976, 4 pp.
1494. Pratt, W. C. and Brown, F. M., "Automated design of micro-
      processor-based controllers", IEEE Trans. Ind. Electron.
      Control Instrum., IECI-22 (3), 273-9 (1975).
1495. Puckett, H. B. and Andrew, F. W., "From pressure switches to
      computers", ASAE Paper 76-3007, 1976, 5 pp.
1496. Rabe, P. R., Greenstein, E. and Hile, J. W., "Analog/digital
      integrated circuit interface for automotive sensors", SAE
      Prepr. No. 760069, 1976, 4 pp.
1497. Raphael, H. A., "Motor control by PLL can be achieved with a
      microprocessor", Electron. Des., 23 (9), 54-7 (1975).
1498. Raphael, H. A., "Low cost A-to-D conversion during micro-
      computer idle time", Comput. Des., 16 (3), 112, 114, 116 (1977).
1499. Rice, J. C., "Base your IC tester on a µP", Electron. Des.,
      24 (1), 88-92 (1976).
1500. Roberts, J. L., "Advanced acoustic position reference system",
      7th Annual Offshore Technology Conference, Proceedings, 1975,
      p. 265-76.
1501. Rose, F. A. and Smith, S. R., "Microprocessor routes data in-
      side programmable slope", Electronics, 49 (8), 148-9 (1976).
1502. Roth, R. M., "Microprocessor simplified design of flexible
      specialized test equipment", EDN, 20 (16), 40-5 (1975).
1503. Runyon, S., "Microprocessors showing promise in test equipment,
      but haven't made it big yet", Electron. Des., 22 (9), 90-5
      (1974).
1504. Saltzman, R. S., Zinn, L. and Sims, R., (Eds.), "Analysis
      instrumentation", 21st Annual ISA Analytical Instrumentation
      Symposium, Vol. 13, 1975, 154 pp.
1505. Santoni, A., "Needed for logic testing: a new breed of instru-
      ments", Electronics, 48 (19), 88-93 (1975).
1506. Sauer, W. J. and Paul, F. W., "Position control of machine
      tools using a microcomputer", Advances in Instrumentation,
      Vol. 31, 1976, Proceedings of 31st Annual ISA Conference and
      Exhibition, Pt. 1, Paper 554, 10 pp.
1507. Sawano, K., "What's new in oscilloscopes? Features and cir-
      cuitry, present and future", JEE J. Electron. Eng., no. 98,
      29-33 (1975).
1508. Schulein, J. M., "Microprocessor converts pot position to
      digits", Electronics, 49 (5), 123 (1976).

1509. Seabury, T., "Microprocessors signal in traffic control",
      EE/Syst. Eng. Today, $\underline{33}$ (1), 76-8 (1974).
1510. Singh, A. and Mekel, R., "Control your analog variables
      digitally", Electron. Des., $\underline{25}$ (5), 68-9 (1977).
1511. Smith, O. B., Donovan, R. L. and Oberg, J. L., "LST power
      system long life design techniques", Am. Astronaut. Soc.,
      21st Annu. Meet., 1975, Paper AAS 75-182, 21 pp.
1512. Steinhauer, H., Jr., "Challenge to those in instrumentation
      related to productivity and measuring power usage", Advances
      in Instrumentation, Vol. 31, 1976, Proceedings of 31st Annual
      ISA Conference and Exhibition,, Pt. 4, Paper 876, 11 pp.
1513. Sverre, P. and Tonning, L. A., "Microprocessors, a new building
      block for electronic equipment", Teknisk Ukeblad, $\underline{121}$ (28),
      24-7 (1974), Norwegian.
1514. Taylor, R. D., "Software links A/D's to computers", Electron.
      Des., $\underline{24}$ (1), 102-5 (1976).
1515. van der Gracht, P. and Mauch, K., "A microprocessor-controlled
      three-phase power inverter", Comput. Des., $\underline{16}$ (5), 120-2 (1977).
1516. Vidal, E., "Multichannel counter design using a microprocessor",
      Mundo Electron., no. 56, 75-83 (1976), Spanish, CCA12-3881.
1517. White, R. M., "Some device technologies applicable to non-
      destructive evaluation", IEEE Trans. Sonics Ultrason., SU-$\underline{23}$
      (5), 306-12 (1976).
1518. Winch, N. R. and Hyde, P. J., "Preprocessor for SPC switching
      systems", International Switching Symposium, Proceedings,
      1974, Paper 147, 6 pp.
1519. Woods, E. L., "Superconducting coil protection", 6th Symposium
      on Engineering Problems of Fusion Research, Proceedings, 1975,
      p. 277-81.
1520. Zuch, E. L., "Op amp complements A-D converter output code",
      Electronics, $\underline{49}$ (5), 123, 125 (1976).

5. Medical/Biological

1521. Arthur, R. M., Wantzelius, D. G., Hernandez, A. and Weiss, A.
      N., "Interactive acquisition of diagnostic electrocardiograms",
      Conference on Computer in Cardiology, 1976, p. 307-11.
1522. Bertrand, M., Guardo, R., Roberge, F. A. and Blondeau, P.,
      "Microprocessor application for numerical ECG encoding and
      transmission", Proc. IEEE, $\underline{65}$ (5), 714-22 (1977).
1523. Bristol, E. H., "Design study of a simple conversational
      language system", ASME Paper No. 76-WA/AUT-11, 1976, 12 pp.
1524. Camp, D. C., Voegele, A. L., Friessen, R. D., Kaufman, L. and
      Hruska, B., "Automated sample changer for X-ray fluorescence
      analysis of bio-medical samples", Chem. Instrum., $\underline{7}$ (2),
      47-63 (1976).
1525. Carolson, R. S., "Consideration in designing a microprocessor
      into a commercial medical instrument", 1975 WESCON Technical
      Papers, Paper 24/2.

1526. Cope, M., "National semiconductor GP/CP 16 bit chip set used
      in a blood sample analyzer", 1975 WESCON Technical Papers,
      Paper 10/2.
1527. Decker, J. R., Hof, P. J. and Phillips, R. D., "Automated
      rodent respiratory monitor and histogram computer: a prelimi-
      nary report", Biomedical Scientific Instruments, Vol. 12,
      1976: Proceedings of the 13th Annual Rocky Mountain Bio-
      engineering Symposium, 1976, p. 7-13.
1528. Drazan, P. J. and Kennett, R., "Opto-pneumatic manipulating
      arm", 4th International Symposium on Industrial Robots, Pro-
      ceedings, 1974, p. 349-58.
1529. Freedy, A., Lyman, J. and Salomonow, M., "Microcomputer aided
      prosthesis control", CISM - IFTOMM Symp., 2nd, Prepr.: On
      Theory and Pract. of Robots and Manipulators, 1976, p. 133-46.
1530. Graupe, D., "Control of upper-limb prostheses in several
      degrees of freedom", U.S. Veterans Adm. Dep. Med. Surg. Bull.
      Prosthet. Res. Bpr. 10-22, 1974, p. 226-36.
1531. Graupe, D. and Monlux, W. J., "Multifunctional control of
      artifical upper limbs based on parameter identification of
      myoelectric signals", Proc. IEEE Conf. Decis. Control Incl.
      Symp. Adapt. Processes, 14th, 1975, p. 105-10.
1532. Hall, E. L., Kruger, R. P. and Turner, A. F., "Optical-digital
      system for automatic processing of chest X-rays", Opt. Eng.,
      $\underline{13}$ (3), 250-7 (1974).
1533. Hansmann, D. R., Sheppard, J. J. and Yeshaya, A., "Ectopic
      beat analysis of continuous ambulatory (Holter) recordings",
      Biomed. Sci. Instrum., Vol. 11, Proc. 12th Annu. Rocky Mt.
      Bioeng. Symp., 1975, p. 117-23.
1534. Harris, W. P., "Microprocessors in an interactive teaching
      machine", Report AD-A017350/0SL, Massachusetts Inst. of
      Tech., Lexington, Lincoln Lab., 1975, 4 pp.
1535. Hathaway, J. C., Cook, A. M. and Smith, W. D., "Versatile
      microprocessor based instrumentation system for use in bio-
      medical engineering instruction", Biomed. Sci. Instrum.,
      Vol. 12, 1976: Proc. of the Annu. Rocky Mt. Bioeng. Symp.,
      13th, 1976, p. 1-5.
1536. Hayes, G. A., "A microprocessor system for automating the
      analysis of blood and urine specimen test results", Report
      AD-A019844/0SL, Air Force Inst. of Tech., Wright-Patterson
      AFB, Ohio, 1975, 137 pp.
1537. Hofstetter, E. M., Tierney, J. and Wheeler, O. C., "Micropro-
      cessor realization of a linear predictive vocoder", Report AD-
      A038241/6SL, Massachusetts Inst. of Tech., Lexington, Lincoln
      Lab., 1977, 176 pp.
1538. Hyman, W. and Lively, W. M., "A proposed study to access the
      impact of microprocessors on health care delivery", Proceedings
      of the 1977 National Computer Conference, AFIPS, Vol. 46,
      p. 309-12.

1539. Jaffa, K. C., "Computation of the mean and variance of light
      scattering with a microcomputer", IEEE 1975 Region 6 - West
      USA - Conference, Proceedings, p. 212.
1540. Johnson, F. and Gibbons, D. R., "Microprocessors in health
      care: Panacea or more effervescent technology?", Bio-Med.
      Eng., 11 (4), 132-6 (1976).
1541. Knudsen, M. J., "Real-time linear-predictive coding of speech
      on the SPS-41 triple-microprocessor machine", IEEE Trans.
      Acoust. Speech Signal Process., ASSP-23 (1), 140-5 (1975).
1542. Krogh, S. C., "A blood gas analyzer using a modular micro-
      computer", In: Microcomputers: Fundamentals and Applications,
      MINICONSULT, London, 1974, Publ. 1975.
1543. Kruger, R. P., Hall, E. L. and Turner, A. F., "Prototype
      optical-digital system for computer aided diagnosis of chest
      radiographs", Proceedings of the National Electronic Conference,
      Vol 29, 1974, p. 364-9.
1544. Lange, A., "Optacon interface permits the blind to 'read'
      digital instruments", EDN, 21 (3), 84-6 (1976).
1545. Lin, W. C., Feng, C. H. and Neuman, M. R., "A microprocessor-
      based data acquisition and processing system for studying the
      kinematics of labor", Proc. IEEE, 65 (5), 722-9 (1977).
1546. Michaels, D. L. and Tole, J. R., "A microprocessor-based
      instrument for nystagmus analysis", Proc. IEEE, 65 (5), 730-5
      (1977).
1547. Miranda, H. and Hatziemmanuel, M., "Blood analyzer tests 30
      samples simultaneously", Electronics, 49 (8), 150-1, 154 (1976).
1548. Moritz, W. E. and Shreve, P. L., "Microprocessor-based spatial-
      locating system for use with diagnostic ultrasound", Proc.
      IEEE, 64 (6), 966-74 (1976).
1549. Pfeiffer, E. A., "Potential applications of microprocessors in
      medical instrumentation", 1975 WESCON Technical Papers, Paper
      24/4.
1550. Power, D. P., "Implementation and prototype testing of a micro-
      processor system for automated blood and urine test analysis",
      Report AD-A023215/7SL, Air Force Inst. of Tech., Wright-
      Patterson AFB, Ohio, 1976, 196 pp.
1551. Ramaswami, R., "The microcomputer in diagnostic health care
      and patient monitoring", 1975 WESCON Technical Papers, Paper
      24/3.
1552. Roberts, M. J., "Microprocessor-controlled water-pollution
      monitoring system", IEEE Trans. Ind. Electron. Control Instrum.,
      IECI-22 (3), 342-5 (1975).
1553. Roesler, H. and Paeslack, V., "Manipulators for tetraplegics
      in adapted environments", CISM-IFTOMM Symposium, 2nd, Prepr.:
      On Theory and Pract. of Robots and Manipulators, 1976,
      p. 411-20.
1554. Rowell, D., Dalrymple, G. F. and McDonald, C. W., "Tension in
      a spinal orthosis for the correction of scoliosis", Proceedings
      of the 4th New England Bioengineering Conference, 1976, p. 345-8.

1555. Schneider, A. J. L., Kreul, J. F. and Zollinger, R. M., Jr., "Patient monitoring in the operating room - an anesthetist's viewpoint", Med. Instrum., 10 (2), 105-9 (1976).

1556. Trautman, E. D., "Microcomputers applied to medical instrumentation", 11th Annual Rocky Mt. Bioeng. Symp. and Int. ISA Biomed. Sci. Instrum. Sympo., 1974, p. 101-4.

1557. Tully, P. D., Moore, A. W. and Dexter, A. L., "Microprocessor-based peripheral circulation analyser", Proceedings of International Symposium and Course on Mini- and Microcomputer and Their Applications, MIMI 76, p. 197-202.

1558. Voss, D. J., Pedersen, P. C., Mahler, G. D. and Barber, F. E., "Microprocessor-based blood flow display for pulsed doppler systems", 10th IEEE Computer Society International Conference, COMPCON 75, Digest of Papers, p. 233-5.

6. Navigation

1559. Abel, J. D., "A proposed microcomputer implementation of an Omega navigation processor", Report N76-19135/2SL, Ohio Univ., Athens, Avionics Engineering Center, 1976, 10 pp.

1560. Anon., "Medium accuracy low cost navigation", AGARD Conference Proceedings No. 176, 1976.

1561. Anon., "µP hooks up with satellite to keep ultra-accurate time", EDN, 21 (19), 43-4 (1976), CCA12-12794.

1562. Anon., "Microcomputer replaces mini in airborne navigation system", EDN, 22 (11), 46-7 (1977).

1563. Bass, S. C., Belter, S. E., Chen, C. L., Findakly, T. and Hwang, H. H., "Application of balanced lines, tone signaling, and microprocessor control technqiues to a category III instrument landing system", Report AD-A022620/9SL, Purdue Research Foundation, Lafayette, Ind., 1976, 114 pp.

1564. Bjerede, B. E., "Unified signal processor for Tacan navigation sets", Navigation, 23 (2), 119-27 (1976).

1565. Boose, E. F. and Husbands, C. R., "Application of microprocessors in Avionics systems", Proceedings, Symposium on Electronics, 1974, p. 287-98.

1566. Busharis, J. G. and Tuppen, A. R., "Low cost navigation processing", National Radio Navigation Symposium, Proceedings, 1973, p. 202-9.

1567. Cotter, G. E., Fitzgerald, F. A. and Driscoll, L. A., "Avionics multiplexing with smart terminals", IEEE Proceedings National Aerospace Electronic Conference, 1975, p. 672-8.

1568. Cox, D. B., Jr., Harringron, E. V., Jr., Lee, W. H. and Stonestreet, W. M., "Digital phase processing for low-cost Omega receivers", Navigation, 22 (3), 221-34 (1975).

1569. De Linhares, T. P. B., "Distributed microcomputer airborne tactical system", Report AD-A021215/9SL, Naval Postgraduate School, Monterey, Calif., 1975, 159 pp.

1570. Dejka, W. J., "Navy opportunities for microcomputers", 11th
Annual IEEE Conf. on Eng. in the Ocean Environ. and Mar.
Technol. Soc., Proc., 1975, p. 246-8.

1571. Delorme, J. F. and Tuppen, A. R., "Low cost airborne Loran-C
navigator", Electr. Commun., 50 (4), 234-9 (1975).

1572. Eloe, E. E. and Scott, R. T., Jr., "Helicopter flight performance system using an LSI microprocessor", Report AD-
765680/4, 1973, 59 pp.

1573. Frazzini, R. and Vaughn, D., "Analysis and preliminary design
of an advanced technology transport flight control system",
Report CR-2490, Honeywell Inc., Hopkins, Minn., 1975, 423 pp.

1574. Fredriksen, P., Tonning, L. and Andersen, O., "Microprocessor
application on board ships", IEEE Trans. Ind. Electron. Control
Instrum., IECI-22 (3), 337-42 (1975).

1575. Gaon, B. N., "Hand held calculator technology applied to an
advanced low cost Omega receiver", AGARD Conference Proceedings
No. 176, 1976, Paper 22, 9 pp.

1576. Hopper, M. R., "Application of mini-processors to navigation
equipment", AGARD Conference Proceedings No. 176, 1976,
Paper 23, 11 pp.

1577. Hughes, J. M., "Multimicroprocessors navigation systems", 13th
IEEE Computer Society International Conference, COMPCON 76,
Digest of Papers, p. 264-8.

1578. Husbands, C. R., "Microprocessors in airborne time division
multiplex terminals", IEEE Proceedings National Aerospace Electronic Conference, 1975, p. 642-7.

1579. Jurison, J. and Marek, V. J., "Avionics processor standardization at microcomputer level", IEEE Proceedings National Aerospace Electronic Conference, 1975, p. 517-22.

1580. Langley, F. J. and Cooney, J. J., "Synchronous microcomputer
system for on-board missile guidance and control", Proceedings,
National Computer Conference, 1975, p. 853-60.

1581. Lilley, R. W., "A memory-mapped output interface: Omega navigation output data from the Jolt (TM) microcomputer", Report
N76-30192/8SL, Ohio Univ., Athens, Dept. of Electrical Engineering, 1976, 3 pp.

1582. Lilley, R. W., "A microprocessor interface for the Ohio University prototype Omega navigation receiver", Report N76-
30193/6SL, Ohio Univ., Athens, Dept. of Electrical Engineering,
1976, 6 pp.

1583. Lilley, R. W., "Demonstration program for Omega receiver prototype microcomputer data processing", Report N77-12029/3SL,
Ohio Univ., Athens, Dept. of Electrical Engineering, 1976,
11 pp.

1584. McCracken, W. L., "Design study of an Avionics navigation microcomputer", Report AD-783868/3WC, 1974, 177 pp.

1585. Mihelich, P. J., Jr., "Automatic fault isolation using a microprocessor", ASSC Conf. Rec., Autom. Support Syst. Symp. for
Adv. Maintainability, 1975, p. 30-6.

1586. Moore, J. P. and Rainsberger, D. B., "The design of a celestial navigation microcomputer with thoughts on an integrated information distribution system", Report AD-A013474/2SL, Naval Postgrade School, Monterey, Calif., 1975, 75 pp.

1587. Napjus, G. A., "The application of microprocessors to strapdown inertial navigation", Proceedings, Symposium on Electronics, 1974, p. 381-91.

1588. Powers, V., "A navigation microcomputer and shipboard information distribution", Report AD-A021829/7SL, Naval Postgraudate School, Monterey, Calif., 1976, 10 pp.

1589. Rose, J. A. and others, "Use of a micro-computer in a missile simulator", Proceedings, National Computer Conference, 1974, p. 821-5.

1590. Smith, R. J., "Users' ingenuity exploits device versatility", Electronics, 49 (8), 136-8 (1976).

1591. Wilson, C. H., "Surface search radar tracking by a microcomputer Kalman filter", Report AD-A028857/1SL, Naval Postgraduate School, Monterey, Calif., 1976, 119 pp.

1592. Zellweger, A., "Computer architectures for advanced air traffic control applications", Proceedings of the International Conference on Parallel Processing, 1976, p. 132-9.

7. Communication

1593. Anon., "Automatic intercept system grows", Bell Lab. Rec., 53 (9), 385 (1975).

1594. Bedzyk, W. L., "Machine translation of Morse code using a microprocessor", Report AD-785130/6, 1974, 125 pp.

1595. Beeforth, T. H., Halsall, F. and Woollons, D. J., "Further work on data communication systems at the University of Sussex", Bull. Sci. Aim, 88 (2), 140-6 (1975).

1596. Ninder, R., Abramson, N., Kuo, F., Okinaka, A. and Wax, D., "Aloha packet broadcasting: a retrospect", AFIPS National Computer Conference, Proceedings, Vol. 44, 1975, p. 203-15.

1597. Bradshaw, M. R., "Solutions to communicating information to port control", Marine Traffic System, Proceedings of an International Symposium, 1976, p. 234-48.

1598. Breinbauer, L., "Data communication on mini-computers for industrial process", Bull. Sci. Aim, 88 (1), 69-72 (1975).

1599. Buedel, C. K. and Wolff, R. W., "No. 1 EAX microprocessor-controlled universal message system", GTE Autom. Electr. Tech. J., 15 (1), 2-10 (1976).

1600. Bursky, D., "Improvements continue in digital and analog communications devices", Electron. Des., 24 (8), 50-5 (1976).

1601. Chen, R. C., Jessel, P. G. and Patterson, R. A., "MININET: a microprocessor-controlled "Mininetwork"", Proc. IEEE, 64 (6), 988-93 (1976).

1602. Cour, J. M., "Microprocessor-controlled implementation of the
      table-ronde distributed communication system", Minicomputer
      and Small Business Systems, Forum, 1976, p. 497-511.
1603. Dromard, D. and Gibergues, O., "Microprogrammed data communi-
      cations procedure controller", 6th Annual Workshop on Micro-
      programming, Conference Record, Preprints, 1973, p. 76-9.
1604. Fralick, S. C. and Brandin, D. H., "The application of micro-
      processors in high speed portable data communications termi-
      nals", Proceedings, Symposium on Electronics, 1974, p. 241-51.
1605. Fraser, A. G., "Spider - an experimental data communications
      system", IEEE International Conference on Communication,
      Record, 1974, Paper 21F, 10 pp.
1606. Fraser, A. G., "A virtual channel network", Datamation, $\underline{21}$ (2),
      51-3, 56 (1975).
1607. Garen, E. R. and Lazar, L., "Microprocessors in telecommunica-
      tions", Telecommunications, $\underline{10}$ (4), 43-6, 48 (1976).
1608. Halsall, F., "Microprocessor-controlled interface for data
      transmission", Radio Electron. Eng., $\underline{45}$ (3), 131-7 (1975).
1609. Holden, J. R. and Valassis, J. G., "Microprocessor techniques
      for telephone-oriented applications", GTE Autom. Electr. Tech.
      J., $\underline{14}$ (6), 284-97 (1975).
1610. Howick, J. F., "Study of message routing within a microcomputer
      network", Report AD-756573, 1972, 45 pp.
1611. Jones, C. L., "C-MOS processor automates motel phone system",
      Electronics, $\underline{49}$ (8), 143-5 (1976).
1612. Kaye, D. N., "Microprocessors help to communicate by voice and
      bit stream", Electron. Des., $\underline{24}$ (8), 58-61 (1976).
1613. Kovar, D., "Communications processor utilizing a Motorola
      M6800", 1975 WESCON Technical Papers, Paper 10/3.
1614. Lake, D. W., "Microprocessors in telephone station equipment",
      9th IEEE Computer Society International Conference, COMPCON 74,
      Digest of Papers, p. 117-8.
1615. Lippman, M. D. and Russo, P. M., "Microprocessor controller for
      international leased data channels", Presented at the IEEE
      International Conference on Communications, June 1974.
1616. Lippman, M. D., Russo, P. M. and Marcantonio, A. R., "A micro-
      processor controlled store-and-forward communications system",
      Proceedings, IEEE International Symposium on Circuit Theory,
      1975, p. 344-7.
1617. Longo, A., Russo, P. M. and Lippman, M. D., "Leased channel
      with microprocessor control", RCA Eng., $\underline{21}$ (3), 60-3 (1975).
1618. Mayo, J. S., "The role of microelectronics in communication",
      Sci. Am., $\underline{237}$ (3), 192-209 (1977).
1619. Mueller, D. J., "Microcomputers decentralize processing in
      data communications network", Comput. Des., $\underline{16}$ (10), 81-8
      (1977).
1620. Padgett, R. S. and Cobb, R. F., "Microprocessor-controlled
      spread-spectrum demodulator", IEEE Electronic and Aerospace
      Systems Convention, EASCON 76, Record, Paper 58, 7 pp.

1621. Parasuraman, B., "LSI microprocessors in telecommunications",
      Bull. Sci. Aim, 88 (1), 23-8 (1975).
1622. Russo, P. M. and Lippman, M. D., "Microprocessor implementa-
      tion of a dedicated store-and-forward data communication
      system", Proceedings, National Computer Conference, 1974,
      p. 439-45.
1623. Schmitz, L. S., Drumbeller, W. D. and Johnston, P. M., "Security
      communications custom microprocessor system", 10th IEEE
      Computer Society International Conference, COMPCON 75, Digest
      of Papers, p. 15-16.
1624. Schoeffler, J. D., Haelsig, M. and Rose, C. W., "Microprocessor
      based communication and instrument control for distributed
      control systems", Proceedings of the IFAC-IFIP Workshop on
      Real-Time Programming, 1976, p. 153-8.
1625. Schutzer, D. M., "Modular approach to the design of a communi-
      cations control processing center: Pros and cons", 11th IEEE
      Computer Society Conference, COMPCON 75, Digest of Papers,
      p. 31-3.
1626. Tescher, A. G., "Advances in image transmission techniques",
      SPIE Seminar Proceedings, Vol. 87, 1976, 267 pp.
1627. Trifari, J., "Speeding traffic in West Germany", Electron.
      Prod., 18 (6), 65-6 (1975).
1628. Watson, P. J., "A microcomputer-based remote data collection
      and transmission system", In: Microcomputers: Fundamentals
      and Applications, MINICONSULT, London, 1974, Publ. 1975.
1629. Weissberger, A. J., "LSI microprocessors in telecommunications",
      IEEE 1975 Region 6 - West USA - Conference, Proceedings,
      p. 186-91.
1630. West, D. G., "Design of a secure data transmission system: An
      implementation of communication theory using a microcomputer",
      Report UCRL-51587, 1974, 40 pp.

8.   Consumer Products

1631. Anon., "Microprocessors in display system", Telecommun. J.,
      43 (12), 708-9 (1976).
1632. Armstrong, L., "Microprocessors steer to Detroit", Electronics,
      47 (8), 65-6 (1974).
1633. Bell, B. and Ogden, D., "Single-chip microprocessor rules the
      roast", Electronics, 49 (25), 105-10 (1976).
1634. Blahuta, T. G., "More complex video games keep player interest
      high", Electronics, 49 (8), 161-2 (1976).
1635. Fineman, H. E., Fitzpatrick, T. E. and Fortin, A. H., "Simpli-
      fying automotive testing with advanced electronics", RCA
      Eng., 20 (6), 40-3 (1975).
1636. Goodchild, K. and Rayner, P., "Nearer the brink - trends in
      overload protection from A.C. motors", Electron. Power, 22
      (2), 102-5 (1976).

1637. Jaasma, E. G., "Latest developments in telephone transmission maintenance", Proceedings National Electronic Conference, Vol. 29, 1974, p. 30.

1638. Johnson, G. R. and Winn, C. B., "Smart thermostats for minimizing energy consumption", 8th Annual Simulation Symposium, Record of Proceedings, 1975, p. 91-104.

1639. Jones, T. O., Schlax, T. R. and Colling, R. L., "Application of microprocessors to the automobile", SAE Special Publication No. 393, 1975, Paper 750432, p. 65-74.

1640. Kim, S. N., "Micro-processors suitable for pocket calculators", Elektron. Ind., 7 (11), 302-4 (1976), German, EEA80-9014.

1641. Laberski, S. E., "Bulk weighting system keeps operator honest", Electronics, 49 (8), 170, 173 (1976).

1642. Michmerhuizen, J. and Gilbert, M., "Digital rhythm and timing generator for electronic music applications", Audio Engineering Society Preprints 1976, 55th Convention, Preprint 1161(E-2), 11 pp.

1643. Moyer, D. F. and Mangrulkar, S. M., "Engine control by an on-board computer", SAE Special Publication No. 393, 1975, Paper 750433, p. 75-7.

1644. Seth, M. K., "Microcomputers in appliance testing", AIIE Systems Engineering Conference, Proceedings, 1976, p. 67-72.

1645. Simanaitis, D. J., "MISAR: An electronic advance", Automot. Eng., 85 (1), 24-9 (1977).

1646. Sonn, E., "Four-bit chip set cuts cash register's cost and size", Electronics, 49 (8), 154, 157 (1976).

1647. Sung, C. H. and Crocetti, M. F., "Air-fuel control in a microprocessor-based system for improving automobile fuel economy", 27th IEEE Vehicular Technology Group Annual Conference, Conference Record, 1977, p. 83-8.

1648. Temple, R. H. and Devlin, S. S., "The use of microprocessors as automobile-on-board controllers", Computer, 7 (9), 33-6 (1974).

1649. Wakui, K., Murakami, K., Kawai, T. and Ohzeki, K., "Color television camera control by a microcomputer", Report N75-19483/7SL, Japan Broadcasting Corp., Tokyo, 1974, 6 pp.

1650. Willis, F. G. and Zimbel, N. S., "The use of microprocessor based electronics in the automobile", Report L750502, Arthur D. Little, Inc., Cambridge, Mass., 1975, 6 pp.

1651. Wise, C. E., "Electronics paces 1977 engine redesign", Mach. Des., 48 (23), 20-2, 24-6 (1976).

9. Others

1652. Anderson, G. E., "The use of microcomputers in DCS Autodin tributaries", Report AD-A035708/7SL, Naval Postgraduate School, Monterery, Calif., 1976, 118 pp.

1653. Anon., "Microcomputer signature-verification unit uses CRT to reduce forgeries", EDN, 22 (10), 39-40 (1977).

1654. Bell, B. A. and Taylor, P. L., "Microprocessor computational
      module for industrial data processing applications", IEEE
      Trans. Ind. Electron. Control Instrum., IECI-22 (3), 329-33
      (1975).
1655. Bernstein, K., "Activation analysis", Ind. Res., 18 (9),
      87-91 (1976).
1656. Bronis, R., "PC profiling: A choice of methods", Electron.
      Packag. & Prod., 15 (6), 44-6, 48 (1975).
1657. Chen, C. F., "Microcomputer application in generating force
      functions for aircraft tire load testing", IEEE Trans. Ind.
      Electron. Control Instrum., IECI-22 (3), 368-71 (1975).
1658. Cushman, R. H., "Getting started in microprocessors on a
      "shorestring" budget", EDN, 20 (19), 64-9 (1975).
1659. Davidow, W., "Impact on the manufacturing, maintenance and
      customer support functions for microprocessor users", Pro-
      ceedings, 1st National Microprocessor Conference, Micropro-
      cessors: Economics, Technology, Applications, 1974.
1660. Davidson, E. S., "The University of Illinois microcomputer
      laboratory", 10th IEEE Computer Society International Con-
      ference, COMPCON 75, Digest of Papers, p. 123-6.
1661. Davison, T. M., "Oscillograph aids digital well-logging
      operations", Oil Gas J., 73 (41), 57-8, 60 (1975).
1662. Demark, A. M., "Distributed control boosts process reliability",
      Electronics, 49 (8), 163-4 (1976).
1663. Dembecki, J. A. and Spalding, B. D., "Power system simulation",
      Natl. Conf. Publ. Inst. Eng. Aust. No. 77/1, 1977: IFAC Symp. -
      Autom. Control and Prot. of Electr. Power Syst., p. 374-8.
1664. Dompe, R. J., "CAMS: A microprocessor-controlled container
      marking system", IEEE Trans. Ind. Electron. Control Instrum.,
      IECI-22 (3), 300-2 (1975).
1665. Drew, B., Jr. and Auyang, J., "A poultry processing application
      using a 16 bit microprocessor", Proceedings of the Technical
      Program, International Microelectronics Conference, 1975,
      p. 213-5.
1666. Goksel, K. and Parrish, E. A., Jr., "Role of microcomputers
      in robotics", Comput. Des., 14 (10), 56-71 (1975).
1667. Gross, J. M., "Architecture for a general purpose transaction
      processing supervisor", European Computer Congress, Conference
      Proceedings, 1974, p. 361-77.
1668. Harman, R. K. and MacKay, N. A. M., "GUIDAR: An intrusion
      detection system for perimeter protection", Ky. Univ. Off.
      Res. Eng. Serv. Bull. No. 110, 1976: Carnahan Conf. on Crime
      Countermeas., Proc., p. 155-9.
1669. Harris, J. E. and Apple, J. H., "Applications of micro and
      mini systems in the automatic fare collection of a modern
      subway system", Symposium on Trends and Applications 1976:
      Micro and Mini Systems, p. 23-6.

1670. Hope, G. S., Malik, O. P. and Dash, P. K., "Differential and
      impedance protection using digital computers", Natl. Conf.
      Publ. Inst. Eng. Aust. No. 77/1, 1977: IFAC Symp. - Autom.
      Control and Prot. of Electr. Power Syst., p. 247-51.

1671. Johnson, B. A., "Microprocessor batching system", IEEE Trans.
      Ind. Electron. Control Instrum., IECI-22 (3), 290-5 (1975).

1672. Johnson, G. R., "Data acquisition, performance evaluation and
      monitoring system for solar heated/cooled residential
      dwellings", ASME Paper No. 76-WA/SOL-13, 1976, 11 pp.

1673. Jordan, B. W., Jr., "A teacher looks at microprocessors", 1974
      WESCON Technical Papers, Paper 19/1.

1674. Kalinowski, J. J. and Brown, J. C., "Microcomputer application
      to transit-passenger counting", 10th IEEE Computer Society
      International Conference, COMPCON 75, Digest of Papers,
      p. 259-62.

1675. Kerns, K. H. and Cooper, R. S., "Microcomputer solution to
      maneuvering board problems", Report AD-767685/1, 1973, 83 pp.

1676. Lambert, A., Di Giacomo, C. and Leroy, C., "Application of
      "acoustic measuring system" to localization and remote control
      of a fish or submarine rescue", 7th Annual Offshore Technology
      Conference, Proceedings, 1975, p. 839-46.

1677. Langballe, M., Tonning, L. and Wiborg, T., "Condition monitor-
      ing of diesel engines", Norw. Marit. Res., 3 (3), 2-16 (1975).

1678. Lomnes, R. K., "Microcomputers applied to underwater diving",
      Can. Electron. Eng., 19 (9), 53-5 (1975).

1679. Mennie, D., "Electronic gamesmanship", IEEE Spectrum, 13 (12),
      26-9 (1976).

1680. Montgomery, R. and Weissberger, A. J., "Automotive applications
      of microprocessors", Proceedings, Symposium on Electronics,
      1974, p. 299-304.

1681. Nicoud, J. D. and Sommer, R., "Modular logic elements, micro-
      processors and peripherals improve efficiency of teaching and
      development", 10th IEEE Computer Society International Con-
      ference, COMPCON 75, Digest of Papers, p. 127-30.

1682. Pease, A. J., "No-drop bomb simulation using micro-computers",
      Report AD-783811/3, 1974, 126 pp.

1683. Pleva, R. M., "A microprocessor controlled interface for burst
      processing", Report R-76-812, Illinois Univ., Urbana, 1976,
      62 pp.

1684. Ranieri, M. A., "Microprocessor applications spark innovation
      in building management systems", Prof. Eng., 46 (9), 16-8
      (1976).

1685. Schultz, G. V , "Which way for storage/order picking?", Factory,
      9 (4), 49-54 (1976).

1686. Seymour, G. W., "On line effective alkali analysis", Tappi
      Alkaline Pulping Conference Preprints, 1976, p. 11-13.

1687. Shavit, G., "A microprocessor based system for energy manage-
      ment system", 13th IEEE Computer Society International Con-
      ference, COMPCON 76, Digest of Papers, p. 126-30.

1688. Solomon, R. D., "A color facsimile system which can be economi-
      cally implemented with a microprocessor", Report, MIT, Cam-
      bridge, Mass., 1976, 5 pp.
1689. Titus, J. A., "Microprocessors in pollution analysis", 9th IEEE
      Computer Society International Conference, COMPCON 74, Digest
      of Papers, p. 241-3.
1690. White, R. I., "ASCII-compatible time-code system for motion-
      picture films using microcomputers", J. SMPTE, 85 (1), 9-15
      (1976).
1691. Wilson, G. H. and Herakovich, C. T., "Application of a micro-
      processor for acquisition of load, strain and acoustic-emission
      data", Exp. Mech., 16 (3), 111-15 (1976).
1692. Woyton, J. T., "Applying microcomputers to power transmission
      tests - 1, 2", Power Transm. Des., 19 (2), 37-40 (1977); ibid:
      19 (4), 53-5 (1977).

SUBJECT INDEX

In this book each chapter is designated by a capital letter and
references are numbered starting with "1". Entries are given in
this index by their number preceded by the letter of the chapter.
When the entry is the title of a chapter or the heading of a
section the page is given.

MIX
Papier aus verantwortungsvollen Quellen
Paper from responsible sources
FSC® C105338

If you have any concerns about our products,
you can contact us on
ProductSafety@springernature.com

In case Publisher is established outside the EU,
the EU authorized representative is:
Springer Nature Customer Service Center GmbH
Europaplatz 3, 69115 Heidelberg, Germany

Printed by Libri Plureos GmbH
in Hamburg, Germany